Development of Functional and Fortified Foods

Development of Functional and Fortified Foods

Guest Editor

Paolo Lucci

MDPI

Basel • Beijing • Wuhan • Barcelona • Belgrade • Novi Sad • Cluj • Manchester

Guest Editor
Paolo Lucci
Department of Agricultural, Food and
Environmental Sciences
Polytechnic University of Marche
Ancona
Italy

Editorial Office
MDPI AG
Grosspeteranlage 5
4052 Basel, Switzerland

This is a reprint of the Special Issue, published open access by the journal *Foods* (ISSN 2304-8158), freely accessible at: https://www.mdpi.com/journal/foods/special_issues/EIOC666P43.

For citation purposes, cite each article independently as indicated on the article page online and as indicated below:

Lastname, A.A.; Lastname, B.B. Article Title. *Journal Name* **Year**, *Volume Number*, Page Range.

ISBN 978-3-7258-7014-1 (Hbk)
ISBN 978-3-7258-7015-8 (PDF)
https://doi.org/10.3390/books978-3-7258-7015-8

Contents

About the Editor

Paolo Lucci

Paolo Lucci attained his PhD in 2008 at SAIFET department of the Polytechnic University of Marche (Italy). Starting from February 2009, he joined the NASCENT European Project as an experienced researcher (Marie Curie Postdoc) at POLYIntell SAS (France), and then, in 2010, he joined the Department of Analytical Chemistry of the University of Barcelona (Spain) as experienced researcher (Marie Curie Postdoc) within the Carbosorb European Project. In April 2011, he joined the School of Sciences of the Pontificia Universidad Javeriana (Colombia), where he was named head of the Research Group of "Foods, Nutrition and Health" in 2012 and then head of the Department of Nutrition and Biochemistry in 2014. Currently, he is an associate professor at the Department of Agricultural, Food and Environmental Sciences, Polytechnic University of Marche (Italy). He has gained expertise in chemical, biological and nutritional characterization of animal and plant matrices by using advanced chromatographic techniques coupled with mass spectrometry as well as high resolution mass spectrometry. His activity research is documented by several publications in international scientific journals present in the database with ISI impact factors. He also published 12 book chapters and 4 books (as editor) present in Scopus database. Paolo Lucci is also a founding member and member of the Italian Society of Food Chemistry "ITACHEMFOOD", a member of the Accademia dei Georgofili, a member of the Italian Olive Academy and a member of the Accademia Agraria in Pesaro. He was the principal investigator and researcher in several financed research projects at regional, national and international levels.

foods

MDPI

Review

Advances in Low-Lactose/Lactose-Free Dairy Products and Their Production

Aili Li [1], Jie Zheng [1], Xueting Han [1], Sijia Yang [1], Shihui Cheng [1], Jingwen Zhao [1], Wenjia Zhou [1] and Yan Lu [2,*]

1 Key Laboratory of Dairy Science, Ministry of Education, College of Food Science, Northeast Agricultural University, 600 Changjiang Road, Harbin 150030, China; ailimail@neau.edu.cn (A.L.); s211001015@neau.edu.cn (J.Z.); s211001003@neau.edu.cn (X.H.); s221001003@neau.edu.cn (S.Y.); s221001007@neau.edu.cn (S.C.); s221001020@neau.edu.cn (J.Z.); s221002143@neau.edu.cn (W.Z.)

2 National Research Center of Dairy Engineering and Technology, Green Food Research Institute of Heilongjiang, Northeast Agricultural University, Harbin 150086, China

* Correspondence: judelily@126.com

Abstract: With increasing health awareness worldwide, lactose intolerance has become a major concern of consumers, creating new market opportunities for low-lactose/lactose-free dairy foods. In recent years, through innovating processes and technologies, dairy manufacturers have significantly improved the variety, and functional and sensory qualities of low-lactose and lactose-free dairy products. Based on this, this paper first covers the pathology and epidemiology of lactose intolerance and market trends. Then, we focus on current advantages and disadvantages of different lactose hydrolysis technologies and improvements in these technologies to enhance nutritional value, and functional, sensory, and quality properties of lactose-free dairy products. We found that more and more cutting-edge technologies are being applied to the production of lactose-free dairy products, and that these technologies greatly improve the quality and production efficiency of lactose-free dairy products. Hopefully, our review can provide a theoretical basis for the marketing expansion and consumption guidance for low-lactose/lactose-free dairy products.

Keywords: lactose intolerance; dairy product production; lactase; fermentation; membrane separation; fortification; nutritional value; sensory properties

1. Introduction

There is improving awareness of lactose intolerance in both research and dairy consumers. Generally, lactose intolerance is caused by primary lactose malabsorption. Lactase activity is highest at birth and declines after weaning. Undigested lactose is metabolized by intestinal microflora and converted into short-chain fatty acids (acetate, propionate, butyrate, lactate, and formate) and gases (hydrogen, methane, and carbon dioxide), causing gastrointestinal discomfort such as diarrhea, bloating and other symptoms. Long-term avoidance of dairy products due to lactose malabsorption or lactose intolerance may lead to malnutrition and skeletal disorders. Epidemiological surveys of lactose intolerance showed that approximately 70% of the world's population was affected by lactose malabsorption, with prevalence as high as 95% to 100% in some Asian and African countries [1].

Current managements for lactose-intolerant people are mainly to replace regular dairy products with low-lactose and lactose-free products [2] and consume dairy products with exogenous lactase or probiotics. Low-lactose/lactose-free dairy products can reduce the incidence of gastrointestinal symptoms in lactose-intolerant patients while ensuring the intake of nutrients in milk. Sharp et al. [3] included 23 previous studies in their systematic review indicating that lactose-free milk and lactose hydrolyzed milk as substitutes for whole milk can reduce the risk of deficiencies and provide important nutrients for lactose-intolerant patients and healthy individuals. In addition, compared with other calcium-rich foods (such as vegetables and mineral water), low-lactose/lactose-free dairy products are a low-cost dietary source of calcium for lactose-intolerant patients [4]. Low-lactose/lactose-free dairy products are recognized as functional foods in the European Union [5]. An increasing variety of low-lactose/lactose-free dairy products, including liquid milk, Greek-style yogurt, kefir, cheese, ice cream and infant formula powder, have a growing appeal to consumers [6].

The requirements for developing low/lactose-free dairy products are avoiding lactose sources, ensuring the nutritional value and sensory properties of the product, and controlling production costs [7,8]. Generally, lactose content is required to be less than 1 g/100 g in low-lactose products and less than 10 mg/100 g in lactose-free products. The main ways to reduce lactose content [9] in dairy products include enzymatic hydrolysis, membrane separation, and fermentation. With the development of lactose removal technologies, the main challenge for manufacturers is to prepare low-lactose/lactose-free dairy products that are consistent with or exceed traditional dairy products in nutrition, flavor, and quality. Processing technologies and conditions are continuously being innovated, for example, studies reported characterization strains [10] of high yield and materials of immobilized lactase [11], coupling membrane, co-fermentation, etc. In this paper, the epidemiology and pathology of lactose intolerance are briefly introduced. Then, we focus on the market, processing technology, fortification, and improvement of low-lactose/lactose-free dairy products, hoping to provide theoretical reference for developing low-lactose/lactose-free dairy products suitable for lactose-intolerant patients.

2. Lactose Intolerance

2.1. Epidemiology

Lactose intolerance is a widespread worldwide problem. The ability of adults to digest lactose (lactase persistence) has become an important topic in genetics, medicine, and evolution. In recent years, the in-depth study of genetic mechanisms of lactase has provided a new theoretical basis for the epidemiological investigation and prevention of lactose intolerance. Storhaug, Fosse and Fadnes [1] presented a systematic review and meta-analysis of lactose intolerance by searching studies from Medline and Embase, including 62,910 participants from 89 countries. The global prevalence of lactose malabsorption estimated was 68%, and when using genotype data only, the estimate was 74%, whereas prevalence was 55% using lactose tolerance test data. Lactase gene polymorphisms have been widely used to predict lactose intolerance [12]. So far, there are 23 SNPs (single-nucleotide polymorphisms) known to be associated with lactase persistence. The most well-known and frequent SNP was identified at −13910 kb (rs4988235). LNP (lactase non-persistent) patients carry a C/C at this site, whereas LP (lactase-persistent) patients carry either a C/T or a T/T.

The distribution of LP genotypes shows significant national and regional differences, e.g., −13910C/T mostly occurrs in European and American countries. For northern Europeans, their endogenous lactase activity is still at a high level with increasing age. Anguita-Ruiz et al. [13] created an online interactive map of the distribution and frequency of LP genotypes worldwide. The prevalence of lactose intolerance increased from northern Europe to Africa and the Middle East and reached its highest in Asia. However, this trend was interrupted in countries with strong cultural admixtures, such as the US and countries in Europe. Chin et al. [14] reported −13910C/T in different ethnicities in the US. Only 17.9% of Caucasians were found to have the LNP genotype, while 96.5% and 69.2% of Asians and African Americans were found to have the LNP genotype, which indicated that there were great differences among different ethnicities within one country. In Africa and the Middle East, −13907:C>G (rs41525747), −14009:T>G (rs869051967), −13915:T>G (rs41380347) and −14010:G>C (rs145946881) were found to be more widespread [15,16]. Priehodová et al. [17] reported that the frequency of the LNP −13910*T variant was only 8.6% in nomadic Arabs, but 84.7% in non-nomads. Although increasing data are becoming available for lactase gene SNPs, studies of some geographical areas are relatively sparse, including Argentina, Uruguay, Cuba and Libya. Additionally, although Asia has a high prevalence of lactose intolerance, many well-known SNPs are difficult to detect, suggesting that there may be exclusive or new SNPs in this region [18]. For example, Peng et al. [19] found three new SNPs (−13838G/A, −13906T/A and −13908C/T) were associated with LNP in the Tibetan population. The creation of a broader database of lactase gene SNPs and collection of different SNP frequency data will be a direction of future analysis and research.

2.2. Lactase Gene and Pathology

It is now clear that lactase gene SNPs plays an important role in lactose metabolism (Figure 1). The MCM6 gene, which is 14 kb chromosomally upstream of lactase gene, contains a highly conserved promoter sequence. Some SNPs enhance transcription factors' binding ability to lactase promoters by binding to them. The molecular mechanism for these SNPs is to create new binding sites for specific transcription factors, especially octamer-binding protein 1 (Oct-1). SNPs associated with LP can combine with transcription factors to promote the start of transcription, while SNPs associated with LNP cannot. Associated with LP, −13915*G was found to be able to interact with Oct-1 in vitro [20]. Other transcription factors involved in lactase activity regulation include caudal type homeobox 2 (Cdx-2), hepatocyte nuclear factor 1α (HNF1-α), and GATA-4, -5, and -6. Jensen et al. [21] found that the −14010*C variant associated with LP had greater binding affinity to Oct-1 than −14010*G. However, due to the stability of DNA sequences, transcriptional mechanisms cannot explain the programmatic decrease in lactase activity with age. The latest studies put forward that epigenetic mechanisms (mainly DNA methylation) could also be involved with LNP. Leseva et al. [22] found that a differential DNA methylation in the lactase gene through epigenome sequencing, which was closely related to activity of lactase and −13910C/T, was associated with the −13910C >T genotype. Moreover, the methylation level of this position combined with gene testing can predict lactase activity more accurately than the hydrogen breath test. LNP haplotypes containing the C(−13910) allele accumulated modified cytosines that silenced the regulatory elements in MCM6 and lactase gene, whereas the LP haplotype containing the T(−13910) allele displayed age-related modification changes that maintained lactase activity [23].

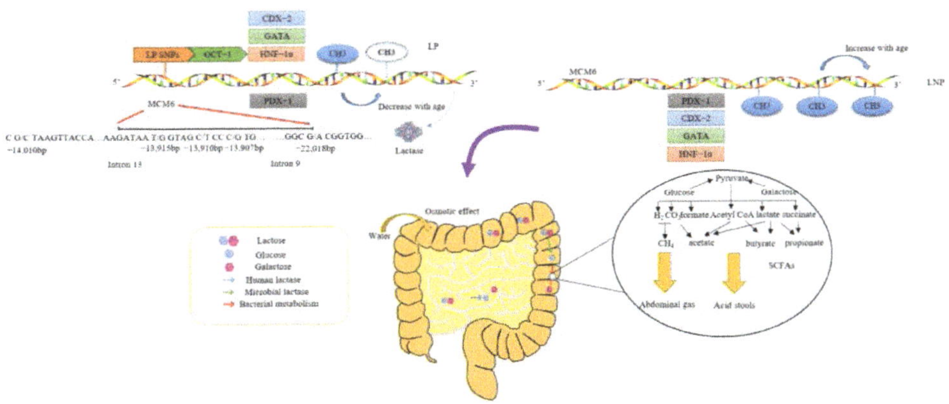

Figure 1. Mechanism of LP and LNP. LP: lactose persistence; LNP: lactose non-persistence; SNPs: single nucleotide polymorphisms; OCT−1: recombinant octamer binding transcription factor 1; PDX−1: pancreatic and duodenal homeobox 1; HNF−1α: hepatocyte nuclear factor 1 homeobox alpha; GATA factors were named after the consensus DNA-binding sequence (A/T)GATA(A/G), which is recognized by the zinc-finger domains common to all family members; CDX−2: caudal type homeobox 2; MCM6: minichromosome maintenance complex component 6; SCFAs: short-chain fatty acids.

Lactase genotypes of hosts are known to influence the gut microbiome, and the influence seems to be modulated by lactose intake [24]. It is reckoned that small amounts of lactase are synthesized by the intestinal flora over a long period of genetic evolution in order to alleviate clinical symptoms of lactose intolerance. Among the colonic microbiota, association of lactase gene SNPs and the *Bifidobacterium* genus is widely identified [25]. Kurilshikov et al. [26] conducted a large-scale association analysis to identify host factors influencing human gut microbiome composition, and found the lactase gene locus reached study-wide significance and showed an age-dependent association with *Bifidobacterium* abundance. Goodrich et al. [27] reported that the association existed between *Bifidobacterium* and the lactase gene locus, and the direction of the genetic association showed lactase non-persisters harbored higher levels of *Bifidobacterium*. In addition, the latest study reported [28] that adults with lactose-tolerant genotype (GG) had higher milk intake, but lower *Bifidobacteria* compared to lactose-intolerant adults (AA/AG). There was a significant interaction between milk intake and LCT variant on *Bifidobacterium*, with a positive association between milk intake, and *Bifidobacterium* was observed only in lactose-intolerant but not in tolerant people.

Generally, lactose intolerance is considered to be a risk factor for reduced milk/calcium intake. According to a survey conducted by Cargill, 61 percent of US consumers avoided dairy products due to lactose intolerance or allergy to milk protein [29]. Obermayer-Pietsch et al. [30] detected lactase genotypes in 258 postmenopausal women and found that individuals with CC (LNP variant) had significantly lower milk calcium intake and decreased hip and lumbar spine bone density. However, recent studies have reassessed the association of lactase genotypes, dairy intake and bone health, and concluded that lactase gene SNPs had a weaker connection with bone health than anticipation. One study reported that LP and LNP genotypes were weak predictors of dairy intake [14]. Considering the geographical variation of lactase gene SNPs, Joslin et al. [31] evaluated the association of LNP and the heredity of numerous diseases in European populations, and found little evidence between LNP and reduction of bone mineral density. Hilliard et al. [32] explored the relationship between LNP and bone health in Africans, and the incidence of hip fracture and less consumption of animal protein was less correlated with LNP. Tolonen et al. [33] selected 882 Finnish women and 669 Finnish men as research objects. A slight difference in trabecular densities at the distal sites of radius and tibia was found in men between different lactase genotypes. Men with the T/T genotype were about 3% higher than those with T/C and C/C. No difference was found in women. More studies are needed to evaluate the association between lactose intolerance and bone health.

3. Lactose-Free Dairy Product Market

Lactose-intolerant consumers need nutritional products that they can utilize. It is known that 25% of consumers in India are motivated to buy or have started to buy lactose-free drinks. The lactose-free dairy products market in Vietnam is also booming, and relevant surveys showed that consumers preferred dairy products that are easy to digest [34]. The large-scale consumption of dairy products in China has just begun in the last 20 years, during which the concept of lactose intolerance has gradually become familiar to consumers. Up to 86.7 percent of Chinese adults are lactose-intolerant, which partly explains why China has the world's largest market for plant-based protein beverages [35]. According to Szabo et al. [5], target group analysis in Hungary showed that currently two-thirds of lactose-sensitive consumers regularly consumed lactose-free products, most of which are female, lactose-intolerant, highly educated and aged under 30 years. Notably, lactose-free products are usually attractive to consumers with digestive problems because the products are always declared as easily digestible.

Lactose-free dairy products have become the fastest-growing part of the dairy industry. The compound annual growth rate of global lactose-free dairy products from 2017 to 2022 was about 7% and accounted for 80% of the total lactose-free products (sales of about 8.8 billion US dollars) [29]. In the US market, lactose-free milk accounted for 4.0% of the total liquid milk sold annually, and sales increased by 12% in 2017 [36]. Over the past five years, the range of lactose-free dairy products has continued to diversify, among which milk and yogurt were the most common lactose-free dairy products, while lactose-free butter, cheese and flavored milk are also on the rise. Świąder et al. [37] analyzed the market of lactose-free dairy products in Poland and categorized 75 lactose-free dairy products, including milk, yogurt, cream, quark, cheese and infant formula.

Figure 2 shows representative lactose-free dairy brands of some countries. Finns have relatively high lactose intolerance prevalence in Europe, at 17 percent, leading to earlier popularity of lactose-free products. Valio, Finland's largest food and dairy company, was the world's first lactose-free dairy company, launching its first lactose-free milk in 2001. Lactalis of France was the world's biggest cheese producer, and its products are low in lactose, often combined with low fat. Finland's Arla Foods owned more than 30 brands, including the well-known Lactofree. Fairlife of America has launched a series of lactose-free ultrafiltered milk products, including shakes and recovery drinks, which are high in calcium and protein and have a strong position in the fast-growing lactose-free dairy market. As early as 2010, China produced only 300,000 t of lactose-free milk, accounting for 1% of liquid milk. In the past five years, while encouraging and guiding lactose-intolerant consumers to make reasonable choices, Chinese dairy enterprises have also increased production of lactose-free products. For example, Monmilk Xinyangdao was certified by the Lactose Intolerance Global Network (LIGN) in 2017, becoming China's first lactose-free dairy brand certified by the international organization. These efforts are all leading lactose-free products to become more mainstream, standardized, and healthier.

Figure 2. Representative lactose-free dairy brands of some countries.

4. Production of Lactose-Free Dairy Products

The development of lactose-free and low-lactose dairy products made it possible for lactose-intolerant people to utilize the rich nutrients in milk, which is one of the effective ways to improve lactose malabsorption and lactase intolerance. Currently, there are three processes commonly used to reduce lactose content: enzymatic hydrolysis of lactose, membrane filtration, and fermentation. Studies have tried to combine several of above processes. Lactose-free milk can be further processed into lactose-free/low-lactose yogurt, cheese, milk powder, ice cream and other dairy products [29] (Figure 3). Many studies have confirmed that most consumers diagnosed with lactose intolerance can tolerate up to 10 g/day lactose and have no observable adverse reactions to 2 g/day lactose. Table 1 lists several regulations that mention lactose threshold. Although there are no common standards on allowable lactose threshold, in most countries, lactose content of low-lactose and lactose-free products are within 1 g/100 g and 10 mg/100 g, respectively.

Table 1. Regulations regarding lactose threshold in different countries have been issued by several authorities.

Threshold	Country/Region	Authorities	References
<1000 mg/L lactose as lactose-free	European countries	EFSA (European Food Safety Authority), 2010	-
<5000 mg/L lactose as lactose-free	China	EFSA (European Food Safety Authority), 2010,	[38]
<10,000 mg/L as low-lactose	India	FSSAI (Food Safety and Standards Authority of India), 2019	[39]
<100 mg/L as lactose-free	India	FSSR (Food Safety and Standards Regulation), 2019	-
0.1% (*w*/*w*) as lactose-free	Italy	Italian Health Ministry	[40]

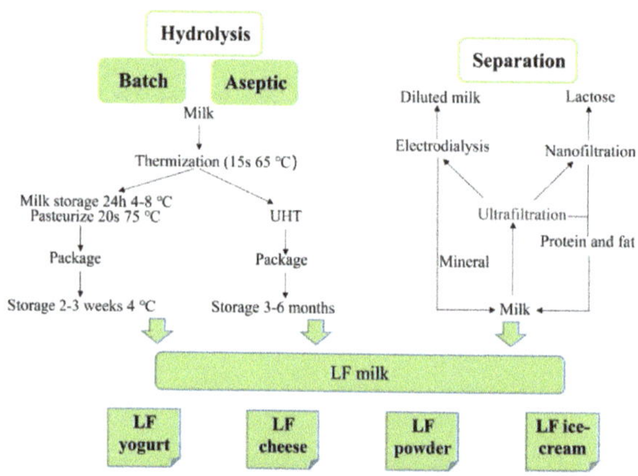

Figure 3. Production of lactose-free dairy products. LF: lactose-free.

4.1. Separation of Lactose

Among methods of separating lactose, the most well-known and widely used are methods of membrane separation, including ultrafiltration, nanofiltration, reverse osmosis, and electrodialysis. In different fields of the dairy industry, membranes are applied to shelf life extension of milk, whey processing, cheese industry, milk protein processing, fractionation of milk fat and desalting or demineralization [41]. The key problem in the removal of lactose from milk is the separation of proteins and lactose, and the broad particle size distribution of proteins in milk reduces the separation efficiency of proteins and lactose [42]. Protein and fat are blocked, while lactose and small molecules are allowed to pass through in ultrafiltration and nanofiltration. Polymer ultrafiltration membranes are usually used in industrial practice because of their ease of preparation and cost-effectiveness. By screening and comparing, Sanchez-Moya et al. [43] found GR60PP (a polysulfone ultrafiltration membrane) to be the most efficient membrane, as 90% of the lactose was separated and 100% protein was recovered. This indicated that significant adsorptive fouling and pore blocking were the main causes of high rejection of lactose. Researchers focused on reducing membrane fouling mainly caused by protein. Attempts were made to achieve high permeation flux and high separation efficiency through ultrasound assistance, optimization of operating conditions, and development of cleaned membranes. Use of ultrasound under optimal power was demonstrated to be able to reduce the fouling by 32% [44]. Sofuwani, Aslina, and Mazlina [45] found the lowest lactose rejection (77.71%) was achieved when adapting 5 kDa cross-flow hollow fiber ultrafiltration membrane and operation parameters of 0.55 bar trans-membrane pressure and 0.74 L min^{-1} feed flow rate. Graphene oxide membrane allowed lactose to diffuse through the nanochannel and exhibit higher lactose permeation flux (2.87 kg m^{-2} day^{-1}) than commercial nanofiltration (0.57 kg m^{-2} day^{-1}) and ultrafiltration (1.61 kg m^{-2} day^{-1}). In addition, the fouling layers on graphene oxide membrane were porous, facilitating higher permeation flux and water flux recovery. Graphene oxide membrane is very promising for lactose separation of lactose-free milk [46].

Generally, ultrafiltration membranes have low permeability and high protein rejection, whereas microfiltration membranes have higher permeability, but higher protein losses [19]. Another common practice in dairy processing is called diafiltration, during which water is added to the concentrates to increase the flux and continue the separation. The trade-off between permeability and selectivity reduces the separation efficiency of proteins and lactose [20]. A microfiltration membrane with smaller pores may be a good choice. Qi et al. [42] selected ceramic microfiltration membranes with high permeation flux and high rejection performance and prepared low-lactose milk with a lactose concentration of less than 5 g/L. The molecular weights of mineral and lactose are very close, so it is difficult to separate them by a single membrane. Coupling membranes are superior, as large molecules such as proteins and fats are first retained using ultrafiltration. Minerals are intercepted by electrodialysis and lactose is recovered by nanofiltration. For example, Zhang et al. [47] obtained low-lactose milk powder with lactose concentration of less than 0.2% and recovered high-purity lactose powder (95.7% lactose content) as a byproduct. Further, membrane separation can be combined with hydrolysis and enzymatic membrane bioreactor function by optimizing GOS production [48].

Several studies explored methods to separate lactose by chromatography and freeze concentration. For example, lactose-free milk produced by Valio Ltd. (Helsinki, Finland) employed chromatography, membrane separation and enzymatic hydrolysis. Freeze concentration is a widely accepted technology and is able to preserve thermosensitive food with high nutritional value. In this process, the temperature of food liquid decreases below its freezing point, and the concentration of liquid is reached by removing water in the form of ice crystals. A new type of lactose-free dairy product was developed by combining progressive freeze concentration with vacuum-assisted block freeze concentration, because carbohydrates accumulated more during progressive freeze concentration, while in block freeze concentration, protein is more likely to enter concentrated liquid [49]. Batista et al. [50] aimed to develop a new system for lactose removal through bioaffinity chromatography. In the research, brosimin, a lactose-binding lectin, was first extracted from *Brosimum gaudichaudii* and immobilized onto polyaniline. The system removed 47% of lactose from skim bovine milk.

4.2. Enzymatic Hydrolysis of Lactose

The process of lactose hydrolysis does minimal damage to the nutritional components in milk and is highly specific. Two processes, batch and aseptic, are used in producing lactose-free milk. In the batch process, neutral lactase is added to milk under slow stirring until lactose is fully hydrolyzed, after which the milk is pasteurized, homogenized, and packaged. The batch process lacks pasteurization during the hydrolysis phase, and enzyme dosage is relatively high because the reaction occurs at low temperatures to prevent microbial spoilage [29]. Research on cold-active β-galactosidase has improved this defect. In the aseptic process, milk is first sterilized using the UHT procedure, after which sterile lactase is injected into the milk just before packaging. Although the lactase dosage decreases compared to the batch process, the aseptic process requires special equipment, and process control is absent since the hydrolysis reaction continues after packaging.

4.2.1. Microorganism Source of Lactase

The most plentiful resource of β-galactosidase (also known as lactase) are microorganisms, which have advantages of a short production cycle and high production yield. Research has been conducted on the optimization of reaction conditions, purification and characterization of β-galactosidase. Commercially available β-galactosidase mainly includes *Aspergillus oryzae, Aspergillus niger, Escherichia coli, Kluyveromyces lactis* and *Kluyveromyces fragilis* [51]. β-galactosidases from microorganisms have different enzymatic characteristics and structure. For example, *Kluyveromyces lactis,* the most reported microorganism in the literature, showed high hydrolytic performance that was relevant with an exclusive insertion in loop 420–443 of its catalytic site, which enhanced affinity of lactase to lactose [52].

Isolation and characterization of β-galactosidase that fulfill the demands of production of low-lactose dairy products remains a hot topic. Genetic engineering technology was used to express the lactase with good enzymatic characteristics in heterologous expression (*E. coli, Lactococcus* and *Pichia pastoris,* et al.). As presented in Table 2, studies showed that the production of recombinant β-galactosidase of different microbial sources increased. Generally, commercial β-galactosidases feature the high lactose affinity (K_M) and low product inhibition (K_I) by galactose, which is adverse for hydrolysis reaction. Engineering technology is used to get satisfactory lactose affinity (K_M) and product inhibition (K_I) [53]. The optimal pH and temperature of commercial β-galactosidases are 7.0 and 35 to 40 °C, which are susceptible to result in the contamination of milk. Thermostable and cold-adapted β-galactosidases had significant advantages in processing, such as higher substrate solubility and reaction rate, as well as lower probability of microbial contamination [54]. Furthermore, lysis of cells and extraction of intracellular lactase increased the costs of production. Ren et al. [55] used the twin-arginine (Tat) signal peptide PhoD to direct the secretion of the β-galactosidase, which is a new pathway to improve the secretion amount of lactase. It provided a new way to improve lactase production.

Galactose oligosaccharides (GOSs) can be applied as prebiotics in a variety of dairy products. The global prebiotics market size exceeded US$ 2.90 billion in 2015, with an expected growth of about 12.7% and a profit of about $10.55 billion by 2025 [56]. Transglycosylation is another property of β-galactosidase. GOSs are synthesized during this process, which reduces concentration of lactose and increases the value of low-lactose milk. In conclusion, the characteristics of lactase are important for the enzymatic production of lactose-free dairy products. The dairy industry aims to produce standard lactose-free milk with low cost and high added value.

Table 2. Studies on strain producing lactase of high activity.

	Enzyme Source	Process	Advantages	References
Strains resistant to low/high temperature and acid environment	*Alteromonas* sp.ML117	*Alteromonas* sp. ML117. β-galactosidases were heterologously expressed in *E. coli* and the recombinant lactase was purified.	Recombinant β-galactosidase was a cold-adapted variant and hydrolyzed 86% lactose of milk after 24 h at 10 °C. The enzyme is NaCl-tolerate.	[57]
	Picrophilus torridus DSM 16176	The enzyme was purified 110-fold and determined.	This enzyme is thermostable. At 70 °C, it retained 76% and 42% activity after 30 and 120 min.	[58]
	Anoxybacillus sp.AH1	The enzyme was purified 10.2-fold.	The purified enzyme was highly stable and retained at 71% of the original activity at 60 °C and 53% at 70 °C within 120 min.	[59]
	Aspergillus niger van Tiegh	Extracellular β-galactosidase was purified to homogeneity using a combination of gel filtration, ion-exchange, chromatography.	The enzyme is highly stable when exposed to simulated gastric conditions in vitro. It retained 68% of original activity. Activity of capsule is some 3.5-fold more than commercial enzyme.	[60]

Table 2. *Cont.*

	Enzyme Source	Process	Advantages	References
Strains with lactose affinity and reduction of product inhibition	*Bifidobacterium adolescentis*	β-galactosidase gene found in *Bifidobacterium adolescentis* and was expressed in *E. coli.*	This enzyme had a Km of 3.7 mM. It exhibited low product inhibition by galactose with a Ki of 116 mM and high tolerance for glucose.	[61]
	Aspergillus candidus	Four amino acid positions (Tyr96, Asn140, Glu142, and Tyr364) were selected for mutation based on their molecular bindings with galactose using site-directed mutagenesis.	β-galactosidase Y364F (Tyr364 mutant) had a galactose inhibition constant (K_I) of 282 mM, which is 15.7-fold greater than that of the wild-type enzyme.	[62]
Strains with high transgalactosylation capacity	*Klebsiella oxytoca* ZJUH1705	Two β-galactosidase genes were isolated from a novel β-galactosidase-producing *Klebsiella oxytoca* ZJUH1705. Two β-galactosidase genes were cloned, expressed in *E. coli* and purified.	β-gal 2 had a high trans-glycosylation capacity. Adding β-gal 2 in lactose with the ratio of 2.5 U/g, a high GOS yield of 45.5%was obtained.	[63]
	Bacillus sp. D1. BglD1	A novel β-glucosidase, BglD1 was screened and cloned from the deep-sea bacterium. a mutant BglD1:E224T was generated based on the semi-rational design.	BglD1 hydrolyzed 88.5% lactose and produced 3.3 g/L GOS when using milk as the substrate. The GOS yield of its mutant was 11.5% higher than that of BglD1.	[64]
	Paenibacillus barengoltzii	β-galactosidase gene was cloned, expressed in *E. coli* and purified.	The recombinant β-galactosidase exhibited high trans-glycosylation activity. Maximum yield of GOS was 47.9% at a lactose concentration of 350 g/L.	[65]
	Alteromonas sp. ANT48	β-galactosidase gene was cloned, expressed in *E. coli.*	90.6% of the lactose was hydrolyzed at 40 °C within 15 min. GOS yield reached 30.9%.	[66]
	Streptococcus thermophilus	Site-directed mutation strategy was attempted to genetically modify β-galactosidase (the enzyme and its mutant were named BagQ and BgaQ-8012 respectively)	The GOS yields increased to 5.8 and 8.3 g/L adding BgaQ or BgaQ-8012. Addition of the β-galactosidases reduced lactose content by 49.3% and 54.4% respectively in yogurt.	[67]

4.2.2. Immobilized Lactase

Immobilization is widely applied in production of lactose-free products because of operational stability, reusability and easy recovery of β-galactosidase in continuous process. Absence of the enzyme in the final product increases its stability. Retained activity of β-galactosidase after recyclable use is critical when deciding whether the immobilized enzyme is suitable for manufacturing. The performance of lactase is greatly influenced by the sources and purity of the enzyme, as well as the type of immobilization method and support materials used. Immobilization methods include adsorption, covalent binding, cross-linking, encapsulation and entrapment. Support materials significantly affect properties (thermal resistance, chemical resistance, mechanical properties and biocompatibility) of immobilized enzymes [68] (Table 3). Classic support materials for β-galactosidase immobilization include alginate, chitosan, silica, resin and so on. New materials are continuing to develop and the particle size of these materials are generally at the nanoscale, which provides large surface-volume ratio, high surface reaction activity and high catalytic efficiency. Nanomaterials mainly applied to the immobilization of β-galactosidase are carbon nanotubes, silicon dioxide nanoparticles, nanodiamonds, silver nanoparticles and zinc oxide nanoparticles [69]. Furthermore, new materials (such as graphene oxide, mesoporous and electrospinning material) are abundant, with a lot of functional groups. For example, many carboxylic (COOH), hydroxyl (–OH) exist in the surface of graphene oxide, which facilitates enzyme–matrix interactions and make it easy for β-galactosidase to be modified [70]. The future direction of development and application lies in reducing costs, sustaining activity of immobilized enzymes, and synthesis of new support materials.

Table 3. Application of immobilized lactase in preparation of low-lactose milk.

Method	Support Material	Activity of Enzyme and Ability of Hydrolysis	Other Advantages	References
Covalent binding	Eupergit CM	The activity of immobilized enzyme decreased after 20 times of repeated use, and reached 99.3% after 15 days of storage. Lactose was completely hydrolyzed within 4 h.	Storage stability and activity of enzyme increase.	[71]
Cross-linking and adsorption	Modified arabic gum-based hydrogel	After 3 cycles, activity of immobilized β-D-galactosidase was 52.79% of the initial enzyme.	Improve the efficiency of lactose hydrolysis and lower costs.	[72]
Adsorption	Fe-chelated cryogel disk	The immobilized lactase lost 29.2% after 70 days and preserved 64.9% of initial activity after 25-runs.	The optimum temperature of immobilized lactase increase.	[73]
Covalent binding	Mesoporous silica/titania with a chitosan coating	Lactase retained approximately 90% of initial activity and achieved full conversion of lactose even after 15 cycles in batch system.	Enzyme is hard to deform and demonstrates high operational stability for application and manufacturing.	[74]
Entrapment	Bacterial cellulose nano crystal	β-galactosidase retained 80% activity after 12 cycles of use.	β-galactosidase showed higher stability to various range of pH and temperature.	[75]
Covalent binding	Gluconic acid coated fullerenes	β-galactosidase was able to be recovered easily and retained 89% activity after 6 repeated uses.	Obvious improvement in lactose hydrolysis was observed at high temperature.	[76]
Entrapment and adsorption	Halloysite nanotubes and cellulose nanocrystals	Enzyme retained 76% activity after 12 cycles.	Enzyme was more thermostable at 55 °C than the free enzyme.	[77]
Covalent binding	Modified gold nanoparticles	β-galactosidase exhibited greater operational activity after 6 reuses.	Stability was significantly enhanced at wider temperature, pH and higher galactose concentrations.	[78]

4.3. Fermentation

Fermentation hydrolyzes 20–30% of the lactose of milk, and protein and fat into peptides, amino acids, fatty acids, which makes it easy for human to digest and absorb nutrients. Traditional yogurt is fermented by *Lactobacillus bulgaricus* and *Streptococcus thermophilus*. The lactose content of the fermented milk decreased and ranged between 4.6% and 3.7%. Moreover, the yogurt strains produce few β-galactolactases, which further hydrolyze lactose. A systematic review by Savaiano et al. [79] concluded that there was a positive correlation between yogurt consumption, improved lactose digestion and improved lactose tolerance symptoms. Greek-style yogurt has become popular in the United States since its first appearance in 2007 and gained 50% market share of yogurt quickly. Greek-style yogurt is made on the basis of yogurt after concentration, centrifugation, isolation of whey or supplementation of protein to achieve the thick and creamy texture [80]. Greek-style yogurt has always been claimed to have high protein (8–12%), low lactose (reduced by half) and low fat. Kefir is another common fermented dairy product. Lactic acid bacteria and yeast in kefir constitutes a complex symbiotic relationship, which is responsible for alcohol and lactic acid fermentation, respectively, giving the product unique flavor [81] and producing functional substances (bioactive peptides, cellular polysaccharides and amino acids, etc.) [82]. The lactose content of kefir is 3.1 g/100 g, and its unique microorganism helps to relieve lactose intolerance by regulating and maintaining the balance of intestinal flora [83]. Cheese is a natural low-lactose dairy product. Most lactose will be excreted with whey during the processing and preparation of cheese. For hard and semihard cheese, the remaining lactose continues to be converted into lactic acid during the cheese ripening process, making them natural lactose-free products. Panseri et al. [84] detected carbohydrates in lactose-free dairy products, and the concentration of lactose in PDO hard cheese was below 0.0001 mg/kg. There are a wide variety of low-lactose/lactose-free cheeses available on the market for lactose-intolerant consumers. The lactose content of Asiago PDO, Gorgonzola PDO, Emmentaler PDO, Pecorino Toscano PDO, Piave PDO, Stelvio PDO, and Montasio PDO ranges from 10 mg/kg to 100–1000 mg/kg [40]. The lactose content of Grana Padano PDO, Parmigiano Reggiano PDO, and Pecorino Romano PDO is less than 10 mg/kg.

The lactose content of traditional fermented dairy products still does not meet the needs of lactose-intolerant people. Lactase can be added to further hydrolyze lactose before (pre-hydrolysis) or during (co-hydrolysis) fermentation. These two hydrolysis methods will affect the fermentation characteristics of yogurt to some extent. Generally, lactose is hydrolyzed into glucose and galactose by lactase, and lactic acid bacteria can directly utilize glucose to produce lactic acid, thus shortening the fermentation time. In the meantime, lactic acid bacteria produce more exopolysaccharides and possibly result in higher viscosity. During storage, a large number of flavor substances such as acetaldehyde and 2,3-butanedione were synthesized at higher levels in low-lactose yogurt than in regular yogurt. For example, Yamamoto et al. [85] found that, compared with unhydrolyzed milk, *Lactobacillus bulgaricus*, extracellular polysaccharide synthesis and viscosity of pre-hydrolyzed milk significantly increased, which may be related to the decrease in dissolved oxygen and the increase in formic acid concentration caused by utilization of glucose. Martins et al. [86] found that the processing time of yogurt co-fermented with *Bifidobacterium animalis*, *Lactobacillus acidophilus* and typical microorganisms of yogurt and co-hydrolyzed reduced from 4.55 h to 3.68 h. Lactose conversion increased from 15.2% to 97.9%.

Pre-hydrolysis to control the lactose content in the final product relatively easily. For example, Raza et al. [87] produced low-lactose cheese containing prebiotics (lactose content 0.8 g/100 g) by converting lactose into galacto-oligosaccharide by adding exogenous lactase. The appearance and overall acceptability of this product are similar to control cheese. Rutkowska et al. [88] added commercial lactase to pre-hydrolyze for 24 h and then added fermentation culture to produce lactose-free kefir. The results showed that the lactose content of the product was as low as 0.1 g/100 g and contained double the ketones (especially 3-hydroxy-2-butanone and 2,3-butanedione), which probably contributed to the high intensity of creamy aroma. Lactose-free kefir is sweeter in taste than traditional kefir and is favored by elderly consumers. The content of lactose in co-hydrolyzed yogurt cannot meet the demand for lactose-free products. Most commercial neutral lactases are completely inactivated when pH is <5.5, and the acidity of yogurt reaches this threshold after 2–3 h of fermentation, which can be solved by adding excess lactase or acidic lactase. However, compared with pre-hydrolysis, co-hydrolysis protects the activity of the fermentation strains and preserves the flavor of fermented milk. Some bacteria may not adapt to the change in the main carbon source from lactose to glucose and the increase in osmotic pressure of milk after rapid hydrolysis of lactose during pre-hydrolysis, which inhibits the activity of lactic acid bacteria to a certain extent. Popescu et al. [89] compared the effects of unhydrolyzed, pre-hydrolyzed and co-hydrolyzed on sensory characteristics of yogurt, and the results showed that the co-hydrolyzed yogurt had the best flavor, which was speculated to be related to the production of more aromatic compounds. Ibrahim et al. [90] evaluated the effects of pre-hydrolysis and co-hydrolysis on the sensory and physicochemical properties of fermented camel milk, and the results showed that the number of bacteria of both the two were higher than unhydrolyzed milk. The decreasing speed of pH and increase in apparent viscosity of hydrolyzed milk was significant, and the co-hydrolyzed camel milk had the best sensory scores. In addition, starter culture highly affected the fermentation characteristics of dairy products. Schmidt et al. [91] researched the changes in the rheological properties of yogurt under pre-hydrolysis and co-hydrolysis conditions, and the results showed that the apparent viscosity of yogurt was affected more by the starter strains than by the hydrolysis method.

5. Detection and Determination of Lactose

Traditionally, detection of lactose in dairy products is carried out using different methods, such as gravimetry, polarimetry, enzymatic methods and high-performance liquid chromatography (HPLC). These methods have been proven to perform well in the un-hydrolyzed milk system. However, when it comes to low-lactose/lactose-free dairy products, they suffer from various drawbacks [39]. The formation of other saccharides during hydrolysis and such low concentration of lactose hinder lactose determination in lactose-free dairy products. There is also a requirement for methods with high sensitivity and precision. Currently available methods are basically improved by enzymatic methods and chromatography. Enzymatic kits and biosensors are based on enzymatic reactions and are used frequently at the commercial level because they are fast and easy to use. Mangan et al. [92] described a novel enzymatic low-lactose determination method, which is based on an optimized glucose removal pre-treatment step followed by a sequential enzymatic assay. Sensitivity was improved through the extension of the typical glucose detection biochemical pathway to amplify the signal response. The limit of detection (LOD) and limit of quantification (LOQ) of this method are 0.13 mg/100 g and 0.44 mg/100 g, respectively.

Chromatography includes a wide range of improved and coupled methods, such as gas chromatography (GC) [93], ultrahigh-performance chromatography coupled with tandem mass spectrometry (UHPLC-MS/MS) [94], high-performance anion-exchange chromatography with pulsed amperometric detection (HPAEC-PAD), high-performance thin-layer chromatography coupled with a fluorescence detector (HPTLC-FLD) [95] and so on. Among all the methods, HPAEC-PAD shows the highest accuracy and specificity without any interference from other disaccharides. Monti et al. [96] reported the obtained LOD and LOQ values were, respectively, 0.25 and 0.41 mg/100 g for lactose, 0.14 and 0.27 mg/100 g for galactose, and 0.16 and 0.26 mg/100 g for glucose in Grana Padano PDO cheese. However, it is quite expensive and therefore not used widely.

Other methods include nuclear magnetic resonance (NMR) [97], capillary zone electrophoresis [98] and depression in freezing point [99]. These methods are applied for the determination of lactose in milk, but still very limited in lactose-free dairy products.

6. Fortification of Lactose-Free Dairy Products

6.1. Function

More and more clinical studies report that probiotics can assist in relieving lactose intolerance [100–102]. Although the specific mechanism is not clear, it is generally believed that changes in microbial flora composition by probiotics and enhancement of immune function are the basis of their effects [103]. Fermented milk is one of the most ideal matrices in the culture of probiotics. *Lactobacillus* (including *Lactobacillus acidophilus*, *Lactobacillus rhamnosus*, *Lactobacillus casei*, etc.), *Bifidobacterium* and *Saccharomyces* and other probiotics are often used in the production and preparation of fermented milk. A systematic review by Oak et al. [104] evaluated the efficacy of eight probiotics in the treatment of lactose intolerance, and the results showed that although the efficacy of different strains in improving intestinal digestion was different, probiotics were generally positively correlated with the alleviation of lactose intolerance. A randomized, double-blind, crossover study by Vitellio et al. [105] showed that formulations of *Bifidobacterium longum* BB536, *Lactobacillus rhamnosus* HN001, and vitamin B6 together significantly improved bloating and constipation in lactose-intolerant patients. Compared with placebo, probiotics drove the enrichment of bacteria involved in lactose digestion in the gut and also produced a few lactases to promote the hydrolysis of lactose in the human body.

Prebiotics are also functionally beneficial for gut health. However, compared with probiotics, there are few studies on improving lactose intolerance using prebiotics alone [106]. At present, only RP-G28 (a galacto-oligosaccharide of more than 95% purity) has been reported. Savaiano et al. [107] reported RP-G28 or placebo was administered to 85 patients with lactose intolerance for 35 days and RP-G28 subjects were six times more likely to claim lactose tolerance post-treatment. Similarly, a randomized, double-blind, placebo-controlled clinical trial by Chey et al. [108] also found that 30 days of treatment with RP-G28 reduced abdominal pain by 50% in lactose-intolerant patients, which was six times more effective than placebo. Presumably, selective utilization of prebiotics by host microorganisms increased the survival of probiotics in the gastrointestinal tract. That is why more studies are focusing on the combination of probiotics and prebiotics. Commonly used prebiotics include galacto-oligosaccharide, fructooligosaccharide (FOS), inulin, etc. For example, Pereira et al. [109] used *Lactobacillus acidophilus* LA-5, *Bifidobacterium lactis* Bb-12 and inulin to produce fermented milk with lactose below 0.1%, and this product had a high sensory score and probiotic activity. Table 4 lists some studies of lactose-free fermented dairy products supplemented with probiotics/prebiotics. The focus of these articles was ensuring the viability of probiotics in the products (more than 10^6 CFU/g at least) to promote lactose digestion and absorption.

Table 4. Lactose-free fermented dairy product supplemented with probiotics/prebiotics.

Product	Study	Conclusion	References
Low-lactose fermented goat milk	Development of low-lactose fermented goat milks with *Bifidobacterium animalis* ssp. lactis Bb-12 and evaluate the effect of prior lactose hydrolysis on the viability of *Bifidobacterium animalis* ssp lactis Bb-12.	The lactose hydrolysis of milk resulted a higher hardness in probiotic fermented goat milk. Moreover, the lactose-free probiotic fermented milk had a more distinct sweet taste than the control one and was characterized by a less sour flavor.	[110]
Lactose-free functional yogurt	Physicochemical, rheological, and microbiological properties of lactose-free functional yogurt supplemented with FOS.	Lactose hydrolysis and FOS supplementation increased acidification rate during fermentation of yogurts. FOS helped to improve syneresis.	[111]
Concentrated lactose-free yogurt	Effect of encapsulated *Bifidobacterium* Bb-12 on the lactose-free yogurt.	Viability of *Bifidobacterium* Bb-12 was found for all spray-dried powders produced with lactose-free skim milk powder, lactose-free skim milk powder and inulin, and lactose-free skim milk powder and oligofructose to be higher than recommended to exert health benefits.	[112]
Lactose-free Greek-style yogurt	Evaluation of potential of lactose-free Greek-style yogurt as probiotic matrix.	Three different microcapsule formulations were produced using gum arabic, inulin and maltodextrin as wall materials. All formulations showed encapsulation yield above 96% and good probiotic viability (>8 log cfu/g) throughout 30 days of storage (4 °C).	[113]
Probiotic Edam cheese	Influence of *Bifidobacterium bifidum* on cheese.	Lactose in control as well as in experimental cheeses (10^7 viable cell) was depleted within 15 days. The free fatty acids increased from 2.23% and 2.31% on 0-day to 2.78% and 2.83% after 3 months, in control and probiotic cheeses, respectively.	[114]
Lactose-free fermented dairy beverages	Influence of co-cultures of *Streptococcus thermophilus* and probiotic lactobacilli on quality and antioxidant capacity parameters of lactose-free fermented dairy beverages containing *Syzygium cumini* (L.) skeels pulp.	Viability of bacteria are above 7 log CFU/g and total phenolic content around 40 mg GAE/100 g. The dairy beverages are good options for functional foods due to its nutritional value, viability of probiotic lactobacilli, phenolic content, and antioxidant capacity, also serving lactose-intolerant people.	[115]

6.2. Nutrition

Milk is a good source of calcium, vitamin B2, vitamin A, and vitamin D. Consumption of 250 mL milk provides 26–40%, 23–52%, 10–24% and about 5% of mineral and vitamins US RDA (recommended dietary allowance) recommendations, respectively. However, lactose-intolerant people often avoid eating dairy products. It is also difficult for them to effectively utilize the rich minerals, vitamins and other nutrients in milk, thus resulting in osteoporosis and other adverse health consequences. Although human clinical trials cannot confirm the effect of lactose on enhancing calcium bioavailability [116], there are numerous animal studies supporting lactose as an enhancer of calcium absorption. Lactose-free products should be fortified to contain at least 20% more calcium than the recommended calcium intake. Lactose-free dairy products should differ from skim dairy products only in lactose content, and can basically meet the nutritional needs of lactose-intolerant patients and other consumers. Within the limits of standard regulations, lactose-free products can be fortified with additional nutrients to satisfy the specific physiological needs of lactose-intolerant individuals. In the US and Canada, milk is mandatorily fortified and is an important source of vitamin D. Vitamin D food fortification policy, started in 2003, recommends that all liquid dairy products, lactose-free milk-, soy-, and cereal-based drinks are to be fortified at a concentration of 0.5 μg/100 g. Jaaskelainen et al. [117] analyzed the health survey data of 6134 and 4051 Finnish adults from 2000 and 2011, and found that the average serum 25-hydroxyvitamin D (S-25(OH)D) concentration increased from 48 nmol/L to 65 nmol/L. This increase is mainly explained by food fortification, especially of fluid milk products.

In addition, many lactose-intolerant consumers turn to plant-based alternatives after finding it hard to digest dairy products. Many plant-based beverages have even higher levels of fortified calcium than dairy products. A review compared nutrient density in milk and 17 plant-based beverages, with milk containing 120 mg/100 mL calcium and fortified plant-based beverages containing 42–197 mg/100 mL calcium [118]. However, the bioavailability of calcium supplements in plant-based beverages was inferior to that in cow's milk, where the absorption of calcium triphosphate was only 75% of calcium in milk. According to Heaney et al. [119], calcium precipitates can also occur in plant-based beverages. The average calcium content of calcium-fortified soy drinks after shaking is only 59% of what is claimed on the label, while the average calcium content of unshaken soy drinks is only 31%.

Apart from direct addition of calcium, phosphorus, and vitamin D, there are other ways to increase the nutritional value of lactose-free dairy products. For example, Dantas et al. [120] increased the carbohydrate and protein content of lactose-free milk by 2.95 and 3.00 times, respectively, through a freezing concentration process. da Silva et al. [121] replaced milk powder with lactose-free whey protein concentrate, which increased the protein and calcium content of Greek yogurt while improving the rheological and sensory properties of the product.

7. Improvement in Sensory Properties and Quality of Lactose-Free Dairy Products

Since consumers are easily driven by the senses, lactose-free dairy products manufacturers attached great importance to sensory properties. Lactose-free dairy products similar to skim milk are believed to be more popular. However, commercially available lactose-free milk typically utilizes lactase to hydrolyze lactose into galactose and glucose, so it tends to taste sweeter than regular milk. Commercial lactase can also have proteolytic activity that releases peptides and free amino acids generating nonenzymic browning and "off" flavors during the shelf life. The extent of proteolysis and product deterioration identified depended on the lactase preparation used and its purity [122]. However, research on the influence of microbial sources on proteolytic activity is very limited. Nielsen et al. [123] compared five commercial lactase preparations from different companies and of different purity. Among them, lactase preparations with the lowest purification gave rise to the highest degree of proteolysis and aggregation. Thus, lactose-free milk was more susceptible to Maillard reaction, and storage increased hexosylation up to elevenfold in lactose-free UHT milk [124]. Lactose-free milk is also often described as having a cooked and eggy flavor. A recent study also showed that the volatile sulfur compounds that create eggy flavor in milk result from Maillard reactions between reducing sugars (lactose) and cysteine and methionine amino acids [125]. When lactose-free milk is produced by a batch process, the lactase and its proteolytic activity disappear after heat treatment. In contrast, under an aseptic process, lactase remain active throughout the shelf life, which could damage the quality of the product during storage. Tossavainen O et al. [126] found that both batch and aseptic processes resulted in proteolytic hydrolysis compared to UHT milk without lactose hydrolysis. However, when the storage temperature was below 5 °C, the proteolytic activity was inhibited. Therefore, the use of commercial lactase preparations with high purity and strict control of storage temperature can ensure the quality of lactose-free milk during the shelf life.

It is generally believed that fermentation can not only improve the flavor of products but also increase functional substances or the content of nutrients in products [127]. The sensory evaluation of many lactose-free yogurts is better than that of unhydrolyzed yogurt. On one hand, this may be because lactic acid bacteria can directly use glucose to produce more aromatic compounds, alcohols, esters, aldehydes, ketones and other flavor substances. On the other hand, more exopolysaccharides are produced and increase the apparent viscosity and improve dehydration characteristics. In addition, some additives can also improve the quality of lactose-free fermented products. For example, Moreira et al. [128] prepared a low-lactose yogurt with added fiber (*Ceratonia siliqua* L.) with content of 1.16–1.44 g/100 g, which had good acceptability. Synthesis of GOS using lactase with high trans-glycosylation activity is another common method to increase the added value and improve the flavor of dairy products. For example, Raza et al. [87] produced prebiotic-enriched cheese using lactase from *Kluveromyces lactis*. Lactose content decreased to 56.25%. The prebiotic cheese has a similar appearance and overall acceptability, as does the control cheese, except for taste and texture, which were improved by trans-glycosylation.

Hydrolyzed milk can further produce lactose-free milk powder, but there are still many problems in the production of lactose-free milk powder. Lactose-free milk powder is more hygroscopic, which is related to the fact that there are more molecules in the amorphous state (glucose and galactose) during drying. Shrestha et al. [129] showed that spray-drying of skim milk with hydrolyzed lactose resulted in very low cyclone recovery of 25% and a large amount of powder remained stuck inside the spray dryer. Torres et al. [130] found that with the increase in lactose hydrolysis rate, adhesion to the drying chamber also increased, due to higher levels of particle agglomeration. In addition, due to the increase of reducing sugar in hydrolyzed milk powder, nonenzymatic browning in high-temperature processing occurs easily, resulting in increased Maillard reaction products and odor. For example, Queiroz et al. [131] found that lactose hydrolysis of goat's milk resulted in greater darkening and increased free fat content. Naranjo et al. [132] researched the kinetics of Maillard reaction in lactose-hydrolyzed milk powder and found that hydrolyzed milk powder was prone to protein deterioration, mainly because galactose reacted faster with lysine than lactose in dairy products. At the same time, temperature is the most important factor. Lower temperatures can reduce the deterioration of nutrients during storage.

In the production of lactose-free ice cream, since the solubility of lactose at room temperature is only 20% of that of sucrose, there are difficulties, including crystallization of α-lactose during freezing and formation of sandy texture. Another problem in lactose-free ice cream production is the low sweetness of lactose, which is about 20% of sucrose. Membrane was combined with enzymatic method to solve the problem of the high sweetness of lactose-free milk [133]. Furthermore, when preparing lactose-free ice cream, the addition of lactase can be used as a sugar-reduction method, because hydrolysis of 70% of the lactose in milk increased the sweetness of milk or yogurt to the same degree as adding 2% sugar [134]. For example, Abbasi et al. [135] showed that with the increase in lactose hydrolysis, the apparent viscosity increased and freezing point decreased, and the sensory properties of ice cream hydrolyzed with 75% lactose were similar to those of ice cream with 25% sugar reduction.

8. Conclusions

With the awakening of consumer health awareness, lactose-intolerant people tend to choose healthier and safer lactose-free dairy products. Dairy enterprises and manufacturers have improved processes and technologies, using immobilized enzymes, genetic engineering, membrane filtration, and fermentation to remove lactose. Different methods have their own advantages and disadvantages. The right choices and combination of several methods can improve sensory properties, decrease the cost of production, and increase the nutritional value and functional effect under the premise of low lactose content. Research in the future will focus on the improvement in production and development of lactose-free dairy products of high quality, so as to provide more choices for lactose-intolerant patients.

Author Contributions: J.Z. (Jie Zheng): conceptualization, writing—original draft, writing—review and editing. X.H., S.Y., S.C., J.Z. (Jingwen Zhao) and W.Z.: conceptualization, writing—original draft. A.L. and Y.L.: conceptualization, writing—review and editing, supervision, and project administration. All authors have read and agreed to the published version of the manuscript.

Funding: This study was supported by the National Natural Science Foundation of China (grant 82008600) and the Platform Project of the Research Institute of Heilongjiang Province for the Study of Standard and Inspection of the National Dairy Technical Innovation Center (CZKYF2020A002).

Data Availability Statement: No new data were created or analyzed in this study. Data sharing is not applicable to this article.

Conflicts of Interest: The authors declare no conflict of interest.

References

1. Storhaug, C.L.; Fosse, S.K.; Fadnes, L.T. Country, regional, and global estimates for lactose malabsorption in adults: A systematic review and meta-analysis. *Lancet Gastroenterol. Hepatol.* **2017**, *2*, 738–746. [CrossRef]
2. Li, A.; Zheng, J.; Han, X.; Jiang, Z.; Yang, B.; Yang, S.; Zhou, W.; Li, C.; Sun, M. Health implication of lactose intolerance and updates on its dietary management. *Int. Dairy J.* **2023**, *140*, 105608. [CrossRef]
3. Sharp, E.; D'Cunha, N.M.; Ranadheera, C.S.; Vasiljevic, T.; Panagiotakos, D.B.; Naumovski, N. Effects of lactose-free and low-lactose dairy on symptoms of gastrointestinal health: A systematic review. *Int. Dairy J.* **2021**, *114*, 104936. [CrossRef]
4. Taeger, M.; Thiele, S. Additional costs of lactose-reduced diets: Lactose-free dairy product substitutes are a cost-effective alternative for people with lactose intolerance. *Public Health Nutr.* **2021**, *24*, 4043–4053. [CrossRef]
5. Szabó, E.; Szakos, D.; Kasza, G.; Ózsvári, L. Analysis of the target group of lactose-free functional foods for product development. *Acta Aliment.* **2021**, *50*, 153–161. [CrossRef]
6. Mygdalia, A.; Sfetsas, T.; Dimitropoulou, G.; Zioupou, S.; Mitsopoulos, T.; Lithoxopoulos, P.; Ioannou, C.; Katsantonis, D. Recipe for Brown Rice Milk-based Vegan Ice Cream. *Asian Food Sci. J.* **2023**, *22*, 33–39. [CrossRef]
7. Suri, S.; Kumar, V.; Prasad, R.; Tanwar, B.; Goyal, A.; Kaur, S.; Gat, Y.; Kumar, A.; Kaur, J.; Singh, D. Considerations for development of lactose-free food. *J. Nutr. Intermed. Metab.* **2019**, *15*, 27–34. [CrossRef]
8. Bartuzi, M.; Szamocka, M.; Ukleja-Sokołowska, N. Social and economic difficulties of patients with food intolerances. *Adv. Dermatol. Allergol./Postępy Dermatol. I Alergol.* **2023**, *40*, 298–307. [CrossRef] [PubMed]
9. López Ruiz, Á.L.; Ruiz Morales, F.d.A.; Ruiz Pérez-Cacho, P.; Galán-Soldevilla, H. Effect of Lactose-Reduction in Murciano-Granadina Semi-Hard Goat Cheese on Physicochemical and Sensory Characteristics. *Foods* **2023**, *12*, 996. [CrossRef]
10. Vasudha, M.; Gayathri, D. Kinetic and modeling analyses of lactose-hydrolyzing β-galactosidase from *Lactiplantibacillus plantarum* GV54. *World Acad. Sci. J.* **2023**, *5*, 11. [CrossRef]
11. Kızıldağ, S.; Işık, C.; Teke, M. Milk lactose removal by β-galactosidase immobilized on eggshell membrane. *Eur. Food Res. Technol.* **2023**, 1–12. [CrossRef]
12. Stouten, K.; Wolfhagen, F.; Castel, R.; van de Werken, M.; Klerks, J.; Verheijen, F.; Vermeer, H.J. Testing for lactase non-persistence in a Dutch population: Genotyping versus the hydrogen breath test. *Ann. Clin. Biochem.* **2023**, 00045632231159288. [CrossRef]
13. Anguita-Ruiz, A.; Aguilera, C.M.; Gil, Á. Genetics of lactose intolerance: An updated review and online interactive world maps of phenotype and genotype frequencies. *Nutrients* **2020**, *12*, 2689. [CrossRef] [PubMed]
14. Chin, E.L.; Huang, L.; Bouzid, Y.Y.; Kirschke, C.P.; Durbin-Johnson, B.; Baldiviez, L.M.; Bonnel, E.L.; Keim, N.L.; Korf, I.; Stephensen, C.B. Association of lactase persistence genotypes (rs4988235) and ethnicity with dairy intake in a healthy US population. *Nutrients* **2019**, *11*, 1860. [CrossRef]
15. Ingram, C.J.; Elamin, M.F.; Mulcare, C.A.; Weale, M.E.; Tarekegn, A.; Raga, T.O.; Bekele, E.; Elamin, F.M.; Thomas, M.G.; Bradman, N. A novel polymorphism associated with lactose tolerance in Africa: Multiple causes for lactase persistence? *Hum. Genet.* **2007**, *120*, 779–788. [CrossRef]
16. Pinto, J.C.; Oliveira, S.; Teixeira, S.; Martins, D.; Fehn, A.M.; Aço, T.; Gayà-Vidal, M.; Rocha, J. Food and pathogen adaptations in the Angolan Namib desert: Tracing the spread of lactase persistence and human African trypanosomiasis resistance into southwestern Africa. *Am. J. Phys. Anthropol.* **2016**, *161*, 436–447. [CrossRef] [PubMed]
17. Priehodová, E.; Abdelsawy, A.; Heyer, E.; Černý, V. Lactase persistence variants in Arabia and in the African Arabs. *Hum. Biol.* **2014**, *86*, 7–18. [CrossRef]
18. Choi, B.J.; Yoon, J.H.; Choi, Y.J.; Han, L.; Park, Y.G.; Park, W.S. No association of LCT-13910 single nucleotide polymorphism with gastroenteritis in Korean children. *Mol. Cell Toxicol.* **2013**, *9*, 23–28. [CrossRef]

19. Peng, M.-S.; He, J.-D.; Zhu, C.-L.; Wu, S.-F.; Jin, J.-Q.; Zhang, Y.-P. Lactase persistence may have an independent origin in Tibetan populations from Tibet, China. *J. Hum. Genet.* **2012**, *57*, 394–397. [CrossRef]
20. Olds, L.C.; Ahn, J.K.; Sibley, E. 13915*G DNA polymorphism associated with lactase persistence in Africa interacts with Oct-1. *Hum. Genet.* **2011**, *129*, 111–113. [CrossRef]
21. Jensen, T.G.K.; Liebert, A.; Lewinsky, R.; Swallow, D.M.; Olsen, J.; Troelsen, J.T. The −14010*C variant associated with lactase persistence is located between an Oct-1 and HNF1α binding site and increases lactase promoter activity. *Hum. Genet.* **2011**, *130*, 483–493. [CrossRef]
22. Leseva, M.N.; Grand, R.J.; Klett, H.; Boerries, M.; Busch, H.; Binder, A.M.; Michels, K.B. Differences in DNA methylation and functional expression in lactase persistent and non-persistent individuals. *Sci. Rep.* **2018**, *8*, 5649. [CrossRef] [PubMed]
23. Labrie, V.; Buske, O.J.; Oh, E.; Jeremian, R.; Ptak, C.; Gasiūnas, G.; Maleckas, A.; Petereit, R.; Žvirbliene, A.; Adamonis, K. Lactase nonpersistence is directed by DNA-variation-dependent epigenetic aging. *Nat. Struct. Mol. Biol.* **2016**, *23*, 566–573. [CrossRef]
24. Lopera-Maya, E.A.; Kurilshikov, A.; van der Graaf, A.; Hu, S.; Andreu-Sánchez, S.; Chen, L.; Vila, A.V.; Gacesa, R.; Sinha, T.; Collij, V. Effect of host genetics on the gut microbiome in 7738 participants of the Dutch Microbiome Project. *Nat. Genet.* **2022**, *54*, 143–151. [CrossRef]
25. Bonder, M.J.; Kurilshikov, A.; Tigchelaar, E.F.; Mujagic, Z.; Imhann, F.; Vila, A.V.; Deelen, P.; Vatanen, T.; Schirmer, M.; Smeekens, S.P. The effect of host genetics on the gut microbiome. *Nat. Genet.* **2016**, *48*, 1407–1412. [CrossRef]
26. Kurilshikov, A.; Medina-Gomez, C.; Bacigalupe, R.; Radjabzadeh, D.; Wang, J.; Demirkan, A.; Le Roy, C.I.; Raygoza Garay, J.A.; Finnicum, C.T.; Liu, X. Large-scale association analyses identify host factors influencing human gut microbiome composition. *Nat. Genet.* **2021**, *53*, 156–165. [CrossRef]
27. Goodrich, J.K.; Davenport, E.R.; Beaumont, M.; Jackson, M.A.; Knight, R.; Ober, C.; Spector, T.D.; Bell, J.T.; Clark, A.G.; Ley, R.E. Genetic determinants of the gut microbiome in UK twins. *Cell Host Microbe* **2016**, *19*, 731–743. [CrossRef] [PubMed]
28. Moon, J.-Y.; Wang, Z.; Usyk, M.; Vazquez-baeza, Y.; Isasi, C.R.; Mossavar-Rahmani, Y.; McDonald, D.; Sotres-Alvarez, D.; Daviglus, M.L.; Cai, J. Abstract P459: Milk Intake, Host LCT Genotype and Gut Bifidobacteria in Relation to Obesity: Results from the Hispanic Community Health Study/Study of Latinos (HCHS/SOL). *Circulation* **2020**, *141*, AP459. [CrossRef]
29. Dekker, P.J.T.; Koenders, D.; Bruins, M.J. Lactose-free dairy products: Market developments, production, nutrition and health benefits. *Nutrients* **2019**, *11*, 551. [CrossRef] [PubMed]
30. Obermayer-Pietsch, B.M.; Bonelli, C.M.; Walter, D.E.; Kuhn, R.J.; Fahrleitner-Pammer, A.; Berghold, A.; Goessler, W.; Stepan, V.; Dobnig, H.; Leb, G. Genetic predisposition for adult lactose intolerance and relation to diet, bone density, and bone fractures. *J. Bone Miner. Res.* **2004**, *19*, 42–47. [CrossRef]
31. Joslin, S.E.; Durbin-Johnson, B.P.; Britton, M.; Settles, M.L.; Korf, I.; Lemay, D.G. Association of the lactase persistence haplotype block with disease risk in populations of European descent. *Front. Genet.* **2020**, *11*, 558762. [CrossRef] [PubMed]
32. Hilliard, C.B. High osteoporosis risk among East Africans linked to lactase persistence genotype. *BoneKEy Rep.* **2016**, *5*, 803. [CrossRef] [PubMed]
33. Tolonen, S. Associations of Lactase and Apolipoprotein E Gene Polymorphisms and Physical Activity with Peripheral Bone Traits. Ph.D. Thesis, University of Helsinki, Helsinki, Finland, 2018.
34. Nguyen, N. An update of the milk market in Vietnam: Lactose-free dairy opportunity. *Theseus* **2014**.
35. Wang, X.; Wang, Y.H.; Yang, H.X.; Chuang, S.C.; Zhou, C.F.; Yu, X.; Zhang, H. *Lysobacter lactosilyticus* sp. nov., a Novel beta-Galactosidase Producing Bacterial Strain Isolated from Farmland Soil Applied with Amino Acid Fertilizer. *Curr. Microbiol.* **2022**, *80*, 43. [CrossRef]
36. Rizzo, P.; Harwood, W.; Drake, M. Consumer desires and perceptions of lactose-free milk. *J. Dairy Sci.* **2020**, *103*, 6950–6966. [CrossRef] [PubMed]
37. Świąder, K.; Kulawiak, M.; Chen, Y.-P. Types of lactose-free products and their availability on the Polish market®. *Postępy Tech. Przetwórstwa Spożywczego* **2020**, *1*, 39–45.
38. EFSA Panel on Dietetic Products, Nutrition and Allergies. Scientific opinion on lactose thresholds in lactose intolerance and galactosaemia. *EFSA J.* **2010**, *8*, 1777.
39. Rao, P.S.; Singh, P.; Sharma, V.; Arora, S. Traditional analytical approaches for lactose residues determination in lactose hydrolysed milks: A review. *LWT* **2021**, *151*, 112069. [CrossRef]
40. Facioni, M.S.; Dominici, S.; Marescotti, F.; Covucci, R.; Taglieri, I.; Venturi, F.; Zinnai, A. Lactose Residual Content in PDO Cheeses: Novel Inclusions for Consumers with Lactose Intolerance. *Foods* **2021**, *10*, 2236. [CrossRef]
41. Kumar, P.; Sharma, N.; Ranjan, R.; Kumar, S.; Bhat, Z.; Jeong, D.K. Perspective of membrane technology in dairy industry: A review. *Asian-Australas. J. Anim. Sci.* **2013**, *26*, 1347. [CrossRef]
42. Qi, T.; Yang, D.; Chen, X.; Qiu, M.; Fan, Y. Rapid removal of lactose for low-lactose milk by ceramic membranes. *Sep. Purif. Technol.* **2022**, *289*, 120601. [CrossRef]
43. Sanchez-Moya, T.; Hidalgo, A.M.; Ros-Berruezo, G.; Lopez-Nicolas, R. Screening ultrafiltration membranes to separate lactose and protein from sheep whey: Application of simplified model. *J. Food Sci. Technol.* **2020**, *57*, 3193–3200. [CrossRef] [PubMed]
44. Khaire, R.A.; Sunny, A.A.; Gogate, P.R. Ultrasound assisted ultrafiltration of whey using dual frequency ultrasound for intensified recovery of lactose. *Chem. Eng. Process.-Process Intensif.* **2019**, *142*, 107581. [CrossRef]
45. Sofuwani, Z.N.; Aslina, H.S.; Mazlina, M.S. Separation of lactose from raw goat's milk by cross-flow hollow fiber ultrafiltration membrane. *Int. Food Res. J.* **2016**, *23*, S209.

46. Morelos-Gomez, A.; Terashima, S.; Yamanaka, A.; Cruz-Silva, R.; Ortiz-Medina, J.; Sánchez-Salas, R.; Fajardo-Díaz, J.L.; Muñoz-Sandoval, E.; López-Urías, F.; Takeuchi, K.; et al. Graphene oxide membranes for lactose-free milk. *Carbon* **2021**, *181*, 118–129. [CrossRef]
47. Zhang, H.; Tao, Y.; He, Y.; Pan, J.; Yang, K.; Shen, J.; Gao, C. Preparation of Low-Lactose Milk Powder by Coupling Membrane Technology. *ACS Omega* **2020**, *5*, 8543–8550. [CrossRef]
48. Cordova, A.; Astudillo, C.; Vera, C.; Guerrero, C.; Illanes, A. Performance of an ultrafiltration membrane bioreactor (UF-MBR) as a processing strategy for the synthesis of galacto-oligosaccharides at high substrate concentrations. *J. Biotechnol.* **2016**, *223*, 26–35. [CrossRef]
49. Dantas, A.; Quinteros, G.J.; Darvishvand, S.Y.; Blanco, M.; Hernandez, E.; Prudencio, E.S.; Samsuri, S. The combined use of progressive and block freeze concentration in lactose-free milk: Effect of process parameters and influence on the content of carbohydrates and proteins. *J. Food Process Eng.* **2021**, *44*, e13867. [CrossRef]
50. Batista, K.A.; Silva, C.N.S.; Fernandes, P.M.; Campos, I.T.N.; Fernandes, K.F. Development of a new bioaffinity stationary phase for lactose removal using a lactose-binding lectin immobilized onto polyaniline. *Sep. Purif. Technol.* **2017**, *185*, 54–60. [CrossRef]
51. Zolnere, K.; Ciprovica, I. The comparison of commercially available β-galactosidases for dairy industry. *Res. Rural. Dev.* **2017**, *1*, 215–222.
52. De Albuquerque, T.L.; de Sousa, M.; Gomes, E.S.N.C.; Girao Neto, C.A.C.; Goncalves, L.R.B.; Fernandez-Lafuente, R.; Rocha, M.V.P. Beta-Galactosidase from *Kluyveromyces lactis*: Characterization, production, immobilization and applications-A review. *Int. J. Biol. Macromol.* **2021**, *191*, 881–898. [CrossRef] [PubMed]
53. Erich, S.; Kuschel, B.; Schwarz, T.; Ewert, J.; Böhmer, N.; Niehaus, F.; Eck, J.; Lutz-Wahl, S.; Stressler, T.; Fischer, L. Novel high-performance metagenome β-galactosidases for lactose hydrolysis in the dairy industry. *J. Biotechnol.* **2015**, *210*, 27–37. [CrossRef]
54. Liu, P.; Xie, J.; Liu, J.; Ouyang, J. A novel thermostable beta-galactosidase from *Bacillus coagulans* with excellent hydrolysis ability for lactose in whey. *J. Dairy Sci.* **2019**, *102*, 9740–9748. [CrossRef]
55. Ren, G.H.; Cao, L.C.; Kong, W.; Wang, Z.J.; Liu, Y.H. Efficient Secretion of the beta-Galactosidase Bgal1-3 via both Tat-Dependent and Tat-Independent Pathways in *Bacillus subtilis*. *J. Agric. Food Chem.* **2016**, *64*, 5708–5716. [CrossRef]
56. Mano, M.C.R.; Neri-Numa, I.A.; da Silva, J.B.; Paulino, B.N.; Pessoa, M.G.; Pastore, G.M. Oligosaccharide biotechnology: An approach of prebiotic revolution on the industry. *Appl. Microbiol. Biotechnol.* **2018**, *102*, 17–37. [CrossRef] [PubMed]
57. Yao, C.; Sun, J.; Wang, W.; Zhuang, Z.; Liu, J.; Hao, J. A novel cold-adapted β-galactosidase from *Alteromonas* sp. ML117 cleaves milk lactose effectively at low temperature. *Process Biochem.* **2019**, *82*, 94–101. [CrossRef]
58. Murphy, J.; Ryan, M.P.; Walsh, G. Purification and Characterization of a Novel beta-Galactosidase from the Thermoacidophile *Alicyclobacillus vulcanalis*. *Appl. Biochem. Biotechnol.* **2020**, *191*, 1190–1206. [CrossRef]
59. Acer, Ö.; Matpan Bekler, F. Characterization of a Thermally Stable β-galactosidase Produced by Thermophilic *Anoxybacillus* sp. AH1. *Türk Doğa Ve Fen Derg.* **2021**, *10*, 130–136. [CrossRef]
60. O'Connell, S.; Walsh, G. A novel acid-stable, acid-active beta-galactosidase potentially suited to the alleviation of lactose intolerance. *Appl. Microbiol. Biotechnol.* **2010**, *86*, 517–524. [CrossRef]
61. Mulualem, D.M.; Agbavwe, C.; Ogilvie, L.A.; Jones, B.V.; Kilcoyne, M.; O'Byrne, C.; Boyd, A. Metagenomic identification, purification and characterisation of the *Bifidobacterium adolescentis* BgaC beta-galactosidase. *Appl. Microbiol. Biotechnol.* **2021**, *105*, 1063–1078. [CrossRef]
62. Zhang, Z.; Zhang, F.; Song, L.; Sun, N.; Guan, W.; Liu, B.; Tian, J.; Zhang, Y.; Zhang, W. Site-directed mutation of beta-galactosidase from Aspergillus candidus to reduce galactose inhibition in lactose hydrolysis. *3 Biotech* **2018**, *8*, 452. [CrossRef]
63. Huang, J.; Zhu, S.; Zhao, L.; Chen, L.; Du, M.; Zhang, C.; Yang, S.T. A novel beta-galactosidase from *Klebsiella oxytoca* ZJUH1705 for efficient production of galacto-oligosaccharides from lactose. *Appl. Microbiol. Biotechnol.* **2020**, *104*, 6161–6172. [CrossRef] [PubMed]
64. Deng, P.; Meng, C.; Wu, Y.; Xu, J.; Tang, X.; Zhang, X.; Xiao, Y.; Wang, X.; Fang, Z.; Fang, W. An unusual GH1 beta-glucosidase from marine sediment with beta-galactosidase and transglycosidation activities for superior galacto-oligosaccharide synthesis. *Appl. Microbiol. Biotechnol.* **2020**, *104*, 4927–4943. [CrossRef] [PubMed]
65. Liu, Y.; Chen, Z.; Jiang, Z.; Yan, Q.; Yang, S. Biochemical characterization of a novel beta-galactosidase from *Paenibacillus barengoltzii* suitable for lactose hydrolysis and galactooligosaccharides synthesis. *Int. J. Biol. Macromol.* **2017**, *104*, 1055–1063. [CrossRef]
66. Li, S.; Zhu, X.; Xing, M. A New beta-Galactosidase from the Antarctic Bacterium *Alteromonas* sp. ANT48 and Its Potential in Formation of Prebiotic Galacto-Oligosaccharides. *Mar. Drugs* **2019**, *17*, 599. [CrossRef]
67. Zhao, J.C.; Mu, Y.L.; Gu, X.Y.; Xu, X.N.; Guo, T.T.; Kong, J. Site-directed mutation of beta-galactosidase from *Streptococcus thermophilus* for galactooligosaccharide-enriched yogurt making. *J. Dairy Sci.* **2022**, *105*, 940–949. [CrossRef]
68. Damin, B.; Kovalski, F.; Fischer, J.; Piccin, J.; Dettmer, A. Challenges and perspectives of the β-galactosidase enzyme. *Appl. Microbiol. Biotechnol.* **2021**, *105*, 5281–5298. [CrossRef] [PubMed]
69. Ureta, M.M.; Martins, G.N.; Figueira, O.; Pires, P.F.; Castilho, P.C.; Gomez-Zavaglia, A. Recent advances in beta-galactosidase and fructosyltransferase immobilization technology. *Crit. Rev. Food Sci. Nutr.* **2021**, *61*, 2659–2690. [CrossRef]
70. Zdarta, J.; Meyer, A.S.; Jesionowski, T.; Pinelo, M. A general overview of support materials for enzyme immobilization: Characteristics, properties, practical utility. *Catalysts* **2018**, *8*, 92. [CrossRef]

71. Aslan, Y.; Taher, A.Y.; Cavidoğlu, İ. Improved catalytic activity of Aspergillus oryzae β-Galactosidase by covalent immobilization on Eupergit CM. *JAPS: J. Anim. Plant Sci.* **2018**, *28*, 1648–1655.
72. Wolf, M.; Gasparin, B.C.; Paulino, A.T. Hydrolysis of lactose using beta-d-galactosidase immobilized in a modified Arabic gum-based hydrogel for the production of lactose-free/low-lactose milk. *Int. J. Biol. Macromol.* **2018**, *115*, 157–164. [CrossRef] [PubMed]
73. Inanan, T. Cryogel disks for lactase immobilization and lactose-free milk production. *LWT* **2022**, *154*, 112608. [CrossRef]
74. Ricardi, N.C.; Arenas, L.T.; Benvenutti, E.V.; Hinrichs, R.; Flores, E.E.E.; Hertz, P.F.; Costa, T.M.H. High performance biocatalyst based on beta-d-galactosidase immobilized on mesoporous silica/titania/chitosan material. *Food Chem.* **2021**, *359*, 129890. [CrossRef] [PubMed]
75. Tizchang, S.; Khiabani, M.S.; Mokarram, R.R.; Hamishehkar, H. Bacterial cellulose nano crystal as hydrocolloid matrix in immobilized β-galactosidase onto silicon dioxide nanoparticles. *LWT* **2020**, *123*, 109091. [CrossRef]
76. Ansari, S.A.; Alshanberi, A.M. Stability studies of β-galactosidase immobilized on gluconic acid coated fullerenes. *Braz. J. Chem. Eng.* **2021**, *39*, 361–367. [CrossRef]
77. Tizchang, S.; Khiabani, M.S.; Mokarram, R.R.; Hamishehkar, H.; Mohammadi, N.S.; Chisti, Y. Immobilization of beta-galactosidase by halloysite-adsorption and entrapment in a cellulose nanocrystals matrix. *Biochim. Biophys Acta Gen. Subj.* **2021**, *1865*, 129896. [CrossRef]
78. Alshanberi, A.M.; Satar, R.; Ansari, S.A. Stabilization of beta-Galactosidase on Modified Gold Nanoparticles: A Preliminary Biochemical Study to Obtain Lactose-Free Dairy Products for Lactose-Intolerant Individuals. *Molecules* **2021**, *26*, 1226. [CrossRef]
79. Savaiano, D.A.; Hutkins, R.W. Yogurt, cultured fermented milk, and health: A systematic review. *Nutr. Rev.* **2021**, *79*, 599–614. [CrossRef]
80. Li, S.; Ye, A.; Singh, H. Effects of seasonal variations on the quality of set yogurt, stirred yogurt, and Greek-style yogurt. *J. Dairy Sci.* **2021**, *104*, 1424–1432. [CrossRef] [PubMed]
81. Dertli, E.; Çon, A.H. Microbial diversity of traditional kefir grains and their role on kefir aroma. *LWT-Food Sci. Technol.* **2017**, *85*, 151–157. [CrossRef]
82. Rosa, D.D.; Dias, M.M.; Grześkowiak, Ł.M.; Reis, S.A.; Conceição, L.L.; Maria do Carmo, G.P. Milk kefir: Nutritional, microbiological and health benefits. *Nutr. Res. Rev.* **2017**, *30*, 82–96. [CrossRef]
83. Malaguarnera, G.; Leggio, F.; Vacante, M.; Motta, M.; Giordano, M.; Biondi, A.; Basile, F.; Mastrojeni, S.; Mistretta, A.; Malaguarnera, M. Probiotics in the gastrointestinal diseases of the elderly. *J. Nutr. Health Aging* **2012**, *16*, 402–410. [CrossRef]
84. Panseri, S.; Pavlovic, R.; Castrica, M.; Nobile, M.; Di Cesare, F.; Chiesa, L.M. Determination of Carbohydrates in Lactose-Free Dairy Products to Support Food Labelling. *Foods* **2021**, *10*, 1219. [CrossRef] [PubMed]
85. Yamamoto, E.; Watanabe, R.; Ichimura, T.; Ishida, T.; Kimura, K. Effect of lactose hydrolysis on the milk-fermenting properties of *Lactobacillus delbrueckii* ssp. *bulgaricus* 2038 and *Streptococcus thermophilus* 1131. *J. Dairy Sci.* **2021**, *104*, 1454–1464. [CrossRef]
86. Martins, A.R.; Monteiro, R.L.; Burkert, J.F.d.M.; Burkert, C.A.V. Simultaneous enzymatic hydrolysis and lactic fermentation to obtain a yogurt with low lactose content. *Ciência E Agrotecnologia* **2012**, *36*, 551–559. [CrossRef]
87. Raza, A.; Iqbal, S.; Shah, F.-u.-H.; Ahmad, Z.; Rehman, M.A.; Waseem, M.; Usman, M. Conversion of milk lactose to galacto-oligosaccharides by enzymes to produce prebiotic enriched cheese. *Future Foods* **2021**, *4*, 100097. [CrossRef]
88. Rutkowska, J.; Antoniewska-Krzeska, A.; Żbikowska, A.; Cazón, P.; Vázquez, M. Volatile composition and sensory profile of lactose-free kefir, and its acceptability by elderly consumers. *Molecules* **2022**, *27*, 5386. [CrossRef] [PubMed]
89. Popescu, L.; Bulgaru, V.; Siminiuc, R. Effects of lactose hydrolysis and milk type on the quality of lactose-free yoghurt. *J. Eng. Sci.* **2022**, *29*, 164–175. [CrossRef]
90. Ibrahim, A.H. Impact of hydrolyzed lactose by β-galactosidase enzyme on the physicochemical and organoleptic properties of fermented camel milk. *Emir. J. Food Agric.* **2018**, *30*, 778–790. [CrossRef]
91. Schmidt, C.; Mende, S.; Jaros, D.; Rohm, H. Fermented milk products: Effects of lactose hydrolysis and fermentation conditions on the rheological properties. *Dairy Sci. Technol.* **2016**, *96*, 199–211. [CrossRef]
92. Mangan, D.; McCleary, B.V.; Culleton, H.; Cornaggia, C.; Ivory, R.; McKie, V.A.; Delaney, E.; Kargelis, T. A novel enzymatic method for the measurement of lactose in lactose-free products. *J. Sci. Food Agric.* **2019**, *99*, 947–956. [CrossRef] [PubMed]
93. Idda, I.; Spano, N.; Ciulu, M.; Nurchi, V.M.; Panzanelli, A.; Pilo, M.I.; Sanna, G. Gas chromatography analysis of major free mono-and disaccharides in milk: Method assessment, validation, and application to real samples. *J. Sep. Sci.* **2016**, *39*, 4577–4584. [CrossRef] [PubMed]
94. Garballo-Rubio, A.; Soto-Chinchilla, J.; Moreno, A.; Zafra-Gómez, A. Determination of residual lactose in lactose-free cow milk by hydrophilic interaction liquid chromatography (HILIC) coupled to tandem mass spectrometry. *J. Food Compos. Anal.* **2018**, *66*, 39–45. [CrossRef]
95. Morlock, G.E.; Morlock, L.P.; Lemo, C. Streamlined analysis of lactose-free dairy products. *J. Chromatogr. A* **2014**, *1324*, 215–223. [CrossRef]
96. Monti, L.; Negri, S.; Meucci, A.; Stroppa, A.; Galli, A.; Contarini, G. Lactose, galactose and glucose determination in naturally "lactose free" hard cheese: HPAEC-PAD method validation. *Food Chem.* **2017**, *220*, 18–24. [CrossRef] [PubMed]
97. Da Cunha, D.A.; Valim, T.C.; Filgueiras, P.R.; Junior, V.L.; Neto, A.C. Lactose quantification in bovine milk by nuclear magnetic resonance without deuterated solvent (No-D qNMR). *Anal. Methods* **2020**, *12*, 4892–4898. [CrossRef]

98. De Oliveira Neves, L.N.; de Oliveira, M.A.L. Quantification of lactose and lactulose in hydrolysed-lactose UHT milk using capillary zone electrophoresis. *Int. Dairy J.* **2020**, *106*, 104710. [CrossRef]
99. Churakova, E.; Peri, K.; Vis, J.S.; Smith, D.W.; Beam, J.M.; Vijverberg, M.P.; Stor, M.C.; Winter, R.T. Accurate analysis of residual lactose in low-lactose milk: Comparing a variety of analytical techniques. *Int. Dairy J.* **2019**, *96*, 126–131. [CrossRef]
100. Cano-Contreras, A.D.; Minero Alfaro, I.J.; Medina López, V.M.; Amieva Balmori, M.; Remes Troche, J.M.; Espadaler Mazo, J.; Perez Lopez, N. Efficacy of i3. 1 probiotic on improvement of lactose intolerance symptoms: A randomized, placebo-controlled clinical trial. *J. Clin. Gastroenterol.* **2022**, *56*, 141–147. [CrossRef]
101. Masoumi, S.J.; Mehrabani, D.; Saberifiroozi, M.; Fattahi, M.R.; Moradi, F.; Najafi, M. The effect of yogurt fortified with *Lactobacillus acidophilus* and *Bifidobacterium* sp. probiotic in patients with lactose intolerance. *Food Sci. Nutr.* **2021**, *9*, 1704–1711. [CrossRef]
102. Gingold-Belfer, R.; Levy, S.; Layfer, O.; Pakanaev, L.; Niv, Y.; Dickman, R.; Perets, T.T. Use of a novel probiotic formulation to alleviate lactose intolerance symptoms—A pilot study. *Probiotics Antimicrob. Proteins* **2020**, *12*, 112–118. [CrossRef]
103. Catanzaro, R.; Sciuto, M.; Marotta, F. Lactose Intolerance—Old and New Knowledge on Pathophysiological Mechanisms, Diagnosis, and Treatment. *SN Compr. Clin. Med.* **2021**, *3*, 499–509. [CrossRef]
104. Oak, S.J.; Jha, R. The effects of probiotics in lactose intolerance: A systematic review. *Crit. Rev. Food Sci. Nutr.* **2019**, *59*, 1675–1683. [CrossRef]
105. Vitellio, P.; Celano, G.; Bonfrate, L.; Gobbetti, M.; Portincasa, P.; De Angelis, M. Effects of *Bifidobacterium longum* and *Lactobacillus rhamnosus* on gut microbiota in patients with lactose intolerance and persisting functional gastrointestinal symptoms: A randomised, double-blind, cross-over study. *Nutrients* **2019**, *11*, 886. [CrossRef] [PubMed]
106. Leis, R.; de Castro, M.-J.; de Lamas, C.; Picáns, R.; Couce, M.L. Effects of prebiotic and probiotic supplementation on lactase deficiency and lactose intolerance: A systematic review of controlled trials. *Nutrients* **2020**, *12*, 1487. [CrossRef] [PubMed]
107. Savaiano, D.A.; Ritter, A.J.; Klaenhammer, T.R.; James, G.M.; Longcore, A.T.; Chandler, J.R.; Walker, W.A.; Foyt, H.L. Improving lactose digestion and symptoms of lactose intolerance with a novel galacto-oligosaccharide (RP-G28): A randomized, double-blind clinical trial. *Nutr. J.* **2013**, *12*, 1–9. [CrossRef] [PubMed]
108. Chey, W.; Sandborn, W.; Ritter, A.J.; Foyt, H.; Azcarate-Peril, M.A.; Savaiano, D.A. Galacto-oligosaccharide RP-G28 improves multiple clinical outcomes in lactose-intolerant patients. *Nutrients* **2020**, *12*, 1058. [CrossRef]
109. Pereira, J.A.; Pinto, S.S.; Dias, C.O.; Vieira, M.P.; Ribeiro, D.H.; Amboni, R.D.; Fritzen-Freire, C.B. Potentially symbiotic fermented milk: A preliminary approach using lactose-free milk. *LWT* **2020**, *118*, 108847. [CrossRef]
110. Pawlos, M.; Znamirowska, A.; Kluz, M.; Szajnar, K.; Kowalczyk, M. Low-lactose fermented goat milks with *Bifidobacterium animalis* ssp. *lactis Bb-12*. *J. Microbiol. Biotechnol. Food Sci.* **2021**, *2021*, 751–755.
111. Pachekrepapol, U.; Somboonchai, N.; Krimjai, W. Physicochemical, rheological, and microbiological properties of lactose-free functional yogurt supplemented with fructooligosaccharides. *J. Food Process. Preserv.* **2021**, *45*, e15017. [CrossRef]
112. Dantas, A.; Verruck, S.; Canella, M.H.M.; Hernandez, E.; Prudencio, E.S. Encapsulated *Bifidobacterium* BB-12 addition in a concentrated lactose-free yogurt: Its survival during storage and effects on the product's properties. *Food Res. Int.* **2021**, *150*, 110742. [CrossRef] [PubMed]
113. Pinto, S.S.; Fritzen-Freire, C.B.; Dias, C.O.; Amboni, R.D. A potential technological application of probiotic microcapsules in lactose-free Greek-style yoghurt. *Int. Dairy J.* **2019**, *97*, 131–138. [CrossRef]
114. Sabikhi, L.; Kumar, M.; Mathur, B. *Bifidobacterium bifidum* in probiotic Edam cheese: Influence on cheese ripening. *J. Food Sci. Technol.* **2014**, *51*, 3902–3909. [CrossRef] [PubMed]
115. Garcia, S.L.A.; da Silva, G.M.; Medeiros, J.M.S.; de Queiroga, A.P.R.; de Queiroz, B.B.; de Farias, D.R.B.; Correia, J.O.; Florentino, E.R.; Buriti, F.C.A. Influence of co-cultures of *Streptococcus thermophilus* and probiotic lactobacilli on quality and antioxidant capacity parameters of lactose-free fermented dairy beverages containing *Syzygium cumini* (L.) Skeels pulp. *RSC Adv.* **2020**, *10*, 10297–10308. [CrossRef]
116. Kwak, H.-S.; Lee, W.-J.; Lee, M.-R. Revisiting lactose as an enhancer of calcium absorption. *Int. Dairy J.* **2012**, *22*, 147–151. [CrossRef]
117. Jääskeläinen, T.; Itkonen, S.T.; Lundqvist, A.; Erkkola, M.; Koskela, T.; Lakkala, K.; Dowling, K.G.; Hull, G.L.; Kröger, H.; Karppinen, J. The positive impact of general vitamin D food fortification policy on vitamin D status in a representative adult Finnish population: Evidence from an 11-y follow-up based on standardized 25-hydroxyvitamin D data. *Am. J. Clin. Nutr.* **2017**, *105*, 1512–1520. [CrossRef]
118. Chalupa-Krebzdak, S.; Long, C.J.; Bohrer, B.M. Nutrient density and nutritional value of milk and plant-based milk alternatives. *Int. Dairy J.* **2018**, *87*, 84–92. [CrossRef]
119. Heaney, R.P.; Rafferty, K. The settling problem in calcium-fortified soybean drinks. *J. Am. Diet. Assoc.* **2006**, *106*, 1753. [CrossRef]
120. Dantas, A.; Orellana-Palma, P.; Kumar, D.; Hernandez, E.; Prudencio, E.S. Block freeze concentration by centrifugation and vacuum increases the content of lactose-free milk macronutrients. *J. Food Sci.* **2022**, *87*, 5317–5329. [CrossRef]
121. Da Silva, A.T.; de Lima, J.J.; Reis, P.; Passos, M.; Baumgartner, C.G.; Sereno, A.B.; Krüger, C.C.H.; Cândido, L.M.B. Application of Lactose-Free Whey Protein to Greek Yogurts: Potential Health Benefits and Impact on Rheological Aspects and Sensory Attributes. *Foods* **2022**, *11*, 3861. [CrossRef]
122. Troise, A.D.; Bandini, E.; De Donno, R.; Meijer, G.; Trezzi, M.; Fogliano, V. The quality of low lactose milk is affected by the side proteolytic activity of the lactase used in the production process. *Food Res Int* **2016**, *89*, 514–525. [CrossRef]

123. Nielsen, S.D.; Zhao, D.; Le, T.T.; Rauh, V.; Sørensen, J.; Andersen, H.J.; Larsen, L.B. Proteolytic side-activity of lactase preparations. *Int. Dairy J.* **2018**, *78*, 159–168. [CrossRef]
124. Milkovska-Stamenova, S.; Hoffmann, R. Influence of storage and heating on protein glycation levels of processed lactose-free and regular bovine milk products. *Food Chem* **2017**, *221*, 489–495. [CrossRef]
125. Jo, Y.; Benoist, D.; Barbano, D.; Drake, M. Flavor and flavor chemistry differences among milks processed by high-temperature, short-time pasteurization or ultra-pasteurization. *J. Dairy Sci.* **2018**, *101*, 3812–3828. [CrossRef]
126. Tossavainen, O.; Kallioinen, H. Proteolytic changes in lactose hydrolysed UHT milks during storage. *Milchwiss.-Milk Sci. Int.* **2007**, *62*, 410–414.
127. Ghosh, K.; Maity, C.; Adak, A.; Halder, S.K.; Jana, A.; Das, A.; Parua, S.; Das Mohapatra, P.K.; Pati, B.R.; Mondal, K.C. Ethnic preparation of haria, a rice-based fermented beverage, in the province of lateritic West Bengal, India. *Ethnobot. Res. Appl.* **2014**, *12*, 39–50.
128. Moreira, T.C.; da Silva, A.T.; Fagundes, C.; Ferreira, S.M.R.; Cândido, L.M.B.; Passos, M.; Krüger, C.C.H. Elaboration of yogurt with reduced level of lactose added of carob (*Ceratonia siliqua* L.). *LWT-Food Sci. Technol.* **2017**, *76*, 326–329. [CrossRef]
129. Shrestha, A.K.; Howes, T.; Adhikari, B.P.; Bhandari, B.R. Water sorption and glass transition properties of spray dried lactose hydrolysed skim milk powder. *LWT-Food Sci. Technol.* **2007**, *40*, 1593–1600. [CrossRef]
130. Torres, J.K.F.; Stephani, R.; Tavares, G.M.; De Carvalho, A.F.; Costa, R.G.B.; de Almeida, C.E.R.; Almeida, M.R.; de Oliveira, L.F.C.; Schuck, P.; Perrone, Í.T. Technological aspects of lactose-hydrolyzed milk powder. *Food Res. Int.* **2017**, *101*, 45–53. [CrossRef] [PubMed]
131. Queiroz, E.S.; Rezende, A.L.L.; Perrone, Í.T.; Francisquini, J.d.A.; de Carvalho, A.F.; Alves, N.M.G.; de Oliveira, L.F.C.; Stephani, R. Spray drying and characterization of lactose-free goat milk. *LWT* **2021**, *147*, 111516. [CrossRef]
132. Naranjo, G.B.; Gonzales, A.S.P.; Leiva, G.E.; Malec, L.S. The kinetics of Maillard reaction in lactose-hydrolysed milk powder and related systems containing carbohydrate mixtures. *Food Chem.* **2013**, *141*, 3790–3795. [CrossRef] [PubMed]
133. Tossavainen, O.; Sahlstein, J. Process for Producing a Lactose-Free Milk Product. U.S. Patent 7,829,130, 9 November 2010.
134. McCain, H.R.; Kaliappan, S.; Drake, M.A. Invited review: Sugar reduction in dairy products. *J. Dairy Sci.* **2018**, *101*, 8619–8640. [CrossRef] [PubMed]
135. Abbasi, S.; Saeedabadian, A. Influences of lactose hydrolysis of milk and sugar reduction on some physical properties of ice cream. *J. Food Sci. Technol.* **2013**, *52*, 367–374. [CrossRef]

foods

MDPI

Article

Assessment of the Quality Attributes of Oat β-glucan Fortified Reduced-Fat Goat Milk Yogurt Supported by Microfluidization

Elif Ayse Anli [1,*], Asuman Gursel [1], Ayse Gursoy [1] and Behic Mert [2]

[1] Department of Dairy Technology, Agricultural Faculty, Ankara University, 06110 Ankara, Turkey; gursel@agri.ankara.edu.tr (A.G.); gursoy@agri.ankara.edu.tr (A.G.)
[2] Department of Food Engineering, Middle East Technical University, 06531 Ankara, Turkey; bmert@metu.edu.tr
* Correspondence: kocaoglu@agri.ankara.edu.tr; Tel.: +90-312-5961775

Abstract: In this study, goat milk blends (1.5% fat) fortified with 0%, 0.25%, and 0.50% oat β-glucan were coded as YC, Y1, and Y2 and MFYC, MFY1, and MFY2. Microfluidization was applied at 103.4 MPa pressure in a 100 μm-process chamber at one stage for MFYC, MFY1, and MFY2 prior to yogurt making. Phase separation occurred due to the casein-β-glucan interaction observed at the oat β-glucan ratio (≥0.25%) but was more distinct at 0.50%. Microfluidization solved the textural instability at all ratios of β-glucan; a creamy and less cohesive structure was maintained in all yogurt samples. Among the samples, Y2 and MFY2 were the least viscous ($p < 0.05$), and syneresis was the highest and the lowest for Y2 and MFY1, respectively ($p < 0.01$). Lightness (L^*) decreased, and yellowness (b^*) and greenness (a^*) increased with oat β-glucan concentration ($p < 0.01$) and MFYC. MFY1 and MFY2 were brighter and less green ($p < 0.05$). Microfluidization enhanced sensory attributes and oat β-glucan suppressed the goaty and salty taste, but the cereal taste became more obvious with the increase in the oat β-glucan ratio. Y1 and MFY1 were generally acceptable, and Y2 was less ($p < 0.01$). A liquid-like structure was observed in Y2 and this affected the sensorial perception in Y2.

Keywords: oat beta-glucan; dietary fiber; goat milk yogurt; high-pressure homogenization; phase separation; thermodynamic incompatibility

1. Introduction

Goat milk is popular due to its dietetic superiorities, especially for children, older people, and lactating women having digestive difficulties mainly associated with α-casein and β-lactoglobulin protein fractions [1,2]. Goat milk contains fewer allergenic proteins (αs1-casein) and fat globules with smaller diameters that help in terms of easy digestion. Milk proteins with different qualities and at different quantities result in different physicochemical properties in goat milk than in cow milk [1–3]. Textural defects, such as a weak texture, a loose gel structure [4–7], and low viscosity and consistency [8,9], are encountered in yogurt [10,11] due to low or deficient αs1-casein levels [1]. Sensorially, goat milk has a unique goaty and salty taste/flavor [12]. The goaty flavor is related to the octanoic acid present in goat milk in higher amounts and generally diminishes its acceptance by consumers.

Fermented dairy products are fortified with dietary fibers for their physiological functionality, prebiotic effect, structural development, and improvement of sensory properties [13–17]. In goat milk yogurt, textural improvement and masking the goaty and salty taste/flavor through fortification are the first possible applications employed for quality enhancement. β-glucans are water-soluble fibers in cereals with many physiological benefits, such as reductions in blood lipid and blood sugar levels, lowering cholesterol, delaying gastric emptying, supplying immune enhancement, and acting on therapeutic effects for coronary heart diseases [13,14,18–20]. According to the American Food and Drug Administration (FDA), the positive effects of β-glucan on health can occur when 0.75 g is consumed per serving [21] and the daily intake of β-glucan recommended by the FDA is 3 g [19,22]. Technologically, β-glucans are also functional depending on their molecular weight (Mw), source of origin, extraction method, and degree of purification [23,24]. Oats are the most important known sources of natural β-glucan; with a content of 3–7% [19,25,26]. Oat β-glucan is a chemically linear polysaccharide of (1 → 3), (1 → 4)-β-D- that encourages the growth and activity of the colonic bacterial population by behaving like a prebiotic [27]. In addition, β-glucans are used as fat replacers in non-fat and low-fat dairy products for their water binding, thickening, gelling, emulsification, and prebiotic properties [19,28–30].

The use of β-glucans in dairy formulations has some limitations because of the phase separation reaction of milk proteins (mainly casein) and β-glucan, known as thermodynamic incompatibility [22,28–31]. This interaction between casein and β-glucan creates two phases, hindering the physical stability of yogurt and dairy beverages, forming a polysaccharide-rich upper phase and a protein-rich lower phase. The molecular weight of β-glucan preparations and the concentration used are the key factors determining the intensity of phase separation [29] and the use of lower molecular weight β-glucans, as well as within the critical β-glucan concentrations, are recommended solutions. Thermodynamic incompatibility is encountered in the aqueous phase of protein and polysaccharide components due to electrostatic repulsive forces. The high-pressure homogenization (HPH) technique attained an approved effect on emulsion stabilization [32]. Microfluidization, a HPH technique, is promising in terms of yogurt texture and is more effective than conventional homogenization [33–35]. HPH causes structural changes in casein micelles that affect the interaction of casein with other constituents and a reduction in the molecular weight of polysaccharides, which is effective for the rheological behavior of polysaccharides in an aqueous system [32]. It provides stable, good-quality viscosity and a creamy structure, especially in low-fat yogurts [33–35].

Based on these, in this study, oat β-glucan and reduced fat goat milk were coupled together for the production of physiologically fortified and sensorially improved goat milk yogurt production, and the effect of the microfluidization technique on some quality characteristics of the final product was also observed. The goaty flavor limits goat milk yogurt consumption. Studies have revealed that sensory improvement is achieved in goat milk yogurt by mixing goat milk and cow milk [36], and also through the use of fruit juices, β-cyclodextrins, etc. As a result, in this study, the oat β-glucan was used in different ratios for sensory improvement of goat milk yogurt. In the literature, only one study used oat flour in goat milk yogurt production, observing the textural and organoleptic characteristics [37]. The β-glucan ratios used for yogurt in the literature are as follows: 0.1, 0.2, 0,3, 0.4, and 0.5% [22], 0.25, 0.50, and 1.0% [19], 0.5, 1.0, 1.5, and 2% [38], 0.1, 0.2, and 0.5 [31], 0.2–0.8% [23], 0.5, 1.0, 1.5., 2.0, and 2.5% [39], and 0.1%, 0.2%, 0.3%, 0.4%, and 0.5% [20]. Among these studies, Brennan and Tudorica, in 2008 [39], found that a 0.5% β-glucan ratio was enough for serum retention; Raikos et al., in 2018 [23], observed an adverse flavor at 0.8% β-glucan; Qu et al., in 2021 [20], declared that yogurt samples were sensorially acceptable at up to oat β-glucan ratios of 0.5% and the texture of yogurt samples was not significantly different up to or equal to 0.4%. Additionally, Vasiljevic et al., in 2007 [28], determined that 0.24% and below was inhibitory for phase separation, while other researchers found the threshold for β-glucan concentrations (%, w/w) phase separation was <0.25% [31], <0.20 [30], and <0.5% [40]. Considering these results and the preliminary experiments carried out; samples with oat β-glucan ratios of more than 0.50% were found to be sensorially unacceptable due to the intense cereal taste and their distinct phase separation. As a result, in the study, the correct oat β-glucan ratios were determined to be 0%, 0.25%, and 0.50%, regarding the sensorial and textural attributes. To the best of our knowledge, in the literature, no other study explored oat β-glucan use in goat milk yogurt and utilized microfluidization for textural improvement in that system. Oat β-glucan-casein interactions are limited in terms of using β-glucans in dairy systems due to thermodynamic incompatibility. From a technological point of view, the results are promising for providing valuable information for developing new-generation dairy systems with β-glucan fortification and providing a new solution with microfluidization for thermodynamic incompatibility reactions faced in β-glucan-casein interactions.

2. Materials and Methods

2.1. Materials

Raw goat milk was freshly obtained from a dairy farmer in Yeni Pecenek Village of Ankara, Turkey. Skim goat milk powder was supplied by Enka Dairy and Food Products Industry and Commerce Ltd., Co., (Konya, Turkey) and used for the enrichment of the non-fat dry matter content of raw milk. The oat β-glucan product used in the fortification of goat milk yogurt samples with dietary fiber was purchased from the producer (PromOat®, Tate & Lyle, Kimstad, Sweden AB). The general characteristics of raw goat milk, skim goat milk powder, and the oat β-glucan product used in yogurt production are presented in Table 1.

Table 1. General characteristics of raw goat milk, skim goat milk powder, and oat β-glucan.

Component (%)	Raw Goat Milk [a]	Skim Goat Milk Powder [b]	Oat β-glucan [c]
Total solid	14.26 ± 0.83	96.61	96.97
Fat	5.13 ± 0.56	0.50	0.75
Total protein	4.51 ± 0.30	- [d]	4.25
Ash	0.78 ± 0.06	8.79	2.09
pH	6.67 ± 0.03	6.66	- [d]
Titratable acidity, % LA	8.37 ± 0.62	0.11	- [d]
β–glucan	- [e]	- [e]	35.00

[a] Values are means ± standard error of parameters (n = 3). [b],[c] Data supplied by the manufacturer. [d] Not determined. [e] Not included.

2.2. Preparation of Milk Blends

Raw goat milk was initially standardized in terms of fat (1.5%) and non-fat solid (16%) for getting physically good quality goat milk yogurt. In the standardization of non-fat solid content, skim goat milk powder (Table 1) was added to approximately 60 °C heated raw goat milk. Milk blends (4000 mL each) were prepared by incorporating different levels of oat β-glucan (0.25% and 0.50% *w*/*v*, on a milk basis), except for control samples. Prepared milk blends were mixed with a laboratory-type Ultra Turax blender (IKA RW 20, Staufen, Germany) at 13,500 rpm for 5 min concerning complete dispersion of the components. Yogurt samples were produced with non-microfluidized and microfluidized milk blends according to the procedure described in the following section. Samples were coded as YC, Y1, and Y2 for non-microfluidized and MFYC, MFY1, and MFY2 for microfluidized yogurt samples (Table 2).

2.3. Microfluidization Process and Yogurt Making

Microfluidization was carried out by Microfluidizer (M-110Y, Microfluidics, Westwood, MA, USA) at 103.4 MPa pressure by passing the milk blends of MFYC, MFY1, and MFY2 through the 100 μm process chamber at one stage. Following this, microfluidized milk blends were processed into yogurt. Non-microfluidized samples (YC, Y1, Y2) were directly processed into yogurt. Yogurt production was carried out starting with heat treatment at 90 °C for 10 min and cooling to 43–45 °C for inoculation with mixed lyophilized commercial starter cultures of *Streptococcus thermophilus* and *Lactobacillus delbrueckii* subsp. *bulgaricus* (Y401, Maysa Food Industry and Commerce Ltd. Co., Istanbul, Turkey) at a ratio of 0.4 g/L for each 4 L-batch. After inoculation, all samples were filled in sterile plastic containers of 100 mL volume and incubated at 43 ± 1 °C until the pH value reached 4.6 ± 0.1. At the end of the incubation period, yogurt samples were left to cool for 10 min at room temperature and then stored in refrigerated conditions (4 ± 1 °C). Yogurt samples were analyzed in their physical, chemical, microbiological, textural and sensory properties every seven days of the 21-day storage period.

Table 2. Details on milk blends for yogurt samples.

Sample Code [1]	Details for Milk Blends			
	Oat β Glucan Ratio (%, *w*/*v*, on Milk Base)			Microfluidization (MF)
	0	0.25	0.50	
YC	✓	-	-	-
Y1	-	✓	-	-
Y2	-	-	✓	-
MFYC	✓	-	-	✓
MFY1	-	✓	-	✓
MFY2	-	-	✓	✓

[1] Sample codes denote: YC: non-microfluidized control yogurt (0% oat β-glucan ratio), Y1: non-microfluidized yogurt with 0.25% oat β-glucan ratio, Y2: non-microfluidized yogurt with 0.50% oat β-glucan ratio, MFYC: microfluidized control yogurt (0% oat β-glucan ratio), MFY1: microfluidized yogurt with 0.25% oat β-glucan ratio, MFY2: microfluidized yogurt with 0.50% oat β-glucan ratio.

2.4. Proximate Composition of Raw Goat Milk

Raw goat milk was analyzed in basic compositional properties, pH, and titratable acidity as Lactic Acid (LA %). Total dry matter (%) and fat (%) and titratable acidity in Lactic Acid (LA %) by the method [41], ash (%) content was determined by [42]. The pH of raw goat milk was measured by a digital pH meter (Mettler Toledo Seven2Go S2; Schwerzenbach, Switzerland). Kjeldahl method, was used in determining the total protein content of yogurt samples [43]. In analysis, samples were initially digested (Büchi K435, Flawil, Switzerland) and distilled (Büchi 323, Flawil, Switzerland). Total nitrogen (%) was converted to Total Protein (TP %) by multiplying with a factor of 6.38.

2.5. Proximate Composition and Physico-Chemical Properties of Yogurt Samples

Yogurt samples were analyzed in total dry matter (%), fat (%), and ash (%) content on the 1st, 7th, 14th, and 21st days of storage according to the methods of [42], Gerber method [41,44,45], respectively. Yogurt samples were diluted with distilled water in a 1:1 ratio, and pH values were measured digitally by a pH meter (Mettler Toledo Seven2Go S2; Schwerzenbach, Switzerland). Titratable acidity values of the yogurt samples were determined and expressed as lactic acid (LA %) [45].

2.6. Oat-β-Glucan Content

The β-glucan content of oat-β-glucan fortified yogurt samples was determined by using an enzymatic kit (Megazyme International Ireland Ltd., Co., Wicklow, Ireland) following the Megazyme method and analysis procedure provided by the company [46]. The amount of β-glucan was calculated according to Equation (1). In the enzymatic reaction, the repeated glucose units in the β-glucan structure were broken down into D-glucose by the action of the enzyme β-glucosidase. The absorbance value of the D-glucose released at the end of the lysis reaction was measured at 510 nm wavelength by spectrophotometer (PerkinElmer Lambda 25 UV/Vis Spectrometer, Republic of Singapore). The β-glucan content of yogurt samples was analyzed on the 1st and 21st days of storage. The values obtained for the samples were expressed in g/100 mL (Equation (1)).

$$\beta - glucan\left(\frac{g}{100}mL\right) = \Delta A \times F \times D \times 0.0027 \quad (1)$$

$$\Delta A = Absorbance_{reaction} - Absorbance_{blank}$$

$$F = 100(\mu g\ D - glucose) \div (Absorbance_{100\mu g D-glucose})$$

$$D = Dilution\ rate\ \ prior\ to\ \beta - glucosidase\ enzyme\ incubation$$

2.7. Color Analysis

Color attributes of yogurt samples were evaluated according to CIELAB color space with the coordinates of L^*, a^*, and b^* denoting lightness/darkness, redness/greenness, and yellowness/blueness, respectively (Konica Minolta CR 410, Sensing Inc., Osaka, Japan). In the coordinates, $L^* = 100$ indicated lightness, and $L^* = 0$ indicated darkness of the sample, (+) a^* values indicated redness and (−) a^* values indicated greenness and (+) b^* values indicated yellowness, and (−) b^* values indicated blueness of the sample. After calibrating the instrument with a standard white plate, the instrument measurements were done by putting yogurt samples in a quartz container. The measurements were performed in triplicate.

Color differences among the samples were given by using calculated indices as the total color difference (ΔE), Chroma (C^*), Hue Angle (h^*), Whiteness Index (WI) and Yellowness Index (YI). In the calculation of these indices the Equations (2)–(6) were used by considering the measured CIE color parameters (L^*, a^* and b^*) of the yogurt samples [47]. The $L_0{}^*$, $a_0{}^*$ and $b_0{}^*$ values were measured values for control samples (YC and MFYC) and considered as the references for calculation of ΔE of non-microfluidized and microfluidized yogurt samples separately (data not given).

$$\Delta E = \sqrt{\Delta a^{*2} + \Delta b^{*2} + \Delta L^{*2}} \quad (2)$$

$$C^* = \sqrt{a{*}^2 + b{*}^2} \quad (3)$$

$$h^* = \tan^{-1}\left(\frac{b^*}{a^*}\right) \quad (4)$$

$$WI = 100 - \sqrt{(100 - L^*)^2 + a^{*2} + b^{*2}} \quad (5)$$

$$YI = \frac{142.86b^*}{L^*} \quad (6)$$

2.8. Evaluation of Yogurt Gel Properties

2.8.1. Syneresis

Robitaille et al. (2009) [48], determined the syneresis of yogurt samples. Twenty-five grams of yogurt samples were centrifuged for 10 min at 2200 rpm and 4 °C by centrifuge (Sigma 3–18K, Osterode am Harz, Germany). At the end of centrifugation, the supernatant (serum) separated at the top of the tubes was weighed, and syneresis (%) was calculated according to Equation (7).

$$Syneresis\ (\%) = (weight\ of\ supernatant \div weight\ of\ yogurt) \times 100 \quad (7)$$

2.8.2. Textural Analysis

Back-extrusion method was used in the characterization of yogurt samples' texture. In the measurements, the texture analyzer (TA. XT Plus, Stable Micro Systems, Surrey, UK) with the equipment of a 5-kg load cell and 35-mm back extrusion disk was used. All samples were analyzed in original containers compatible with the method and equipment. The test conditions were adjusted as: 1.5 mm/s probe test speed and 20 mm for probe penetration distance. Sample temperature is a critical parameter in texture analysis, so samples were removed from the refrigerator (4 ± 1 °C) immediately before the measurement. Quaternary measurements were performed for each sample in the 1st, 7th, 14th, and 21st days of storage. The derived texture attributes from the force-time curve obtained by the back-extrusion method were firmness, consistency, cohesiveness, and index of viscosity.

2.9. Culture Viability of Yogurt Bacteria

The growth of yogurt bacteria (*Streptococcus thermophilus* and *Lactobacillus delbrueckii* subsp. *bulgaricus*) in oat β-glucan fortified goat milk yogurt was evaluated during the storage period according to procedures expressed in International Dairy Federation Standard [42,43]. For enumeration of *Lactobacillus delbrueckii* subsp. *bulgaricus* MRS agar medium was used, and incubation was carried out in anaerobic conditions with jars at 37 °C for 72 h for *Streptococcus thermophilus* in M17 agar medium was used. Petri plates were incubated at 37 °C for 48 h aerobically. Yogurt samples were also observed for yeast and mold in PDA (Potato dextrose agar) agar medium at 25 °C for 72–120 h of incubation. Before pouring into the plates, the PDA medium was initially acidified by sterile tartaric acid (10%), 1 mL for each 100 mL PDA agar medium. Plates with 30–300 colonies were recorded, and bacterial counts were evaluated as log CFU × g^{-1}. Duplicate examinations were done for all yogurt samples.

2.10. Sensory Analysis

Sensory evaluation was conducted with 7-experienced panelists who regularly consumed dairy products in the Department of Dairy Technology, Ankara, Turkey. A scoring guide was prepared with some additional modifications related to criteria encountered in goat milk yogurt and oat-beta glucan (Table 3) [49,50]. In Table 3, panelists were asked to evaluate the yogurt samples over 5 points in terms of color and appearance, structure and texture attributes, degree of liking, and in terms of odor and flavor 10-point rating scale was used. In the evaluation, approximately 50 mL of randomly-coded yogurt sample (in 3-digit numbers) was presented to the panelists with a glass of drinking water in their original containers.

Table 3. Scoring guide for sensory evaluation of yogurt samples.

Quality Criteria	Possible Defects in Goat Milk Yogurt
Color and appearance The score ranges from 1–5 points 5: defines no defect	Non-uniform color Free whey Unnatural color Shrinkage Surface growth of bacteria
Structure and texture attributes The score ranges from 1–5 points 5: defines no defect	Too thin or too firm Creamy Drinkable Syneresis Lumpy Granular Ropy

Table 3. *Cont.*

Quality Criteria	Possible Defects in Goat Milk Yogurt
Flavor and Odor The score ranges from 1–10 points 10: defines no defect	Lack of yogurt taste and aroma Cooked flavor Creamy Cereal flavor Goaty flavor and odor Fermented Sour Salty Sweet Yeasty/Fruity Foreign

2.11. *Statistical Analysis*

The study was organized by considering the three different levels of oat β-glucan ratio (0%, 0.25% and 0.50%), four different levels of storage time (1, 7, 14 and 21 days) and two different levels of microfluidization conditions (either non-microfluidized or microfluidized) as factors in oat β-glucan fortified reduced fat goat milk yogurt production. This experiment, conducted using the repeated measurement ANOVA technique with oat β-glucan and storage time factors, was replicated in non-microfluidized and microfluidized conditions. The experiment is a repeated-trial conducted using the factorial design with the repeated measurement analysis of variance (ANOVA) method. Statistical analysis of the data was performed using Minitab statistical software (version 16.0, Minitab Inc., State College, PA, USA) and multiple comparisons were made by Duncan test at significance levels of $p < 0.05$ and/or $p < 0.01$.

3. Results and Discussion

The proximate composition, physicochemical, textural, and microbiological properties of non-microfluidized (YC, Y1, Y2) and microfluidized (MFYC, MFY1, MFY2) oat β-glucan fortified goat milk yogurt samples in the storage days (1st, 7th, 14th and 21st) were discussed in the whole manuscript by covering the individual effect of oat β-glucan use and its concentration, the individual effect of using microfluidization and the individual effect of storage time on yogurt sample characteristics, and also the effect of interactions of these factors in statistical significance level ($p < 0.01$ and $p < 0.05$) as given in Table 4. For this reason, the results of the parameters for yogurt samples were grouped by factor issues as given in Table 4 and discussed in the relevant sections.

Fat (%) content of yogurt samples was determined as 1.60% and remained un-changed during 21 days of storage. In the manufacture raw goat's milk was initially standardized to 1.5% fat but fat content of oat β-glucan also contributed to the fat content of the yoghurt samples (Table 1).

Table 4. Demonstration of p values; individual effect of oat β-glucan use, storage time, microfluidization and their mutual interactions.

Parameter	ANOVA						
	P-Oat β-Glucan	*P*-Storage Time	*P*-MF	*P*-MF × *P*-β-Glucan	*P*-MF × *P*-Storage Time	*P*-Oat β-Glucan × *P*-Storage Time	*P*-Oat β-Glucan × *P*-Storage Time × *P*-MF
Dry matter (%)	**	NS	NS	NS	**	NS	NS
Lactic Acid (%)	*	NS	NS	NS	**	NS	NS
pH	*	NS	NS	NS	*	NS	NS
Ash (%)	NS	NS	NS	NS	**	NS	NS
Oat β-glucan	**	NS	NS	NS	NS	NS	NS
Syneresis (%)	NS	NS	NS	**	*	NS	NS
Firmness (N)	NS	**	NS	NS	NS	NS	NS
Consistency (N×s)	NS	*	NS	NS	NS	NS	NS
Cohesiveness (N)	NS	NS	NS	NS	*	NS	NS
Index of viscosity (N×s)	*	NS	NS	NS	NS	NS	NS
*L**	NS	NS	NS	**	*	NS	NS
*a**	NS	NS	NS	*	**	NS	NS
*b**	NS	*	NS	**	NS	NS	NS
S. thermophilus (log CFU/g)	NS	**	NS	NS	NS	NS	NS
Lb. delbrueckii subsp. *bulgaricus* (log CFU/g)	NS	**	NS	NS	NS	NS	NS
General acceptability	NS	NS	NS	**	NS	NS	NS

* $p < 0.05$, ** $p < 0.01$; NS, not significant $p > 0.05$, *P*-oat β-glucan, individual effect of oat β-glucan, *P*-storage time, individual effect of storage time, *P*-MF, individual of effect of microfluidization, *P*-MF × *P*-β-glucan, interaction between microfluidization and oat β-glucan, *P*-MF × *P*-storage time, interaction between microfluidization and storage time, *P*-oat β-glucan × *P*-storage time, interaction between oat β-glucan and storage time, *P*-oat β-glucan × *P*-storage time × *P*-MF, interaction between oat β-glucan, storage time and microfluidization.

3.1. Oat β-glucan Content and Effect of Oat-β Glucan Concentration on Some Characteristics of Yogurt Samples

In the inclusion of oat β-glucan, the oat β-glucan with 35% β-glucan were used (Table 1). The amounts incorporated to yogurt samples in ratios of 0.25% and 0.50% were calculated as 0.09 g and 0.18, respectively. However, according to Table 5, yogurt with oat β-glucan 0.25% and oat β-glucan 0.50% contained β-glucan in amounts of 0.05 ± 0.0190 and 0.11 ± 0.0030. These values were lower than inclusion amounts so some amount of β-glucan might have been consumed during fermentation since no change was observed from day 1 to day 21 (data not given). In literature, some researchers declared the utilization of β-glucan by probiotic bacteria and yogurt bacteria resulting in a loss of β-glucan that may weaken the product's functional properties. So, Gee et al. (2007) [51] and Snart et al. (2005) [52], suggested that the addition of β-glucan should be done after fermentation. Some studies show that selected probiotics and strains of yogurt bacteria can benefit from β-glucan [53]. However, it is not completely clear whether the strains digest β-glucan or whether it has mediating effects for culture growth [28]. Vasiljevic et al. (2007) [28] also concluded that it was unclear whether yogurt starter cultures and probiotics can utilize β-glucan as a prebiotic. However, it was evident that their growth and viability were enhanced. In the studies, no adverse effect of β-glucan use on yogurt bacteria growth rate was declared by researchers [51,54]. Researchers stated that the addition of β-glucan did not affect the growth of yogurt bacteria [28,55–57].

Table 5. The individual effect of oat β-glucan concentration in some characteristics of both non-microfluidized and microfluidized goat milk yogurt samples.

Parameter	$\overline{X} \pm S_x$ [1]		
	Control	**Oat β-Glucan 0.25%**	**Oat β-Glucan 0.50%**
Dry matter (%)	15.91 ± 0.1100 ^{B**}	16.21 ± 0.1290 ^{A**}	16.44 ± 0.1300 ^{A**}
LA (%)	1.51 ± 0.0155 ^{B*}	1.56 ± 0.0127 ^{A*}	1.49 ± 0.0170 ^{B*}
pH	4.29 ± 0.0196 ^{AB*}	4.25 ± 0.0181 ^{B*}	4.33 ± 0.0180 ^{A*}
Oat β-glucan content (g/100 mL) [2]	-	0.05 ± 0.0190 ^{B**}	0.11 ± 0.0030 ^{A**}
Index of viscosity (N × s)	1.13 ± 0.0668 ^{A*}	1.17 ± 0.0460 ^{A*}	0.70 ± 0.0442 ^{B*}

[1] Data are the means ± SE (n = 24) of parameters for both goat milk yogurt samples (non-microfluidized and microfluidized) with the same oat β-glucan content during the 21-day storage period; [2] The values are the means ± SE (n = 12) of oat β-glucan content of both non-microfluidized and microfluidized goat milk yogurt samples on days of 1 and 21. Control: Means of both non-microfluidized and microfluidized control yogurts (YC and MFYC). Oat β-glucan 0.25%: Means of both non-microfluidized and microfluidized yogurts with 0.25% oat β-glucan ratio (Y1 and MFY1), Oat β-glucan 0.50%: Means of both non-microfluidized and microfluidized yogurt with 0.50% oat β-glucan ratio (Y2 and MFY2), different capital letters in the same row indicate significant difference (** $p < 0.01$, * $p < 0.05$).

According to Table 5, oat β-glucan use and its concentration had a significant effect on dry matter ($p < 0.01$) and similarly on LA %, pH, and index of viscosity values of all samples ($p < 0.05$). Similarly declared as; the dry matter content of oat β-glucan fortified yogurt samples increased relevant to the amount of oat β-glucan used in the samples [19]. The dry matter content of yogurt samples fortified with 0.25% oat β-glucan and 0.50% oat β-glucan (both non-microfluidized and microfluidized) were significantly higher than that of control yogurts ($p < 0.01$) but not found significantly different among yogurt samples fortified with 0.25% oat β-glucan and 0.50% oat β-glucan.

Oat β-glucan use, and its concentration created a significant difference in LA % and pH of all samples ($p < 0.05$). Among the samples, oat β-glucan fortified yogurt with 0.25% concentration was significantly different from control and 0.50% oat β-glucan fortified samples in terms of LA % and again significantly different than 0.50% oat β-glucan fortified samples for pH (Table 5). Sahan et al. (2008) [19] found that using β-glucan composite did not significantly change non-fat yogurt samples' pH and titratable acidity.

The index of viscosity measures the sample's resistance to flow off through the disk during instrumental texture analysis (back extrusion method) [33]. An oat β-glucan ratio-dependent variation was observed for the index of viscosity of yogurt samples. The index of viscosity of yoghurt samples increased when the increase in oat β-glucan ratio was at the level of 0.25%, but a significant decrease was observed at the level of 0.50% ($p < 0.05$) (Table 5). From Table 5, for 0.25% oat β-glucan fortified goat milk yogurt samples an insignificant increase was observed in index of viscosity when compared with control yogurts. In dairy systems with β-glucans above the critical β-glucan level, the casein-β-glucan interaction causes a 2-layer structure resulting in drinkable, low-viscosity yogurts [28,58]. If the emulsion is unstable, the components compromising the system will separate [59]. Stability, viscosity, and water holding capacity are yogurt quality parameters, which are enhanced by homogenization enabling uniform spreading of milk components and ingredients. High-pressure homogenization by microfluidization supplies stable emulsions [33], as it was in our samples (Figure 1).

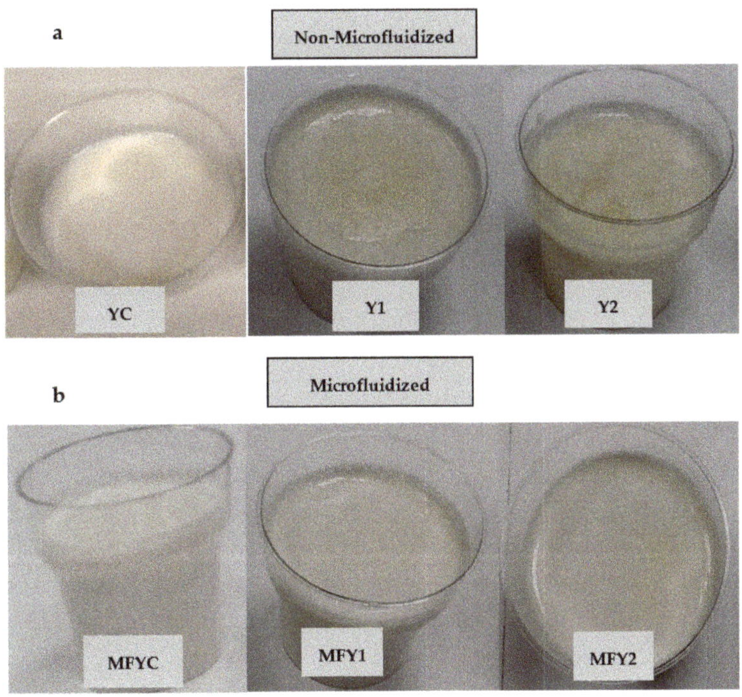

Figure 1. The visual appearance of non-Microfluidized (**a**) and Microfluidized (**b**) oat β-glucan fortified goat milk yogurt samples. YC: non-microfluidized control yogurt (0% oat β-glucan ratio), Y1: non-microfluidized yogurt with 0.25% oat β-glucan ratio, Y2: non-microfluidized yogurt with 0.50% oat β-glucan ratio, MFYC: microfluidized control yogurt (0% oat β-glucan ratio), MFY1: microfluidized yogurt with 0.25% oat β-glucan ratio, MFY2: microfluidized yogurt with 0.50% oat β-glucan ratio.

3.2. Impact of Microfluidization and Oat-β Glucan Concentration on Color Attributes, Syneresis, and General Acceptability of Yogurt Samples

Color parameters (L^*, a^*, and b^*) and syneresis (%) of goat-milk yogurt samples were significantly affected by the interaction of oat β-glucan and the microfluidization interaction and given in Table 6. Non-microfluidized samples differed in L^* and b^* ($p < 0.01$). Among the microfluidized samples, goat-milk yogurt with oat β-glucan 0.50% ratio was significantly different from the others in terms of L^* value (less bright) ($p < 0.01$). Lightness decreased, and oppositely yellowness increased (higher b^* values) in both non-microfluidized and non-microfluidized samples with increasing amounts of oat β-glucan. As the concentration of oat β-glucan increases b^* value increases and the L^* value decreases due to the slightly yellowish color of the oat β-glucan product used. Microfluidized control yogurt was significantly different in b^* than microfluidized goat-milk yogurt with oat β-glucan 0.50% ratio ($p < 0.01$). In non-microfluidized samples, the intense mass accumulation and serum separation occurred due to an increase in β-glucan concentration, which might decrease L^* values and increase yellowness (b^*). In oat-based fermented products, control yogurts were whiter than samples with oat β-glucan [60]. Similar results were observed in probiotic yogurts with oat-based and barley-based β-glucan [61]. Including oat β-glucan up to 0.3% had no significant effect on yogurt samples' lightness (L^*). However, a darker color was formed with lower L^* values by increasing the concentration of oat β-glucan from 0.3% to 0.5% [22].

Table 6. Color attributes, syneresis (%), and general acceptability of oat β-glucan fortified goat milk yogurt samples depend on microfluidization and oat β-glucan concentration.

Parameter	Yogurt Sample	$\overline{X} \pm S_x$ [1]	
		Non-Microfluidized	Microfluidized
L^*	Control	69.63 ± 0.3980 ^{Ab**}	70.30 ± 0.0832 ^{Aa**}
	Oat β-glucan 0.25%	68.76 ± 0.3830 ^{Bb**}	70.08 ± 0.0989 ^{Aa**}
	Oat β-glucan 0.50%	67.26 ± 0.3590 ^{Ca**}	69.46 ± 0.1540 ^{Ba**}
a^*	Control	−2.12 ± 0.1050 ^{Aa*}	−2.08 ± 0.0814 ^{Aa*}
	Oat β-glucan 0.25%	−2.23 ± 0.1720 ^{Aa*}	−2.00 ± 0.0551 ^{Aa*}
	Oat β-glucan 0.50%	−2.70 ± 0.1160 ^{Bb*}	−2.15 ± 0.0786 ^{Aa}
b^*	Control	7.90 ± 0.3750 ^{Ca**}	8.31 ± 0.0889 ^{Ba**}
	Oat β-glucan 0.25%	8.52 ± 0.4060 ^{Ba**}	8.53 ± 0.0535 ^{ABa**}
	Oat β-glucan 0.50%	9.26 ± 0.3300 ^{Aa**}	8.80 ± 0.0484 ^{Ab**}
Syneresis (%)	Control	12.53 ± 0.434 ^{Ba**}	12.81 ± 0.969 ^{Aa**}
	Oat β-glucan 0.25%	21.88 ± 0.853 ^{Aa**}	10.34 ± 0.785 ^{Ab**}
	Oat β-glucan 0.50%	24.59 ± 2.280 ^{Aa**}	15.08 ± 1.400 ^{Ab**}
General acceptability	Control	3.62 ± 0.125 ^{Aa**}	3.27 ± 0.163 ^{Aa**}
	Oat β-glucan 0.25%	3.71 ± 0.132 ^{Aa**}	3.83 ± 0.122 ^{Aa**}
	Oat β-glucan 0.50%	1.80 ± 0.210 ^{Bb**}	3.49 ± 0.168 ^{Aa**}

[1] Data are the means ± SE (n = 12) of parameters for oat β-glucan fortified goat milk yogurt samples during a 21-day storage period; Control: Means of goat-milk yogurt with 0% oat β-glucan ratio either non-microfluidized or microfluidized, Oat β-glucan 0.25%: Goat-milk yogurt fortified with 0.25% oat β-glucan ratio either non-microfluidized or microfluidized, Oat β-glucan 0.50%: Goat-milk yogurt fortified with 0.50% oat β-glucan ratio either non-microfluidized or microfluidized, different capital letters in the same column indicate significant difference (* $p < 0.05$, ** $p < 0.01$), different lower case letters in the same row indicate significant difference (* $p < 0.05$, ** $p < 0.01$).

The negative a^* values indicated that all goat milk yogurt samples (with or without oat β-glucan) were in green color space, and non-microfluidized samples were more greenish than microfluidized ones with higher (−) a^* values. As the concentration of oat β-glucan increased, higher a^* values were obtained in the negative area. No significant difference was found among microfluidized samples, but in non-microfluidized samples, goat-milk yogurt fortified with 0.50% oat β-glucan ratio was significantly different and more greenish than control and goat-milk yogurt with 0.25% oat β-glucan ratio ($p < 0.05$).

The interaction of oat β-glucan concentration and microfluidization treatment indicated that the lightness (L^*) of the samples increased ($p < 0.01$), but a^* (greenness) decreased with microfluidization ($p < 0.05$). The increase in lightness among non-microfluidized and microfluidized yogurt groups can be explained by whey protein denaturation and its association with casein micelle induced by pressures of 100–200 MPa applied during homogenization. However, higher pressures (300 MPa) reduce lightness [62]. Similar to our results (Table 6), an increase in lightness was associated with microfluidization application at different pressures for Cheddar cheese [63]. Oppositely, Gervilla et al. (2001) [64], declared that a decrease in L^* was caused by the dispersion of casein micelles into smaller pieces resulting from high pressures applied and milk becoming transparent. Microfluidization significantly changed L^* ($p < 0.01$) but not for a^* and b^* for control yogurts. The b^* (yellowness) value increased with microfluidization in control yogurt. In goat-milk yogurt fortified with 0.25% oat β-glucan ratio, L^* differed statistically ($p < 0.01$) among non-microfluidized and microfluidized samples. Additionally, the difference was not significant for L^* but was significant for a^* ($p < 0.05$) and b^* ($p < 0.01$) among non-microfluidized and microfluidized yogurt samples with 0.5% oat β-glucan (Table 6). Microfluidized samples were less green (a^*) than non-microfluidized ones. Lower b^* values were measured for only microfluidized goat-milk yogurt fortified with 0.50% oat β-glucan ratio primarily related to homogenization supplied by microfluidization.

According to Table 6, using oat β-glucan in varying amounts and microfluidization treatment created a significant difference in the syneresis of samples ($p < 0.01$). In non-microfluidized samples, syneresis values increased gradually with oat β-glucan concentration and were higher than the control sample. Like ours, syneresis increased at β-glucan levels higher than 0.3% [54]. However, no linear relationship between the amount of β-glucan composite and the syneresis and this time oppositely, syneresis decreased in yogurt samples with β-glucan composite addition [19]. The difference was significant with non-microfluidized control and non-microfluidized oat β-glucan fortified samples ($p < 0.01$) but insignificant among samples with 0.25% and 0.50% oat β-glucan ratio. For microfluidized samples, syneresis values were lower than non-microfluidized couples with similar oat β-glucan content (Table 6). In microfluidized samples, the highest syneresis values were obtained in yogurt samples with 0.5% oat β-glucan ratio, and the lowest was in yogurt samples with 0.25% oat β-glucan ratio, probably due to the more homogeneous distribution of oat β-glucan in the system and the improved water-holding capacity of oat β-glucan (dietary fiber) by microfluidization. However, this behavior did not continue parallel with the increase in oat β-glucan concentration. Although insignificant, higher syneresis was observed for microfluidized yogurt with a 0.50% oat β-glucan ratio (Table 6). According to Lazaridou and Biliaderis (2007) [58], this was due to phase separation due to the interaction between oat β-glucan and casein above a critical concentration level and the thermodynamic incompatibility between these two molecules. Oat and barley β-glucan addition resulted in higher syneresis values obtained in yogurts compared to the control sample and recommended using a β-glucan concentration not higher than 0.4% [28]. At levels up to 0.3% of β-glucan addition, a significant increase in hardness values was observed in yogurt gels with the increase in concentration, and this was found concerning decreased water in the gel structure due to increased syneresis [22].

Oat β-glucan concentration and microfluidization interaction created no significant difference in terms of syneresis among both non-microfluidized and microfluidized control yogurts ($p > 0.05$). However, they were significant ($p < 0.01$) among oat β-glucan fortified non-microfluidized and microfluidized groups (Table 6). In general, in the samples, syneresis (%) decreased by microfluidization (Table 6). The denaturation degree of serum proteins is effective on water holding capacity and the syneresis values of yogurt samples as the denaturation degree of serum proteins increases water holding capacity increases and meaning that syneresis decreases [65,66]. In addition, high-pressure technology causes the casein micelles to break into smaller fragments, dissolution of colloidal calcium phosphate, and denaturation of serum proteins [67,68]. Yogurt texture is improved with increased amounts of reacting denatured serum proteins and the increased number of fat globules and casein micelles [33].

The difference in syneresis of yogurt samples with similar oat β-glucan ratio was correlated with the effect of microfluidization on the functionality of dietary fibers and the increased denaturation ratio of serum proteins. Chen et al., 2013 [69] found that microfluidization treatment increased the solubility of insoluble dietary fibers (peach and oat) by changing the structures of the regions acting on water binding and improving their water-holding capacity. Similarly, Ronkart et al. (2010) [70] stated that microfluidization decreases the particle size of inulin in inulin-water systems independent of the inulin concentration, and the more porous structure is formed by microfluidization and water holding capacity is increased. In our samples (Table 6), the joint effect of microfluidization on denaturation of serum proteins and oat beta-glucan functionality could result in lower syneresis values in microfluidized oat β-glucan fortified yogurt samples ($p < 0.01$).

In sensorial evaluation, the interaction of oat β-glucan concentration and microfluidization effect was found statistically significant in the general acceptability scores of the samples ($p < 0.01$). Non-microfluidized control (YC) and oat β-glucan fortified yogurt with a ratio of 0.25% (Y1) did not differ in terms of general acceptability, but Y2 differed significantly from YC and Y1 ($p < 0.01$). Y1 got the highest score and was followed by YC and Y2 orderly. Increasing oat β-glucan brings many obstacles, especially in dairy systems, due to the interaction of casein and β-glucan. Textural defects are considered significant determinants in yogurt quality, dominating the other favorable properties. As a result, the definite result of using 0.50% oat β-glucan in non-microfluidized samples had an adverse effect on the total quality of goat milk yogurt samples. In microfluidized samples, no significant difference was observed in the general acceptability scores of samples (MFYC, MFY1, and MFY2) (Table 6). MFY1 got the highest score and was followed by MFY2 and MFYC orderly. Fortification concentration of oat β-glucan was found to be statistically significant among the non-microfluidized sample (Y2) and microfluidized sample (MFY2) ($p < 0.01$).

3.3. *Derived Color Indices of non-Microfluidized and Microfluidized Oat β-Glucan Fortified Goat Milk Yogurt Samples during Storage Period*

In Table 7, the results of calculated color indices obtained by the Equations (2)–(6) were given. ΔE values were derived by measured values as L^*, a^* and b^* (Equation (2)). The ΔE, indicates the color difference among samples and their controls. The degree of the difference was analytically scaled as very distinct difference when $\Delta E > 3$, distinct difference when $1.5 < \Delta E < 3$ and a small difference when $\Delta E < 1.5$ [47]. From Table 7, in non-microfluidized samples a small difference was observed between YC and Y1 in all storage days with ΔE values smaller than 1.5. At ratios of oat β-glucan 0.50%, samples a distinct difference was observed on days of 1st, 7th and 14th (Table 7). In case of microfluidized samples, all samples (MFY1 and MFY2) were had small difference than MFYC. These results can be explained by more homogeneous structure maintained by microfluidization and also more successful dispersion of oat β-glucan and milk constituents in microfluidized samples.

Higher Chroma values meaning that the higher the color in quantity and more intense the color sensed by human eye [47]. The Chroma values increased gradually with oat β-glucan ratio in non-microfluidized and microfluidized yogurt samples but lower values observed for the former. The Chroma values of microfluidized samples increased gradually with oat β-glucan ratio but were nearly constant over time for MFY1 and MFY2. Practically a steady increase was observed for YC, Y1 and Y2 during storage period (Table 7).

Hue angle, defines the qualitative measure of color comparing a certain color with grey color as reference at the same lightness. Angles of 0° and/or 360° defines red hue, 90°, 180°, and 270° defines yellow, green and blue hues in order [47]. The hue angle values of all oat β-glucan fortified samples were in yellow/green color region but mostly yellow with angles very close to 90°. The hue angle of non-microfluidized samples were higher but with a very small margin than microfluidized samples. The hue angles of microfluidized samples were very close to each other presented nearly a constant trend in storage period (Table 7).

Table 7. Change in color indices of non-Microfluidized and Microfluidized oat β-glucan fortified goat milk yogurt samples during storage period.

Color Indices	Sample	Non-Microfluidized				Sample	Microfluidized			
		Storage Time (Day)					Storage Time (Day)			
		1	7	14	21		1	7	14	21
ΔE	YC	-	-	-	-	MFYC	-	-	-	-
	Y1	1.13 ± 0.17	1.02 ± 0.13	1.13 ± 0.08	1.34 ± 0.09	MFY1	0.38 ± 0.06	0.33 ± 0.06	0.42 ± 0.08	0.46 ± 0.07
	Y2	2.12 ± 0.20	2.15 ± 0.04	1.75 ± 0.05	1.21 ± 0.26	MFY2	0.89 ± 0.24	0.94 ± 0.17	0.63 ± 0.04	0.44 ± 0.13
Chroma	YC	7.06 ± 1.42	8.45 ± 0.08	8.55 ± 0.16	8.69 ± 0.21	MFYC	8.44 ± 0.08	8.63 ± 0.11	8.67 ± 0.19	8.54 ± 0.17
	Y1	7.66 ± 1.51	9.19 ± 0.02	9.09 ± 0.12	9.39 ± 0.13	MFY1	8.66 ± 0.03	8.76 ± 0.08	8.84 ± 0.11	8.84 ± 0.10
	Y2	8.73 ± 1.25	9.99 ± 0.09	10.07 ± 0.16	9.85 ± 0.11	MFY2	8.97 ± 0.13	9.04 ± 0.05	9.17 ± 0.11	9.05 ± 0.03
Hue Angle	YC	107.55 ± 4.40	103.79 ± 0.65	103.46 ± 0.20	106.98 ± 1.72	MFYC	104.24 ± 1.52	104.00 ± 1.29	103.95 ± 1.67	104.20 ± 1.78
	Y1	107.52 ± 5.70	102.55 ± 1.45	103.33 ± 1.79	107.42 ± 2.94	MFY1	103.32 ± 0.96	103.16 ± 0.94	103.09 ± 0.91	103.02 ± 1.07
	Y2	107.43 ± 3.16	105.03 ± 0.92	104.65 ± 0.55	108.64 ± 2.09	MFY2	103.94 ± 0.96	103.63 ± 0.91	103.55 ± 1.28	103.69 ± 1.56
Whiteness Index	YC	67.64 ± 1.18	68.83 ± 0.24	68.46 ± 0.45	69.12 ± 0.09	MFYC	69.16 ± 0.06	69.16 ± 0.28	69.09 ± 0.27	68.96 ± 0.16
	Y1	66.59 ± 1.10	68.02 ± 0.16	67.53 ± 0.28	67.88 ± 0.11	MFY1	68.97 ± 0.15	68.87 ± 0.25	68.74 ± 0.26	68.70 ± 0.26
	Y2	64.77 ± 0.79	65.92 ± 0.19	65.92 ± 0.24	66.74 ± 0.34	MFY2	68.15 ± 0.39	67.99 ± 0.45	68.15 ± 0.22	68.29 ± 0.16
Yellowness Index	YC	13.99 ± 2.85	16.74 ± 0.10	17.05 ± 0.24	16.84 ± 0.29	MFYC	16.61 ± 0.27	16.98 ± 0.37	17.09 ± 0.55	16.84 ± 0.50
	Y1	15.37 ± 3.19	18.45 ± 0.16	18.34 ± 0.40	18.42 ± 0.05	MFY1	17.14 ± 0.09	17.37 ± 0.26	17.56 ± 0.35	17.57 ± 0.34
	Y2	18.01 ± 2.59	20.43 ± 0.17	20.64 ± 0.35	19.51 ± 0.43	MFY2	17.90 ± 0.08	18.11 ± 0.12	18.33 ± 0.37	18.04 ± 0.21

YC: non-microfluidized control yogurt (0% oat β-glucan ratio), Y1: non-microfluidized yogurt with 0.25% oat β-glucan ratio, Y2: non-microfluidized yogurt with 0.50% oat β-glucan ratio, MFYC: microfluidized control yogurt (0% oat β-glucan ratio), MFY1: microfluidized yogurt with 0.25% oat β-glucan ratio, MFY2: microfluidized yogurt with 0.50% oat β-glucan ratio.

Whiteness index, indicates the degree of whiteness of a product in general [47]. The highest whiteness index values were observed for microfluidized samples. Among the samples YC and MFYC got highest whiteness index values. Inclusion of oat β-glucan resulted in a gradual decrease but with nearly 2 units among Y1 and Y2 (more than MFY1 and MFY2) possibly related to effect of homogeneous distribution of oat β-glucan by microfluidization (Table 7).

Yellowness index, changes by processing, exposure to light and chemicals and explains the degree of yellowness [47]. A gradual increase was observed in yellowness index values of non-microfluidized samples with nearly 2 units. Similar trend was observed in microfluidized samples but this time with small increments. Yellowness increased was initially thought to be associated with oat β-glucan inclusion and its ratio first, but microfluidized samples also presented a homogeneity in values in general that were lower than Y1 and Y2 (Table 7).

3.4. *Effect of Microfluidization and Storage Time on Some Physico-Chemical, Proximate Composition, and Textural Characteristics of Yogurt Samples*

The interaction of microfluidization and storage time was found significant in terms of dry matter, lactic acid, ash, and a^* ($p < 0.01$) and for pH, L^*, syneresis, and cohesiveness ($p < 0.05$) (Table 8). Non-microfluidized samples did not differ significantly in dry matter and ash in all storage time ($p > 0.05$), but non-microfluidized samples were different from microfluidized samples in terms of dry matter and ash (means ± SE) on all storage days ($p < 0.01$). In non-microfluidized and microfluidized samples, pH was significantly changed during the storage period, with no change on day 7 and day 14 (Table 8). However, among the groups (non-microfluidized and microfluidized), the pH changes on the 7th, 14th, and 21st days was significant ($p < 0.05$). In the samples on the 1st day, means ± SE of titratable acidity and pH values of non-microfluidized samples were lower than (less acidic) microfluidized ones. A gradual increase was observed in LA % of all yogurt samples, and from day 1 to day 21, these changes were 0.18 and 0.06 units for non-microfluidized and microfluidized samples, respectively (Table 8). During storage, yogurt samples' pH values gradually decreased [71,72]. According to Table 8, the pH drop in the samples was more pronounced between 1st and 7th days of storage and similar to results obtained it was associated with the high rate of lactose consumption and higher production of lactic acid and galactose in these days [73]. Prebiotics in apple pomace powder nourished lactic acid bacteria and acted in additional lactic acid production [72]; similar can be concluded for oat β-glucan fortified yogurt samples. High-pressure homogenization has an inhibitory effect on the growth of yogurt bacteria and *Lactobacillus delbrueckii* subsp. *bulgaricus* and stimulate the *Streptococcus thermophilus* [34]. The exposure of microfluidized milk blends to an additional treatment in which a slight temperature increase occurred during microfluidization could be responsible for this difference. The initial pH of both groups of yogurt samples ranged from 4.3–4.41, similar to Herrero and Requena (2006) [74], which ranged from 4.31–4.48. The activity of yogurt starter cultures *L. delbrueckii* subsp. *bulgaricus* and *Streptococcus thermophilus* continues during refrigerated conditions, and acidity increases [75]. The post-acidification reaction of yogurt starter cultures is responsible for the pH drop during the storage period [76–78] and related to β-galactosidase activity that is active at 0–5 °C is very closely linked to the drop of pH values lower than 4.2. At this stage survival degree of lactic acid bacteria is adversely affected, and syneresis is observed [75]. From Table 8, the pH values of all samples were higher than this limit indicating over-acidification.

Table 8. Some physicochemical, proximate composition, and textural characteristics of goat milk yogurt samples depend on microfluidization and storage time.

Parameter	Storage Period (day)	Sample $\overline{X} \pm S_x$ [1] Non-Microfluidized	Sample $\overline{X} \pm S_x$ [1] Microfluidized
Dry matter (%)	1	16.73 ± 0.0923 ^{Aa**}	15.58 ± 0.1620 ^{Bb**}
	7	16.62 ± 0.1160 ^{Aa**}	15.92 ± 0.2050 ^{Ab**}
	14	16.63 ± 0.1010 ^{Aa**}	15.60 ± 0.1300 ^{Bb**}
	21	16.66 ± 0.1300 ^{Aa**}	15.75 ± 0.1630 ^{ABb**}
Lactic Acid (%)	1	1.41 ± 0.0266 ^{Ca**}	1.46 ± 0.0125 ^{Ba**}
	7	1.53 ± 0.0250 ^{Ba**}	1.53 ± 0.0090 ^{Aa**}
	14	1.56 ± 0.0195 ^{ABa**}	1.54 ± 0.0096 ^{Aa**}
	21	1.59 ± 0.0175 ^{Aa**}	1.52 ± 0.0277 ^{Ab**}
pH	1	4.41 ± 0.0186 ^{Aa*}	4.37 ± 0.0247 ^{Aa*}
	7	4.33 ± 0.0287 ^{Ba*}	4.21 ± 0.0147 ^{BCb*}
	14	4.29 ± 0.0166 ^{BCa*}	4.24 ± 0.0250 ^{Bb*}
	21	4.26 ± 0.0173 ^{Ca*}	4.19 ± 0.0187 ^{Cb*}
Ash (%)	1	1.40 ± 0.0104 ^{Aa**}	1.09 ± 0.0482 ^{Bb**}
	7	1.39 ± 0.0105 ^{Aa**}	1.25 ± 0.0080 ^{Ab**}
	14	1.40 ± 0.0047 ^{Aa**}	1.28 ± 0.0044 ^{Ab**}
	21	1.40 ± 0.0074 ^{Aa**}	1.28 ± 0.0062 ^{Ab**}
L	1	67.34 ± 0.7740 ^{Bb*}	69.99 ± 0.1940 ^{Aa*}
	7	68.93 ± 0.4020 ^{Ab*}	69.93 ± 0.2400 ^{Aa*}
	14	68.64 ± 0.3660 ^{Ab*}	69.95 ± 0.1650 ^{Aa*}
	21	69.29 ± 0.3320 ^{Aa*}	69.91 ± 0.1210 ^{Aa*}
a	1	−2.14 ± 0.1170 ^{Aa**}	−2.08 ± 0.0901 ^{Aa**}
	7	−2.21 ± 0.1310 ^{Aa**}	−2.07 ± 0.0751 ^{Aa**}
	14	−2.22 ± 0.1190 ^{Aa**}	−2.08 ± 0.0836 ^{Aa**}
	21	−2.84 ± 0.2100 ^{Bb**}	−2.08 ± 0.1010 ^{Aa**}
Syneresis (%)	1	18.30 ± 2.390 ^{Ba*}	12.50 ± 1.460 ^{ABb*}
	7	18.43 ± 2.050 ^{Ba*}	11.58 ± 1.320 ^{Bb*}
	14	22.77 ± 2.970 ^{Aa*}	12.23 ± 1.090 ^{ABb*}
	21	19.16 ± 2.040 ^{Ba*}	14.66 ± 1.600 ^{Ab*}
Cohesiveness (N)	1	0.68 ± 0.0456 ^{Ba*}	0.68 ± 0.0727 ^{Aa*}
	7	0.72 ± 0.039 ^{Aa*}	0.68 ± 0.0746 ^{Ab*}
	14	0.72 ± 0.0452 ^{Aa*}	0.70 ± 0.0739 ^{Aa*}
	21	0.73 ± 0.0347 ^{Aa*}	0.67 ± 0.0721 ^{Ab*}

[1] Data are the means ± SE (n = 9) of all non-microfluidized goat milk yogurt samples (YC, Y1, Y2) and *microfluidized* (MFYC, MFY1, MFY2) goat milk yogurt samples for the parameters in storage days of 1, 7, 14 and 21, different capital letters in the same column indicate significant difference (* $p < 0.05$, ** $p < 0.01$), different lowercase letters in the same row indicate significant difference (* $p < 0.05$, ** $p < 0.01$).

The L^* values increased in non-microfluidized samples, and the first day's L^* values differed significantly from the other storage days ($p < 0.05$) (Table 8). L^* of microfluidized samples remained almost constant ($p > 0.05$), possibly due to supplied homogeneity and stability in the samples. Both non-microfluidized and microfluidized samples got negative a^* values throughout the storage period, indicating they exist in green color space. The a^* values increased greenness and became more intense from day 1 to day 21 in non-microfluidized samples, with a significant increase on the 21st day. Non-microfluidized samples were more greenish than microfluidized ones and significant only on the 21st day ($p < 0.01$). In microfluidized samples, similar to L^*, a^* remained nearly constant during storage ($p > 0.05$) (Table 8).

According to microfluidization and storage time interaction, the syneresis of non-microfluidized samples was higher than microfluidized samples on all storage days ($p < 0.05$). Microfluidization showed superiority in preventing syneresis in oat β-glucan fortified goat milk yogurt systems (Table 8). Syneresis is a textural defect considered a problem in set-type yogurt. It is especially possible in non-fat, low-fat, and low-dry matter yogurt systems [79]. Casein content is known to be more effective on syneresis than serum proteins. In the samples, casein-β-glucan interaction supported syneresis and presented a significant two-layered structure in increasing β-glucan concentrations. However, all yogurt samples got similar fat content but different dry matter content due to oat β–glucan fortification (Table 5). This time microfluidization created a difference among yogurt samples. Syneresis decreased by microfluidization and increased by storage time, related to increased acidity (Table 8). Microfluidization served uniform distribution of oat β-glucan and enhanced water holding capacity of oat β-glucan that resulted in lower syneresis (Table 8). The researchers observed a decrease in syneresis in yogurt during storage [80,81]. Syneresis decreased by storage, especially in cow's milk yogurt samples with β-glucan composite [19]. However, the reverse was valid for the samples and can be explained by decreased pH values (Table 8).

In non-microfluidized samples, cohesiveness increased from day 1 to day 21 and was statistically significant on the first day than the other days ($p < 0.05$). In microfluidized samples, cohesiveness remained nearly the same during storage ($p > 0.05$). A significant difference was observed between the non-microfluidized and microfluidized samples on the 7th and 21st day (Table 8). In the back-extrusion method on a force-time graph, cohesiveness is the maximum negative force, and it determines the resistance of the samples to the moving disk going away from the sample [33]. As the strength of the internal bonds increases, internal stickiness increases [82]. Microfluidization maintains a more homogeneous and creamy texture, but this is not related to the fat content; it defines the smoothness of the product [83]. In low-fat yogurts, microfluidization improves the texture in that creamy, softness, enhanced density, and viscosity in mouthfeel are supplied [35]. Lower cohesiveness values in microfluidized yogurt samples can depend on the increased smoothness and creaminess (Table 8). Opposite to ours, firmness, elasticity, adhesiveness, and gumminess of milk gels can be enhanced, but cohesiveness can be diminished by β-glucan with specific molecular characteristics [84]. Cohesiveness values of microfluidized yogurts with 0.50% oat β-glucan were higher than non-microfluidized candidates due to the homogeneous structure formed by microfluidization since in non-microfluidized samples with 0.50% oat β-glucan concentration a distinct phase separation was observed due to thermodynamic incompatibility of casein and oat β-glucan. Although further research is required, high-pressure homogenization mainly causes differences in gel microstructure rather than texture in yogurt samples [85].

3.5. *Effect of Storage Time on Color, Microbiology, and Texture Attributes of Yogurt Samples*

Storage time effect was found significant on the parameters for all non-microfluidized and microfluidized goat milk yogurt samples and given in Table 9. The b^* value of the samples differed significantly on the first day of storage ($p < 0.05$) and increased gradually up to day 14 with no significant change in the following days of storage ($p > 0.05$). Similarly, b^* values of apple fiber-added strained yogurts increased significantly during the 21-day storage period [13].

Table 9. The individual effect of storage time on b^* value, the viability of yogurt bacteria, and some texture characteristics of goat milk yogurt samples.

Parameter	Storage Time (Day)	$\overline{X} \pm S_x$ [1]
b^*	1	7.96 ± 0.3960 ^{B*}
	7	8.75 ± 0.1180 ^{A*}
	14	8.80 ± 0.1280 ^{A*}
	21	8.70 ± 0.1030 ^{A*}
Firmness (N)	1	1.43 ± 0.147 ^{B**}
	7	1.50 ± 0.151 ^{AB**}
	14	1.51 ± 0.161 ^{A**}
	21	1.53 ± 0.159 ^{A**}
Consistency (N×s)	1	14.45 ± 1.68 ^{B*}
	7	14.96 ± 1.72 ^{A*}
	14	15.12 ± 1.73 ^{A*}
	21	15.15 ± 1.76 ^{A*}
S. thermophilus (log CFU/g)	1	9.93 ± 0.209 ^{B**}
	7	10.98 ± 0.369 ^{A**}
	14	10.23 ± 0.276 ^{B**}
	21	10.13 ± 0.293 ^{B**}
Lb. delbrueckii subsp. *bulgaricus* (log CFU/g)	1	8.62 ± 0.183 ^{A**}
	7	8.08 ± 0.127 ^{B**}
	14	7.36 ± 0.142 ^{C**}
	21	6.85 ± 0.217 ^{D**}

[1] Data are the means ± SE (n = 18) of both non-microfluidized and microfluidized goat milk yogurt samples for the parameters in storage days of 1, 7, 14, and 21, different capital letters in the same column indicate a significant difference (* $p < 0.05$, ** $p < 0.01$).

Firmness values changed significantly on the 1st and 14th, and 21st days of storage ($p < 0.01$) and increased from day 1 to day 21. In dietary fiber-added strained yogurt samples, firmness increased with storage time. It was explained by the increased water-holding capacity of milk proteins and dietary fibers during storage [13]. Similar to firmness values, the consistency of yogurt samples increased slightly during the storage period, and it was significantly different on the 1st day of storage than in the remaining periods ($p < 0.05$) (Table 9). Consistency is a measure of the thickness of samples [33]. Protein rearrangement occurred during storage, increasing firmness and consistency [71].

The viability of the yogurt starter culture presented a storage time-dependent trend with statistical significance ($p < 0.01$). The health benefits of yogurt can be observed when the live bacteria level is 10^8–10^9 CFU/g [65]. According to Table 9, *Streptococcus thermophilus* increased in number of colonies until day 7 significantly ($p < 0.01$), but no significant difference was observed on the 14th day and 21st day compared to the 1st day. Table 9 shows a slight decrease in colonies from day 7 to 21. *Lb. delbrueckii* subsp. *bulgaricus* showed a gradual decrease in the number of colonies during storage ($p < 0.01$) (Table 9). The inhibitory effect of high-pressure homogenization on the growth of yogurt bacteria was determined by Lanciotti et al. (2004) [34]. Consistent with the results of Gee et al. (2007) [51] and Valerie (2009) [54], Lanciotti et al. (2004) [34] declared that high-pressure homogenization stimulates the growth of *S. thermophilus* than *Lb. delbrueckii* subsp. *bulgaricus* independent of the applied pressure values. Similarly, during the cold storage of yogurt, the number of *L. delbrueckii* subsp. *bulgaricus* colonies decreased over time; a more pronounced decrease was observed in high-pressure processed milk used in yogurt production (at 200 MPa pressure (30 °C or 40 °C), and this situation was found in relation with the accumulation of pyruvate having a toxic effect on the starter bacteria [86].

3.6. Sensory Evaluation

The sensory evaluation results of non-microfluidized (YC, Y1, and Y2) and microfluidized (MFYC, MFY1, and MFY2) oat β-glucan fortified goat milk yogurt samples were shown in Figure 2 for the storage period of 21 days. In terms of color/appearance (top left), among the non-microfluidized group YC (control) and Y1 got very close scores (almost 5 points) except Y2 during the 1st, 7th, and 14th days of storage. On day 21, YC was evaluated with the highest and Y2 with the lowest score. In sensory evaluation, no change was pronounced in the color/appearance of the control sample during the storage period. According to panelists' verbal definitions, the control sample (YC) was whiter in color than the others, and yellowness increased in the order of Y1 and Y2 depending on the oat β-glucan concentration.

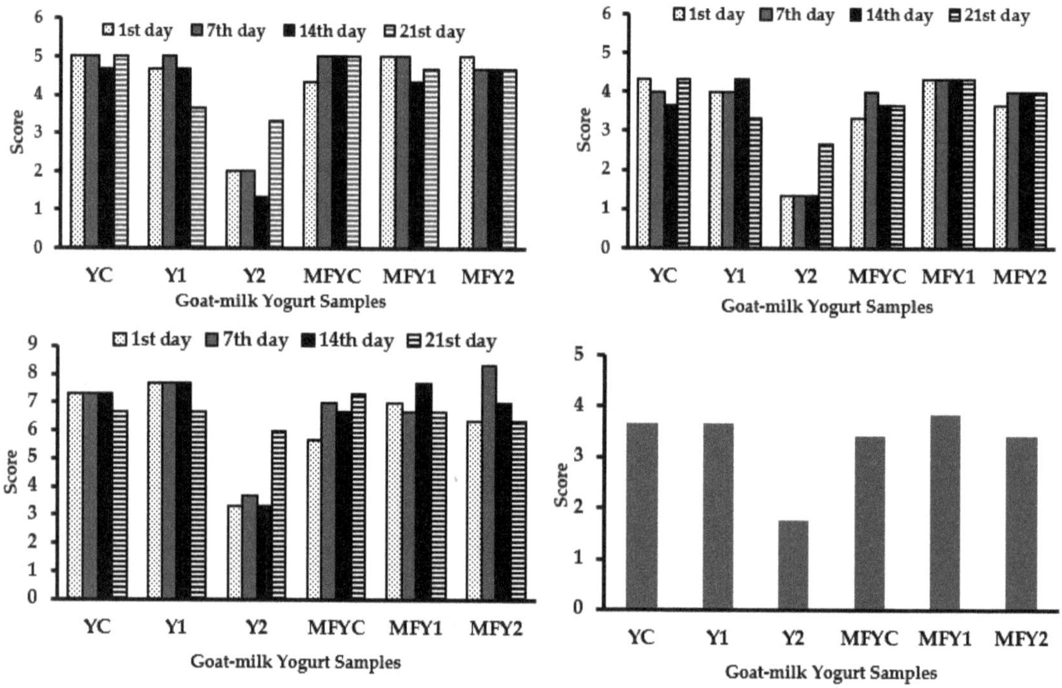

Figure 2. Scores of sensory parameters (Color/appearance (top left), body/texture (top right), taste/flavor (bottom left), general acceptability (bottom right) for non-microfluidized and microfluidized oat β-glucan fortified goat milk yogurt samples (values are the means of the treatments), (n = 3). YC: non-microfluidized control yogurt (0% oat β-glucan ratio), Y1: non-microfluidized yogurt with 0.25% oat β-glucan ratio, Y2: non-microfluidized yogurt with 0.50% oat β-glucan ratio, MFYC: microfluidized control yogurt (0% oat β-glucan ratio), MFY1: microfluidized yogurt with 0.25% oat β-glucan ratio, MFY2: microfluidized yogurt with 0.50% oat β-glucan ratio.

No apparent texture defect was detected in Y1, and very little serum accumulation on the surface of the yogurt was observed. Y2 was the least favored sample on all storage days regarding all sensory parameters (Figure 2). From the 1st day of storage, a definite phase separation (a two-layered structure) was observed in Y2. The serum separation and viscosity of yogurt are essential in its sensorial acceptability [87]. Syneresis increased in yogurts above the 0.3% β-glucan level (critical β-glucan level), and the color darkened compared to the control sample [22]. Y1 had a lower amount of oat β-glucan than the reported critical concentration, so the phase separation was not so apparent. However, in Y2, a very distinct two-layered structure was observed (Figure 1).

In general, microfluidized samples were superior to non-microfluidized samples with more homogeneous scores (Figure 2, top left). Microfluidization treatment improved the samples both texturally and also in color/appearance properties (Figure 2). The microfluidized samples MFYC, MFY1, and MFY2 were improved in color and appearance by homogenization effect acting on increased lightness (L^*). In particular, the two-layered phase separation problem occurred due to increasing oat β-glucan concentration was avoided entirely by microfluidization (Figure 1). In terms of color/appearance, MFY2 was evaluated with higher scores than Y2 in all days of storage (Figure 2).

YC and Y1 got higher scores in texture among non-microfluidized samples, and Y2 was the least liked sample in terms of texture (Figure 2). YC got a typical yogurt gel structure but was slightly weak due to goat milk characteristics. Goat's milk is similar to cow's milk in total dry matter, protein, fat, and lactose levels [3,81]. However, casein micelles differ significantly in structure, composition, and size, leading to differences in physicochemical properties [81]. Goat milk contains higher levels of non-protein nitrogen and mineral substances and a higher proportion of small-diameter fat globules [3,81]. Qualitative and quantitative differences exist in proteins, especially αs1-casein [1]. Fat content and fat globule size affect the gel firmness. Goat milk yogurts are characterized by soft consistency, low viscosity, low firmness, weak gel structure, and syneresis susceptibility when compared with cow's milk [5–9,88]. In non-microfluidized samples, the panelists reported phase separation defects becoming intense with increasing concentrations of oat β-glucan (Figure 1). By the way, this resulted in lower scores in Y2 (Figure 2, top right). Y1 exhibited an excellent mouth feel, a non-granular but lumpy texture with a pudding-like structure. According to sensorial texture, Y1 was not thin as YC; besides that, firmer than YC. Panelists reported a significant phase separation in Y2 due to thermodynamic incompatibility between milk proteins and β-glucan. Excessive serum separation, weak gel structure, granular texture, drinkable consistency, and an untypical gel structure far from yogurt texture are some verbal definitions of defects observed for Y2 during storage (Table 3). Similar to that, researchers determined significant syneresis in yogurt samples with 0.50% oat β-glucan product, so they gathered lower texture scores [28].

In general, microfluidization improved the textural properties of all samples, also measured instrumentally. However, as expected MFYC sample was thinner in structure than MFY1 and MFY2 related to goat milk's weak textural characteristics. A creamy structure was observed in MFYC, MFY1 and MFY2. This creaminess created lower firmness values in MFYC than in YC (data not given). The creamy texture was regarded as a positive consequence of microfluidization on texture. In a study conducted by Brennan and Tudorica (2008) [39], it was revealed that the creamy texture of the samples containing 0.5% and 1.5% β-glucan increased with microfluidization. Similarly, in high-pressure applied samples, the creamy structure became dominant [35,83,89]. Microfluidized oat β-glucan fortified goat milk yogurt samples got higher scores than the control of this group, and MFY1 was evaluated with the highest and identical scores during storage days (Figure 2). Increasing concentrations of oat β-glucan created a structure different than yogurt gel and defined as pudding-like structure and besides that phase separation defect was entirely prevented by microfluidization. However, in microfluidized samples, firmness decreased in the order of MFY1, MFY2, and MFYC (data not given). MFY2 samples were firmer than MFYC due to the joint effect of water-holding capacity oat β-glucan as a dietary fiber and microfluidization.

Regarding taste/flavor, yogurt samples were evaluated over 10 points. Among the non-microfluidized samples, Y1 was the most liked sample with constant scores on the 1st, 7th, and 14th days, and Y2 was the least liked on almost all days. In non-microfluidized oat β-glucan fortified samples, it was advised to consume after stirring due to phase separation defect, especially in Y2, since the two-layered structure affected the taste perception. In oat β-glucan dairy systems, a two-layered structure is formed due to thermodynamic incompatibility, with the upper phase being polysaccharide-rich and the lower phase being protein-rich [28]. A goaty and salty taste was generally pronounced in YC due to goat milk origin. Goat's milk is characterized by a unique goaty and salty taste/flavor [12] and in the samples of Y1 and Y2, cereal flavor suppressed the salty and goaty taste and the typical yogurt flavor. Typical yogurt flavor and acidity were detected mainly in YC. Grain flavor dominated with increasing concentrations of oat β-glucan.

Like other sensory attributes, the taste/flavor of oat β-glucan fortified goat milk yogurt samples were improved by microfluidization. A noticeable improvement in taste and flavor was maintained in the 0.50% oat β-glucan fortified sample (MFY2). Taste/flavor scores indicated fluctuation during shelf life; the highest scores were obtained for MFYC on the 1st day, for MFY2 on the 14th day, and MFY2 on the 7th day (Figure 2). Oat fiber use in plain yogurt improved texture but decreased flavor attributes in plain yogurt [90] According to a study, in the samples with 0.5% and 1.5% β-glucan, a creamy texture and a soft feeling were maintained on the palate with no election among the samples [39]. The use of β-glucan hydrocolloidal composite in ratios of 0.25% or 0.50% in non-fat yogurts was satisfying, and these were evaluated with scores close to the control that was the most preferred sample [19]. As previously discussed in the related section, in terms of general acceptability, Y2 was the most significantly different sample in the non-microfluidized group and also in its microfluidized couple ($p < 0.01$) (Table 6). To sum up, since the microfluidization process improved the overall quality of the samples and eliminated the physical defects that occurred in the yogurt sample due to oat β-glucan existence, it is possible to say that the general acceptability of microfluidized oat β-glucan fortified samples was higher than their non-microfluidized candidates.

4. Conclusions

The study was designed to observe the compatibility of oat β-glucan and goat milk in yogurt system and together with observing the effect of microfluidization treatment in this system. Oat β-glucan fortification can suppress goaty and salty taste of goat milk yogurt so could increase the acceptability depending on the inclusion ratio. Sensorial and textural improvement can be maintained by microfluidization in goat milk yogurt samples independent on oat β-glucan ratio but oat β-glucan ratio could be determinative on sensorial and textural properties of non-microfluidized yogurt samples. It could be concluded that microfluidization can be successfully used in oat β-glucan fortified dairy systems for preventing phase separation occurred due to thermodynamic incompatibility reaction between casein and oat β-glucan.

Author Contributions: Investigation, E.A.A. and A.G. (Asuman Gursel); methodology, E.A.A., A.G. (Asuman Gursel), A.G. (Ayse Gursoy) and B.M.; formal analysis, E.A.A.; data curation, E.A.A. and A.G. (Asuman Gursel); conceptualization, E.A.A. and A.G. (Asuman Gursel); validation, E.A.A., A.G. (Asuman Gursel), A.G. (Ayse Gursoy) and B.M.; writing—reviewing and editing, E.A.A. and A.G. (Asuman Gursel); supervision, A.G. (Asuman Gursel), A.G. (Ayse Gursoy) and B.M.; funding acquisition, A.G. (Asuman Gursel). All authors have read and agreed to the published version of the manuscript.

Funding: This work was prepared from Ph. D. Thesis of E.A.A. that was funded by Ankara University Scientific Research Projects Coordination Unit, Ankara, Turkey (Project number: 14L0447004) by the supervision of Asuman Gursel. The authors thank the organization for financial support for the project.

Data Availability Statement: All data are contained within the article.

Acknowledgments: We would like to thank the farmer Sudi Pınar (Yeni Pecenek Village, Ankara, Turkey) supplying raw goat's milk for the project.

Conflicts of Interest: The authors declare no conflict of interest.

References

1. Haenlein, G.F.W. Goat milk in human nutrition. *Small Rumin. Res.* **2004**, *51*, 155–163. [CrossRef]
2. Raynal-Ljutovac, K.; Lagriffoul, G.; Paccard, P.; Guillet, I.; Chilliard, Y. Composition of goat and sheep milk products: An update. *Small Rumin. Res.* **2008**, *79*, 57–72. [CrossRef]

3. Attaie, R.; Richter, R.L. Size distribution of fat globules in goat milk. *J. Dairy. Sci.* **2000**, *83*, 940–944. [CrossRef]
4. Nguyen, H.T.; Afsar, S.; Day, L. Differences in the microstructure and rheological properties of low-fat yoghurts from goat, sheep and cow milk. *Food Res. Int.* **2018**, *108*, 423–429. [CrossRef]
5. Zee, J.A.; Tirand-Collet, P.; Lavigne, C. Amelioration de la qualite du yogourt fabriqué à partir de lait de chèvre. In Proceedings of the 23rd International Dairy Congress: Brief Communications and Poster Abstracts, Montreal, QC, Canada, 6–12 October 1990; Volume 1, p. 23.
6. Agnihotri, M.K.; Prasad, V.S.S. Biochemistry and processing of goat milk and milk products. *Small Rumin. Res.* **1993**, *12*, 151–170. [CrossRef]
7. Kalantzopoulos, G. Etat de la recherche sur le lait de chevre en Grece. *Lait* **1993**, *73*, 431–441. [CrossRef]
8. Ambrosoli, R.; Di Stasio, L.; Mazzoco, P. Content of αs1-casein and coagulation properties in goat milk. *J. Dairy. Sci.* **1988**, *71*, 24–28. [CrossRef]
9. Addeo, F.; Masi, P.; Rubino, R. The aptitude of caprine milk to cheese making. Relationship between αs1-casein content and rheological properties of curd and cheese. *Dairy. Sci. Abstr.* **1992**, *54*, 472.
10. Guo, M. Goat's milk. In *Encyclopedia of Food Sciences and Nutrition*; Caballero, B., Trugo, L., Finglas, P., Eds.; Academic Press: London, UK, 2003; pp. 2944–2949.
11. Park, Y.W.; Juarez, M.; Ramos, M.; Haenlein, G.F.W. Physico-chemical characteristics of goat and sheep milk. *Small Rumin. Res.* **2007**, *68*, 88–113. [CrossRef]
12. Pal, M.; Dudhrejiya, T.P.; Pinto, S.; Brahamani, D.; Vijayageetha, V.; Reddy, Y.K.; Kate, P. Goat milk products and their significance. *Beverage Food World* **2017**, *44*, 21–25.
13. Seçkin, A.K.; Baladura, E. Effect of using some dietary fibers on color, texture and sensory properties of strained yogurt. *J. Food* **2012**, *37*, 63–69.
14. Butt, S.M.; Tahir-Nadeem, M.; Iqbal Khan, M.K.; Shabir, R. Oat: Unique among the cereals. *Eur. J. Nutr.* **2008**, *47*, 68–79. [CrossRef]
15. Yang, Y.Y.; Ma, S.; Wang, X.X.; Zheng, X.L. Modification and application of dietary fiber in foods. *J. Chem.* **2017**, *2017*, 9340427. [CrossRef]
16. Dello-Staffolo, M.; Bertola, N.; Martino, M.; Bevilacqua, A. Influence of dietary fiber addition on sensory and rheological properties of yogurt. *Int. Dairy. J.* **2004**, *14*, 263–268. [CrossRef]
17. Mousavi, M.; Heshmati, A.; Garmakhany, A.D.; Vahidinia, A.; Taheri, M. Optimization of the viability of Lactobacillus acidophilus and physico-chemical, textural and sensorial characteristics of flaxseed-enriched stirred probiotic yogurt by using response surface methodology. *LWT-Food Sci. Technol.* **2019**, *102*, 80–88. [CrossRef]
18. Ahmad, A.; Munir, B.; Abrar, M.; Bashir, S.; Adnan, M.; Tabassum, T. Respective of β-glucan as functional ingredient for food industry. *J. Nutr. Food Sci.* **2012**, *2*, 133–139.
19. Sahan, N.; Yasar, K.; Hayaloglu, A.A. Physical, Chemical and Flavour Quality of Non-fat Yogurt as Affected by a β-glukan Hydrocolloidal Composite during Storage. *Food Hydrocoll.* **2008**, *22*, 1291–1297. [CrossRef]
20. Qu, X.; Nazarenko, Y.; Yang, W.; Nie, Y.; Zhang, Y.; Li, B. Effect of oat β-glucan on the rheological characteristics and microstructure of set-type yogurt. *Molecules* **2021**, *26*, 4752. [CrossRef]
21. U.S. Food and Drug Administration. Food labeling: Health claims; oats and coronary heart disease. Interim Final rule. *Fed. Regist.* **1997**, *62*, 3583–3601.
22. Singh, M.; Sanghoon, K.; Sean, X.L. Effect of Prufied Oat β-Glukan on Fermentation of Set-Style Yogurt Mix. *J. Food Sci.* **2012**, *77*, 195–201. [CrossRef]
23. Raikos, V.; Grant, S.B.; Hayes, H.; Ranawana, V. Use of β-glucan from spent brewer's yeast as a thickener in skimmed yogurt: Physicochemical, textural, and structural properties related to sensory perception. *J. Dairy. Sci.* **2018**, *101*, 5821–5831. [CrossRef] [PubMed]
24. Rosa-Sibakov, N.; de Oliveira Carvalho, M.J.; Lille, M.; Nordlund, E. Impact of enzymatic hydrolysis and microfluidization on the techno-functionality of oat bran in suspension and acid milk gel models. *Foods* **2022**, *11*, 228. [CrossRef] [PubMed]
25. Charalampopoulos, D.; Wang, R.; Pandilla, S.S.; Webb, C. Application of cereals and cereal components in functional foods: A review. *Int. J. Food Microbiol.* **2002**, *79*, 131–141. [CrossRef] [PubMed]
26. Lee, S.; Inglett, G.E.; Palmquist, D.; Warner, K. Flavor and texture attributes of foods containing β-glucan-rich. *LWT-Food Sci. Technol.* **2009**, *42*, 350–357. [CrossRef]
27. Daou, C.; Zhang, H. Oat beta-glucan: Its role in health promotion and prevention of diseases. *Compr. Rev. Food Sci. Food Saf.* **2012**, *11*, 355–365. [CrossRef]
28. Vasiljevic, T.; Kealy, T.; Mishra, V.K. Effects of β-Glucan Addition to a Probiotic Containing Yogurt. *Food Chem. Toxicol.* **2007**, *72*, 405–411. [CrossRef]
29. Sarantis, S.D.; Eren, N.M.; Kowalcyk, B.; Jimenez–Flores, R.; Alvarez, V.B. Thermodynamic interactions of micellar casein and oat β-glucan in a model food system. *Food Hydrocoll.* **2021**, *115*, 106559. [CrossRef]
30. Sharafbafi, N.; Tosh, S.M.; Alexander, M.; Corredig, M. Phase behaviour, rheological properties, and microstructure of oat β-glucan-milk mixtures. *Food Hydrocoll.* **2014**, *41*, 274–280. [CrossRef]
31. Zhao, Y.; Fu, R.; Li, J. Effects of the β-glucan, curdlan, on the fermentation performance, microstructure, rheological and textural properties of set yogurt. *LWT-Food Sci. Technol.* **2020**, *128*, 109449. [CrossRef]

32. Ye, R.; Harte, F. High pressure homogenization to improve the stability of casein–hydroxypropyl cellulose aqueous systems. *Food Hydrocoll.* **2014**, *35*, 670–677. [CrossRef]
33. Ciron, C.I.E.; Gee, V.L.; Kelly, A.L.; Auty, M.A.E. Comparison of the effects of high-pressure microfluidization and conventional homogenization of milk on particle size, water retention and texture of non-fat and low-fat yoghurts. *Int. Dairy. J.* **2010**, *20*, 314–320. [CrossRef]
34. Lanciotti, R.; Vannini, L.; Pittia, P.; Elisabetta, M.; Guerzoni, M.E. Suitability of high-dynamic-pressure- treated milk for the production of yoghurt. *Food Microbiol.* **2004**, *21*, 753–760. [CrossRef]
35. Ciron, C.I.E.; Gee, V.L.; Kelly, A.L.; Auty, M.A.E. Effect of microfluidization of heat-treated milk on rheology and sensory properties of reduced fat yoghurt. *Food Hydrocoll.* **2011**, *25*, 170–1476. [CrossRef]
36. Vargas, M.; Cháfer, M.; Albors, A.; Chiralt, A.; González-Martínez, C. Physicochemical and sensory characteristics of yoghurt produced from mixtures of cows' and goats' milk. *Int. Dairy J.* **2008**, *18*, 1146–1152. [CrossRef]
37. Alqahtani, N.K.; Darwish, A.A.; El-Menawy, R.K.; Alnemr, T.M.; Aly, E. Textural and organoleptic attributes and antioxidant activity of goat milk yoghurt with added oat flour. *Int. J. Food Prop.* **2021**, *24*, 433–445. [CrossRef]
38. Kaur, R.; Riar, C.S. Sensory, rheological and chemical characteristics during storage of set type full fat yoghurt fortified with barley β-glucan. *J. Food Sci. Technol.* **2020**, *57*, 41–51. [CrossRef]
39. Brennan, C.S.; Tudorica, C.M. Carbohydrate-based fat replacers in the modification of the rheological, textural and sensory quality of yoghurt: Comparative study of the utilisation of barley beta-glukan, guar gum and inulin. *Int. J. Food Sci. Technol.* **2008**, *43*, 824–833. [CrossRef]
40. Lazaridou, A.; Serafeimidou, A.; Biliaderis, C.G.; Moschakis, T.; Tzanetakis, N. Structure development and acidification kinetics in fermented milk containing oat β-glucan, a yogurt culture and a probiotic strain. *Food Hydrocoll.* **2014**, *39*, 204–214. [CrossRef]
41. *TS 1018*; İnek Sütü-Çiğ. (Raw Cow's Milk). Turkish Standards Institute: Ankara, Turkey, 2002.
42. *No. 149A*; Dairy Starter Cultures of Lactic Acid Bacteria (LAB). International Dairy Federation (IDF): Brussels, Belgium, 1997.
43. *No. 20B*; International Dairy Federation (IDF). Determination of Nitrogen Content. International Dairy Federation: Brussels, Belgium, 1993.
44. Renner, E. *Milchpraktikum*; Justus-Liebig-Universität: Giessen, Germany, 1993.
45. Bradley, R.L., Jr.; Arnold, E., Jr.; Barbano, D.M.; Semerad, R.G.; Smith, D.E.; Vines, B.K. Vines. Chemical and physical methods. In *Standard Methods for the Examination of Dairy Products*; Marshall, R.T., Ed.; American Public Health Association: Washington, DC, USA, 1993; Volume 70, pp. 433–531.
46. Anonymous. Mixed-Linkage Beta-Glucan Assay Procedure (McLeary Method). 2015. Available online: https://www.megazyme.com/beta-glucan-assay-kit (accessed on 21 November 2015).
47. Pathare, P.B.; Opara, U.L.; Al-Said, F.A.J. Color measurement and analysis in fresh and processed foods: A review. *Food Bioprocess Technol.* **2013**, *6*, 36–60. [CrossRef]
48. Robitaille, G.; Tremblay, A.; Moineau, S.; St-Gelais, D.; Vadeboncoeur, C.; Britten, M. Fat-free yogurt made using a galactose-positive exopolysaccharide-producing recombinant strain of Streptococcus thermophilus. *J. Dairy. Sci.* **2009**, *92*, 477–482. [CrossRef]
49. Bodyfelt, F.W.; Tobias, J.; Trout, G.M. *The Sensory Evaluation of Dairy Products*; Van Nostrand, R., Ed.; Springer Science & Business Media: New York, NY, USA, 1988.
50. Tribby, D. Yogurt. In *The Sensory Evaluation of Dairy Products*; Clark, S., Costello, M., Drake, M.A., Bodyfelt, F., Eds.; Springer Science Business Media LLC: New York, NY, USA, 2009; pp. 191–224.
51. Gee, V.L.; Vasanthan, T.; Temelli, F. Viscosity of model systems enriched with β-glucan as influenced by starter cultures. *Int. Dairy. J.* **2007**, *17*, 1083–1088. [CrossRef]
52. Snart, J.; Bibilonil, R.; Grayson, T.; Lay, C.; Zhang, H.; Allison, G.E.; Laverdiere, J.K.; Temelli, F.; Vasanthan, T.; Bell, R.; et al. Supplementation of the Diet with High-Viscosity Beta-Glucan Results in Enrichment for Lactobacilli in the Rat Cecum. *Appl. and Environ. Microbiol.* **2005**, *72*, 1925–1931. [CrossRef]
53. Donkor, O.N.; Tsangalis, D.; Shah, N.P. Viability of probiotic bacteria and concentrations of organic acids in commercial yoghurts during refrigerated storage. *Food Aust.* **2007**, *59*, 121–126.
54. Valerie, A.R. Viability of Bifidobacteria in Yogurts Containing Oat Beta-Glucan and/or Corn Starch during Cold Storage. Master's Thesis, Iowa State University, Ames, IA, USA, 2009, *unpublished*.
55. Angelov, A.; Gotcheva, V.; Kuncheva, R.; Hristozova, T. Development of a new oat-based probiotic drink. *Int. J. Food Microbiol.* **2006**, *112*, 75–80. [CrossRef]
56. Rosburg, V.; Boylston, T.; White, P. Viability of Bifidobacteria Strains in Yoghurt with Added Oat Beta-Glucan and Corn Starch during Cold Storage. *J. Food Sci.* **2010**, *75*, 439–444.
57. Kearney, N.; Stack, H.M.; Tobin, J.T.; Chaurin, V.; Fenelon, M.A.; Fitzgerald, G.F.; Ross, R.P.; Stanton, C. Lactobacillus. paracasei NFBC 338 producing recombinant beta-glucan positively influences the functional properties of yoghurt. *Int. Dairy. J.* **2011**, *21*, 561–567. [CrossRef]
58. Lazaridou, A.; Biliaderis, C.G. Molecular aspects of cereal β-glucan functionality: Physical properties, technological applications and physiological effects. *J. Cereal Sci.* **2007**, *46*, 101–118. [CrossRef]
59. Ketenoglu, O.; Mert, B.; Tekin, A. Effects of microfluidized dietary fibers on stability properties of emulsions. *J. Texture Stud.* **2014**, *45*, 295–306. [CrossRef]

60. Martensson, O.; Andersson, C.; Andersson, K.; Öste, R.; Holst, O. Formulation of an oat-based fermented product and its comparison with yoghurt. *J. Sci. Food Agric.* **2001**, *81*, 1314–1321. [CrossRef]
61. Kurtuldu, O.; Ozcan, T. Effect of β-glucan on the properties of probiotic set yoghurt with Bifidobacterium animalis subsp. lactis strain Bb-12. *Int. J. Dairy Technol.* **2018**, *71*, 157–166. [CrossRef]
62. Pedras, M.M.; Tribst, A.A.; Cristianini, M. Effects of high-pressure homogenisation on physicochemical characteristics of partially skimmed milk. *Int. J. Food Sci. Technol.* **2014**, *49*, 861–866. [CrossRef]
63. Lemay, A.; Paquin, P.; Lacroix, C. Influence of microfluidization of milk on cheddar cheese composition, color, texture, and yield. *J. Dairy Sci.* **1994**, *77*, 2870–2879. [CrossRef]
64. Gervilla, R.; Ferragut, V.; Guamis, B. High Hydrostatic Pressure Effects on Color and Milk- Fat Globule of Ewew's Milk. *J. Food Sci.* **2001**, *66*, 880–885. [CrossRef]
65. Trujillo, A.J.; Capellas, M.; Saldo, J.; Gervilla, R.; Guamis, B. Applications of high-hydrostatic pressure on milk and dairy products: A Review. *Innov. Food Sci. Emerg. Technol.* **2002**, *3*, 295–307. [CrossRef]
66. Martin-Diana, A.B.; Janer, C.; Pelaez, C.; Requena, T. Development of a fermented goat's milk containing probiotic bacteria. *Int. Dairy. J.* **2003**, *13*, 827–833. [CrossRef]
67. Iordache, M.; Jelen, P. High pressure microfluidization treatment of heat denatured whey proteins for improved functionality. *Innov. Food Sci. Emerg. Technol.* **2003**, *4*, 367–376. [CrossRef]
68. Penna, A.L.B.; Gurram, S.; Barbosa-Ca'novas, G.V. High hydrostatic pressure processing on microstructure of probiotic low-fat yogurt. *Food Res. Int.* **2007**, *40*, 510–519. [CrossRef]
69. Chen, J.; Gao, D.; Yang, L.; Gao, Y. Effect of microfluidization process on the functional properties of insoluble dietary fiber. *Food Res. Int.* **2013**, *54*, 1821–1827. [CrossRef]
70. Ronkart, S.N.; Paquot, M.; Deroanne, C.; Fougnies, C.; Besbes, S.; Blecker, C.S. Development of gelling properties of inulin by microfluidization. *Food Hydrocoll.* **2010**, *24*, 318–324. [CrossRef]
71. Anli, E.A.; Kıral, A.G.; Sert, D. Physicochemical and Textural Properties of Ready-to-eat yogurt fortified with dried Purslane (*Portulaca oleracea* L.). *J. Food* **2021**, *46*, 229–242. [CrossRef]
72. Popescu, L.; Ceșco, T.; Gurev, A.; Ghendov-Mosanu, A.; Sturza, R.; Tarna, R. Impact of apple pomace powder on the bioactivity, and the sensory and textural characteristics of yogurt. *Foods* **2022**, *11*, 3565. [CrossRef]
73. Béal, C.; Skokanova, J.; Latrille, E.; Martin, N.; Corrieu, G. Combined effects of culture conditions and storage time on acidification and viscosity of stirred yogurt. *J. Dairy. Sci.* **1999**, *82*, 673–681. [CrossRef]
74. Herrero, A.M.; Requena, T. The effect of supplementing goat's milk with whey protein concentrate on textural properties of set-type yoghurt. *Int. J. Food Sci. Technol.* **2006**, *41*, 87–92. [CrossRef]
75. Kailasapathy, K. Survival of free and encapsulated probiotic bacteria and their effect on the sensory properties of yoghurt. *LWT-Food Sci. Technol.* **2006**, *39*, 1221–1227. [CrossRef]
76. Espírito-Santo, A.P.; Lagazzo, A.; Sousa, A.L.O.P.; Perego, P.; Converti, A.; Oliveira, M.N. Rheology, spontaneous whey separation, microstructure and sensorial characteristics of probiotic yoghurts enriched with passion fruit fiber. *Food Res. Int.* **2013**, *50*, 224–231. [CrossRef]
77. Mani-López, E.; Palou, E.; López-Malo, A. Probiotic viability and storage stability of yogurts and fermented milks prepared with several mixtures of lactic acid bacteria. *J. Dairy. Sci.* **2014**, *97*, 2578–2590. [CrossRef]
78. Rudra, S.G.; Nath, P.; Kaur, C.; Basu, S. Rheological, storage stability and sensory profiling of low-fat yoghurt fortified with red capsicum carotenoids and inulin. *J. Food Process. Preserv.* **2017**, *41*, e13067. [CrossRef]
79. Bruzantin, F.P.; Daniel, J.L.P.; Da Silva, P.P.M.; Spoto, M.H.F. Physicochemical and sensory characteristics of fat-free goat milk yogurt with added stabilizers and skim milk powder fortification. *J. Dairy. Sci.* **2016**, *99*, 3316–3324. [CrossRef]
80. Barrantes, E.; Tamime, A.Y.; Muir, D.D.; Sword, A.M. The effect of substitution of fat by microparticulate whey protein on the quality of set-type yoghurt. *J. Soc. Dairy. Technol.* **1994**, *47*, 61–68. [CrossRef]
81. Domagala, J. Instrumental texture, syneresis and microstructure of yoghurts prepared from goat, cow and sheep milk. *Int. J. Food Prop.* **2009**, *12*, 605–615. [CrossRef]
82. Özcan, T.; Yıldız, E. Sebze püresi ile üretilen yoğurtların tekstürel ve duyusal özelliklerinin belirlenmesi. *Turk. J. Agric.-Food Sci. Technol.* **2016**, *4*, 579–587. [CrossRef]
83. Cayot, P.; Schenker, F.; House, G.; Sulmont-Rosse, C.; Colas, B. Creaminess in relation to consistency and particle size in stirred fat-free yogurt. *Int. Dairy. J.* **2008**, *18*, 303–311. [CrossRef]
84. Lazaridou, A.; Vaikousi, H.; Biliaderis, C.G. Impact of mixed-linkage (1→3, 1→4) β-glucans on physical properties of acid-set skim milk gels. *Int. Dairy. J.* **2008**, *18*, 312–322. [CrossRef]
85. Levy, R.; Okun, Z.; Shpigelman, A. High-pressure homogenization: Principles and applications beyond microbial inactivation. *Food Eng. Rev.* **2021**, *13*, 490–508. [CrossRef]
86. Serra, M.; Trujillo, J.; Guamis, B.; Feragut, V. Flavour profiles and survival of starter cultures of yoghurt produced from high-pressure homogenized milk. *Int. Dairy. J.* **2009**, *19*, 100–106. [CrossRef]
87. Lee, W.J.; Lucey, J.A. Formation and Physical Properties of Yogurt. *Asian-Australas. J. Anim. Sci.* **2010**, *23*, 1127–1136. [CrossRef]
88. Joon, R.; Mishra, S.K.; Brar, G.S.; Singh, P.K.; Panwar, H. Instrumental texture and syneresis analysis of yoghurt prepared from goat and cow milk. *Pharma Innov.* **2017**, *6*, 971.

89. Needs, E.C.; Capellas, M.; Bland, P.; Manoj, P.; MacDougal, D.B.; Gopal, P. Comparison of heat and pressure treatments of skimmed milk, fortified with whey protein concentrate, for set yoghurt preparation: Effects on milk proteins and gel structure. *J. Dairy. Res.* **2000**, *67*, 329–348. [CrossRef]
90. Fernández-Garía, E.; McGregor, J.U.; Traylor, S. The addition of oat fiber and natural alternative sweeteners in the manufacture of plain yogurt. *J. Dairy Sci.* **1998**, *81*, 655–663. [CrossRef]

foods

MDPI

Article

Inulin Functionalized "Giuncata" Cheese as a Source of Prebiotic Fibers

Diego Romano Perinelli, Agnese Santanatoglia, Giovanni Caprioli, Giulia Bonacucina, Sauro Vittori, Filippo Maggi * and Gianni Sagratini

Chemistry Interdisciplinary Project (ChIP) Research Center, School of Pharmacy, University of Camerino, Via Madonna delle Carceri 9, 62032 Camerino, MC, Italy; diego.perinelli@unicam.it (D.R.P.); agnese.santanatoglia@unicam.it (A.S.); giovanni.caprioli@unicam.it (G.C.); giulia.bonacucina@unicam.it (G.B.); sauro.vittori@unicam.it (S.V.); gianni.sagratini@unicam.it (G.S.)

* Correspondence: filippo.maggi@unicam.it

Abstract: The development of functional foods in the dairy sector represents a flourishing field of technological research. In this study, an Italian fresh cheese as "giuncata" was enriched with inulin, a dietary fiber, with the aim of developing a product with improved nutritional properties in terms of prebiotic action on intestinal microbiota. An inulin concentration of ~4% w/w was determined in the fresh cheese after the fortification process, enabling the claim of being a "source of dietary fiber" (inulin > 3 g/100 g) according to the European regulation. The addition of inulin has no effect on the pH of cheese and does not relevantly influence its color as well as the total fat content (fat reduction ~0.61%) in comparison to the control. Mechanical properties of the cheese were also not markedly affected as evidenced from rheological and tensile testing analyses. Indeed, the incorporation of inulin in "giuncata" only exerts a slight "softening effect" resulting in a slightly lower consistency and mechanical resistance in comparison to the control. Overall, this study demonstrates the feasibility of producing a fiber-enriched dairy functional food from a large consumed fresh and soft cheese as "giuncata".

Keywords: inulin; prebiotic; fresh cheese; functional food; derivatization; UV spectroscopy; water content; rheology; physicochemical characteristics

1. Introduction

Inulin is a polysaccharide composed of D-fructose molecules, usually ending with a glucose residue, and occurs as a storage carbohydrate in many plants [1]. For example, it is found in onions (1–5% of fresh weight), garlic (4–12%), bananas (0.2%), and chicory roots (15–20%) [2]. The inulin contained in chicory has a degree of polymerization (DP) of 3 to 60 monosaccharide units, on average about 10; the product of partial enzymatic hydrolysis is oligofructose (OF), which has a DP of 2 to 8 (averagely 4) [3]. This indigestible soluble dietary fiber is now used in an increasing number of foods, including dairy and bakery products, beverages, cereals and granola bars, low-fat spreads, ice cream, and packaged products [4]. Its wide use in the food industry is based on its nutritional and technological properties. As regards the nutritional properties, the prebiotic properties of inulin are important since this fructan can induce a specific change in the composition of the colonic microbiota with beneficial effects on the human host [5]. This increase in beneficial bacteria is found in people of all ages and is associated with several positive physiological effects. These include improved intestinal habits, increased calcium absorption with positive effects on bone health, a reduction in serum lipids with relevance to heart health, a positive effect on satiety with potentially positive consequences for weight management, and a significant effect of increasing resistance to infection and stimulating the immune system [6]. In addition, the technological use of inulin is based on its properties as a sugar substitute (especially in combination with high-intensity sweeteners) [7], as a fat substitute [8], and as a texture modifier [9,10]. Lourencetti et al. conducted experiments with cookies. The partial replacement of fat with inulin in the production of cookies resulted in a decrease in the percentage of total fats in the final product, but also the consistency/texture of the final enrichment product did not change with this substitution [11]. Therefore, the inulin ingredient can be considered as a fat replacer, which is efficient and feasible for the formulation of cookies with good technological and nutraceutical properties. Instead, Mansouripour et al. [12] developed a formulation and produced an acceptable prebiotic ketchup using inulin and GOSs (galactooligosaccharides). For fat replacement in low-fat dairy products, inulin seems to be particularly suitable as it can contribute to a better mouthfeel [13]. Instead, some studies have shown that the addition of long-chain inulin to low-fat yogurt improves its creaminess. Others have shown that this effect also occurs in low-fat cheese [14–18]. When inulin is added to foods at low concentrations, the rheological properties and sensory quality of the product are not greatly affected due to the neutral or slightly sweet taste, and the effects on the viscosity of this ingredient are limited. However, to obtain nutrition and health claims, inulin must be added in amounts greater than 3 g per 100 g or 100 mL [19,20], and a level of 3–8 g per serving should be added to ensure its prebiotic effect. The physicochemical properties of inulin are related to the degree of polymerization [21,22]. The short-chain fraction, oligofructose, is much more soluble and sweeter than native and long-chain inulin and may contribute to improved mouthfeel because its properties are closely related to those of other sugars [23]. Long-chain inulin is less soluble and more viscous than the native product and may act as a texture modifier. Other physicochemical properties influenced by DP include melting transition temperature, the ability to form gels and resulting gel strength, and interaction with other food ingredients such as starch or hydrocolloids [24]. These properties are also important for technical applications of inulin, especially for its ability to modify and improve texture. In the context of the current increase in obesity, there is a strong demand in the food industry for low-calorie, good-tasting foods. Reducing the fat (or sugar) content of a food is, therefore, an important starting point for product development. However, eliminating or reducing fat in foods not only changes the composition and structure, but also the interactions between the different components, resulting in clearly perceptible changes [23]. Highly valued and frequently consumed dairy products can be optimal food matrixes for inulin fortification, due to their impact on the diet of many consumers. Giuncata cheese is an Italian dairy product made from cow's, sheep's or goat's milk. It is a soft and fresh cheese with a texture similar to that of "ricotta" cheese and a higher moisture content that makes it tender. The traditional process for making giuncata cheese is to curdle the milk with rennet or a natural coagulant, and then the curd is collected and placed in molds made of juniper leaves ("giunco" in Italian language), which give the cheese its name. In addition, the juniper leaf molds also give the cheese its unique appearance and flavor. Nowadays for the production of giuncata, firstly pasteurized raw milk is heated and kept at 37 °C. After the addition of salt and rennet, the resulting curd is cut by the cheese maker to help separate the cheese from the whey. The curd is then collected in perforated baskets to allow the whey to drain. This yields the final cheese curd, which can be formed into cylindrical or cuboid shapes using special molds. After molding, the finished cheese is kept in cold water (10 °C) for 2 h before being packaged and stored [25,26].

In the present study, it was chosen to add inulin to the milk, before adding the rennet, to obtain a homogeneous system. The addition of the fiber was quick before the curd forms. As a fresh cheese, giuncata has a soft and creamy consistency, which makes it ideal for spreading, mixing into dishes, or using as a filling for various culinary preparations. For all these interesting reasons, this cheese was chosen for the first time for the development and validation of the inulin fortification design. Therefore, the present study aims to develop an inulin-fortified giuncata and assess whether it can be claimed as a "source of dietary fiber", without altering its physicochemical and mechanical properties. To our knowledge, this cheese type has never been considered in studies of fortification with inulin, despite its high consumption.

2. Materials and Methods

2.1. Chemicals and Reagents

The inulin powder (INULIN 90%, 90% of fibers content), obtained from the chicory root (*Cichorium intybus*) was purchased from ACEF (Fiorezuola d'Arda, Italy). The analytical standard for resorcinol (purity $\geq$ 99%, $C_6H_6O_2$, molecular weight 110.11 g/mol CAS No. 108-46-3) and the Carrez clarification kit were purchased from Sigma-Aldrich (St. Louis, MO, USA). The extract from the liquid rennet concentrates was produced by Caglificio Clerici spa (Cadorago, Italy). The fresh pasteurized milk was purchased from the company TreValli Cooperlat (Jesi, Italy).

2.2. Fortified Giuncata Cheese Manufacturing

As mentioned before in this case, it was chosen to add inulin to the milk, before adding the rennet, in order to obtain a homogeneous system. The addition of inulin was quick and easy before the curd forms. In addition, it does not require the use of machinery other than that used in traditional production, and the subsequent stirring results in a homogeneous product. However, this method did not provide homogeneity of inulin distribution and recoveries in the cheese after coagulation.

To investigate the inulin recoveries in giuncata cheese and to develop a new functional cheese, the giuncata production process was replicated in the laboratory, in triplicate: first, 500 g of fresh pasteurized milk was weighed, and then the milk was heated to 37 °C in a water bath under constant monitoring with a thermometer. After that, 6 g of salt (1.2 g/100 g of milk), inulin, and finally 0.5 g of rennet were added. After stirring, it took about 30 to 40 min for the milk to curdle properly, and then the resulting gelatinous mass was broken to facilitate the draining of the whey. The drainer used had holes in both the bottom and sides to ensure optimal drainage. After waiting 5 to 10 min to allow the whey to settle, the curd was placed in cheese molds, which gives the giuncata its shape but at the same time allows the whey to settle further (Figure 1). Finally, we cooled the obtained rushes in water at 10 °C for 2 h. Different enrichment experiments were carried out at two concentrations of inulin: 10 g/500 mL and 15 g/500 mL.

Figure 1. (**A**) Heated fresh milk to 37 °C in a water bath under constant monitoring with a thermometer, and then salt and rennet were added. (**B**) Coagulation. (**C**) Whey removal and separation of whey and curds.

2.3. Inulin Extraction Procedure

Several extraction tests were preliminarily performed on cheese samples (2 g) fortified with 100 mg of the ingredient starting from the method of [26] to optimize the procedure. Two grams of cheese were weighed and placed in a 100 mL flask, and afterwards 50 mL of boiling water was poured over the sample in the flask. Then, the sample was kept at 85 °C for 10 min, under constant stirring with a magnetic stirrer. Next, the sample was allowed to cool down (20 min) at room temperature and Carrez reagents were added: 5 mL of Carrez reagent I (aqueous solution of $K_4Fe(CN)_6 \times 3H_2O$, 15 g/100 mL) and 5 mL of Carrez reagent II (aqueous solution of $Zn(CH_3COO)_2 \times 2H_2O$, 30 g/100 mL) (Figure S1).

These solutions were used to precipitate the proteins, whose presence interferes with inulin quantification. Finally, the sample was moved to a 50 mL tube, vortexed for 4 min, and then centrifuged at room temperature for 15 min at 5000 rpm. Subsequently, the supernatant was filtered over a 100 mL flask with filter paper and made up to 100 mL with distilled water.

2.4. Derivatization and Spectrophotometric Method for Inulin Quantification

The derivatization procedure was firstly performed using 1 mL of standard solution (1 mg/mL inulin solution in distilled water) for the validation of the method. To this solution, 5 mL of a resorcinol reagent (1 mg/mL in EtOH) and 10 mL of a 30% *v*/*v* HCl were added. Subsequently, the samples were then brought to 20 mL through the addition of distilled water, under agitation. Finally, after being covered with foil, they were left in the water bath at 80 °C for 30 min. A temperature of 80 °C was required for the reaction between resorcinol and inulin and the subsequent staining of the sample. Finally, after being cooled down, they were analyzed through a spectrophotometer (Agilent Technologies, Woburn, MA, USA) using 480 nm as the wavelength of quantification.

The spectrophotometric method was based on the formation of a colored compound by the interaction of inulin with resorcinol in an acidic environment, as described in the well-known Seliwanoff test for ketoses [27] (Figure 2).

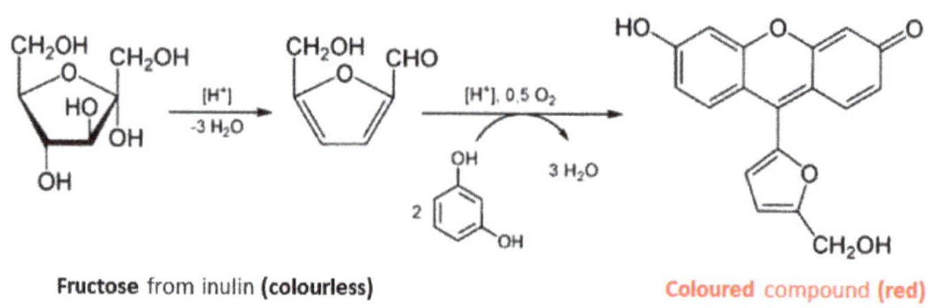

Figure 2. Mechanism of the Seliwanoff test for ketoses. The spectrophotometric method was based on the formation of a colored compound by the interaction of inulin with resorcinol in an acidic environment.

This derivatization reaction is specific for ketoses, and the presence of aldoses does not interfere with inulin analysis. The developed method was validated by studying the linearity, repeatability, detection limits, and quantification of inulin. The repeatability (Table S1) of the analytical method was determined by monitoring the variations in instrumental response (relative standard deviation—RSD %) after analysis of a standard sample (5 ppm) three times in a row on the same day (intraday repeatability) and on three consecutive days (interday repeatability) [28]. The intraday RSD % was 0.2%, while the interday RSD % was 0.7%, confirming very low instrumental variation and repeatability of the analytical method. Subsequently, the developed analytical method was performed to determine inulin concentration in fortified giuncata. A total of 5 mL of the cheese extract containing inulin was diluted up to 50 mL with distilled water and analyzed by UV spectroscopy after derivatization as described above.

2.5. Characterization of Inulin-Fortified Giuncata Cheese

2.5.1. Physical–Chemical Composition

A digital pH meter (Mettler Toledo, Columbus, UK) was used to determine pH, while a chroma Meter CR—400 (Konica Minolta®, Tokyo, Japan) was used to determine color. The measurement was performed according to the L*a*b* system of the Commission International de l'Eclairage (CIE). In this way, the instrument calculates three parameters simultaneously for each measurement: Brightness (L*); Green to Red component (a*); Blue to Yellow component (b*). By convention, the L* component ranges from 0 (zero brightness, black) to 100 (maximum brightness, a specific white chosen as reference and calibrator). The a* and b* components can take any value except in the cases L* = 100 and L* = 0, where they can only be 0. Positive values of a* indicate a color tending to red, and negative values indicate a color tending to green. Positive values of b* indicate a yellow color, and negative values indicate a blue color.

The fats were extracted according to two methodologies: continuous single-phase extraction and biphasic solvent extraction (Folch extraction). According to the first method, fats extractions were performed using 250 mL of n-hexane as solvent [29]. Extractions were performed using a finely ground 50 g sample of the cheese matrix, which was lyophilized, for one night, and extracted with the selected solvent (n-hexane) in a Soxhlet apparatus (Universal Extractor—Buchi, Mod. E-800, Uster, Switzerland) for 2 h. Then, the extracts were evaporated, and the solvent was removed with rotavapor. At the same time, according to the Folch extraction, 90 mL of chloroform-methanol 2:1 (v/v) was added to 5 g of ground cheese in a 250 mL beaker and the mixture was extracted with stirring (Ultra-Turrax, IKA, Darmstadt, Germany) for 2 min and then transferred to a separating funnel by washing the beaker walls with a total of 10 mL of solvent mixture [30]. A total of 20 mL of saturated aqueous sodium chloride solution was added; the two phases of the system were shaken, and the organic phase was collected and dried over anhydrous sodium sulphate. Finally, the extracts were evaporated under vacuum and then dried and weighed.

2.5.2. Viscoelastic Properties

Oscillatory measurements (stress sweep test and frequency sweep test) were performed using a rotational rheometer (Kinexus Lab+, Malvern, UK), equipped with a 20 mm plate–plate geometry. For the stress sweep test, rheological moduli (elastic modulus G' and viscous modulus G'') were recorded at the frequency of 1 Hz in the range of applied stress from 0.1 Pa to 100 Pa at 25 °C. For the frequency sweep test, G' and G'' were recorded at the applied stress of 1 Pa and in the range of frequencies 0.1–10 Hz. The gap was set at 3 mm. Analyses were performed after one day from the preparation of the cheese in triplicate.

2.5.3. Texture Analysis

Texture analysis on the cheese was carried out using a tensile tester apparatus (5543, Instron, Pianezza, Italy), equipped with a 20 mm diameter cylindrical flat probe. The test speed was 1.0 mm s^{-1}, and the depth of penetration was half of the height of the cheese (20 mm). Force/displacement curves were recorded and the parameters "maximum force" (F_{max}, N) corresponding to the hardness of the cheese, positive area (N, mm) corresponding to the stiffness of the cheese, negative maximum force (negative F_{max}, N), and negative area (N, mm) corresponding to the adhesiveness of the cheese were calculated [31]. Analyses were performed at room temperature after one day from the preparation of the cheese in at least six repetitions.

2.5.4. Thermo Gravimetric (TGA) and Differential Thermal (DTA) Analyses

The weight (%) and heat flow (mW) variations over temperature for the cheese samples were recorded in a nitrogen environment using a simultaneous thermal analyzer STA 6000 (Perkin Elmer Inc., Waltham, MA, USA) at a heating rate of 10 °C/min from 35 °C to 400 °C. The weight loss of samples was determined using the software STA 6000 data analysis (Perkin Elmer Inc., Waltham, MA, USA) using Delta-Y function. Analyses were performed after one day from the preparation of the cheese in triplicate.

2.5.5. Total Water Content

The total water content (%) of the cheese was determined using a moisture analyzer (HE53, Mettler Toledo). A total of 0.7 g of the cheese was heated at 120 °C in an automatic mode until no weight variation was recorded by the instrument. The weight loss of the samples after applying the thermal program was considered as the water content (%) of the samples. Analyses were performed after one day after the preparation of the cheese in triplicate.

2.6. Statistical Analysis

All the analyses were performed at least in triplicate ($n \geq 3$) and data are presented as mean ± standard deviation. Statistical comparisons between control giuncata and inulin-enriched giuncata cheese were performed using Student's *t*-test and $p < 0.05$ was considered statistically significant.

3. Results

3.1. Fortified Giuncata Manufactoring

On average, 122–137 g of the cheese was obtained from 500 g of milk, with a yield of 21–24.4%, depending on the amount of inulin added to the milk. Table 1 gives an overview of the inulin content in the cheese at different levels of enrichment (10 and 15 g/500 mL). To confirm the concentration of this ingredient in giuncata cheese, extraction, derivatization, and spectrophotometric analysis were performed according to the methods described previously. The highest inulin contents in giuncata were obtained at a degree of enrichment of 15 g/500 mL of milk (4 g/100 g of the cheese), and the difference with the enrichment of 10 g/500 mL was found to be statistically significant (Table 1). Nutrition and health claims state that a food is a "source of dietary fiber" if the product contains at least 3 g of dietary fiber/100 g [19,20]; therefore, it can be assumed that all enrichment levels tested satisfied this claim. According to this, 15 g/500 mL was chosen as the enrichment inulin concentration in the milk for further studies.

Table 1. Concentrations of inulin in giuncata cheese at different fortification levels (10 g or 15 g of inulin added to milk).

Inulin Added to Milk (g)	Milk (g)	Produced Cheese (g)	Concentration of Inulin (g/100 g)
10	500	137.1 ± 12.0 [a]	3.6 ± 0.3 [a]
15	500	122.6 ± 4.5 [b]	4.0 ± 0.1 [b]

[a,b] Values with different uppercase letters are significantly different ($p < 0.05$).

3.2. Inulin Fortification and Extraction Procedure

An efficient extraction method was developed to study inulin levels in the fortified cheese (Figure S2). Considering the complexity of the fortified food matrix (giuncata), which is rich in protein (11% *w/w*) and fat (15% *w/w*), it was necessary to study a selective extraction method for inulin with high recoveries [32].

Among all the preliminary screened procedures, the highest recoveries of inulin from the cheese after derivatization were obtained with the optimized method of [27] performed with some modifications as reported previously (Section 2.3) (Figure S2). This extraction method (Figure S2) achieved recoveries of 79.8 ± 2.8%. The high extraction temperature (85 °C) improved the solubility of inulin in water, thus promoting the extraction.

3.3. Study of Giuncata Enrichment

3.3.1. Physical–Chemical Composition

The enriched giuncata samples (Giuncata IN) (Table 2) were analyzed to study the effects of enrichment on pH, color, and fat content compared to the non-fortified cheese as the control (Giuncata CTR). It was found that the addition of inulin has no effect on the pH of giuncata cheese. Therefore, the measured pH was not significantly different (pH 6.6 vs. pH 6.5) between the Giuncata IN and Giuncata CTR cheeses, as previously reported by [33]. The color of a food product is an important purchase criterion for the consumer. The data regarding color showed that the addition of inulin has no significant effect on the cheese color. However, Giuncata IN cheese tends to show higher brightness (L*: 95.6 ± 0.6) than Giuncata CTR cheese (L*: 93.9 ± 0.2) and a stronger tendency toward blue in functionalized cheese (b*: 7.5 ± 0.6) than in the control, which tends more toward yellow (b*: 7.8 ± 0.7). This positive trend could be explained by the white color of the inulin component. In fact, the white color and less tendency to the yellow color of fresh cheese is perceived by consumers as a quality parameter [17]. Therefore, the enrichment with inulin has a little effect on the color of the cheese but tends to increase the typical white color of giuncata (Figure 3).

Table 2. Comparison of the chemical–physical characteristics (pH, Color, Fats percentage) between functionalized (Giuncata IN) and conventional (Giuncata CTR) cheese.

	Giuncata IN	Giuncata CTR
pH	6.6 ± 0.0 [a]	6.5 ± 0.0 [a]
Color		
- L*	95.6 ± 0.6 [a]	93.9 ± 0.2 [b]
- a*	−2.0 ± 0.3 [a]	−1.8 ± 0.2 [a]
- b*	7.5 ± 0.6 [a]	7.8 ± 0.7 [b]
Fats (Soxhlet extraction)	11.72 ± 0.6% [a]	12.28 ± 0.7% [b]
Fats (Folch extraction)	11.91± 0.2% [a]	12.52± 0.3% [b]

[a,b] Values with different uppercase letters are significantly different ($p < 0.05$).

Figure 3. Visual appearance of the cheese Giuncata, produced in the laboratory. **Left** normal (CTR), **right** fortified with inulin (IN).

The fats were also extracted from Giuncata IN and Giuncata CTR cheeses in order to study the effects of enrichment with inulin on the fat content of fresh cheese and to compare the results (Table 2). Thus, we applied two conventional methods (Folch and Soxhlet extraction), with the combination of different solvents, in order to improve the extraction yield and obtain a more reliable fat profile. The fat content obtained from Soxhlet extraction was 11.72 ± 0.6% and 12.28 ± 0.7% for the Giuncata IN and Giuncata CTR, respectively. The percentage reduction in fat content by Soxhlet was 0.56% (from 12.28% to 11.72%). In the case of Folch extraction, the yields obtained were 11.91 ± 0.2% for the enriched cheese and 12.52 ± 0.3% for the control cheese. The percentage decrease in yield with Folch was 0.61% (from 12.52% to 11.91%). The decrease in lipid content was similar between the two methodologies and these results confirmed the minimal effect of inulin on lipid content, at least for the prepared giuncata cheese.

3.3.2. Viscoelastic Properties of Giuncata Cheese

The viscoelastic properties of giuncata cheese were assessed by measuring the rheological moduli (G′ and G″) as a function of the applied stress (stress sweep test) and the applied frequency (frequency sweep test). The stress sweep test (Figure 4A) showed the variation of G′ modulus over the applied stress that can be considered as a measure for the consistency of a semi-solid material as giuncata cheese. Specifically, the measured G′ values were almost constant, indicating that the cheese does not destructure over the experimental conditions of applied stress. Moreover, the obtained G′ values were similar between the giuncata control (Giuncata CTR) and inulin-fortified giuncata (Giuncata IN) around 5000–6000 Pa with mean values only slightly lower for Giuncata IN than Giuncata CTR. The frequency sweep test (Figure 4B) highlighted the solid-like behavior of both giuncata cheese, since the G′ values were higher than the G″ values over all the applied frequencies. Indeed, when G′ > G″ the material shows a prevalence of the elastic properties over the viscous ones. Even more, according to the frequency sweep test, it can be noticed that both G′ and G″ moduli are slightly higher for Giuncata CTR than Giuncata IN, confirming the not marked effect of the incorporation of inulin in both the consistency and viscoelastic properties of giuncata cheese. Moreover, both cheeses show a dependency of both moduli over the applied frequency (i.e., the values of both moduli increase over frequency) denoting them as "weak" solid-like samples. This behavior over frequencies has been explained through a not complete relaxation of the network structure of the cheese over the frequency increase [34], and it has been observed for other soft cheese such as mozzarella [35] and others [36,37].

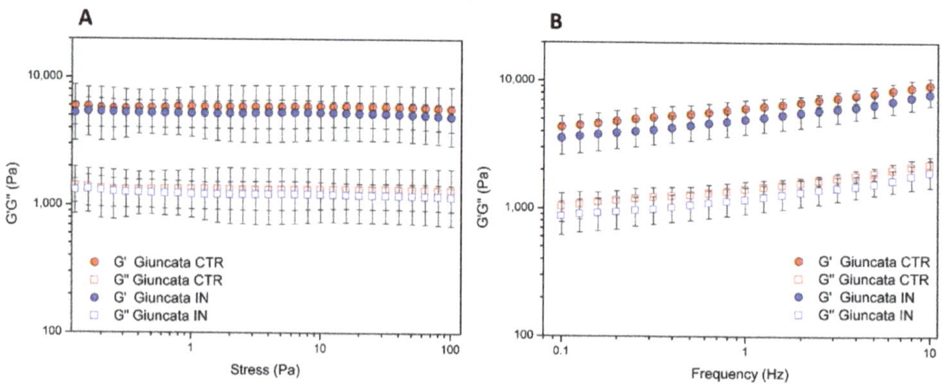

Figure 4. (**A**) Variation of G′ modulus over the applied stress (stress sweep test) and (**B**) variation of G′ and G″ moduli over applied frequency (frequency sweep test) for control giuncata cheese (Giuncata CTR) and inulin-fortified giuncata (Giuncata IN) measured at 25 °C.

3.3.3. Texture Analysis

The mechanical properties of giuncata cheese can be further investigated using texture analysis. Figure 5 shows the displacement vs. force plots obtained from the compression analysis of Giuncata CTR and Giuncata IN both during the penetration of the probe inside the cheese and its release up. For all samples, the maximum force was recorded at the highest penetration (20 mm) of the probe inside the cheese (set as the half distance of the height of the cheese). The increase in the registered force is not linear, but it shows two main slopes. Specifically, a sharper increase in the force was registered in the first 2 mm of compression, related to the elasticity of the material. At any displacement value, the registered force for Giuncata IN was lower than for Giuncata CTR, confirming that as for the consistency (measured by rheology), the mechanical resistance to compression is also affected by the incorporation of inulin into the cheese. These results are comparable to others reported in the literature in which inulin was incorporated into the cheese with the aim to act as a "fat replacer" [33,38]. The softening effects exerted on the cheese by the incorporation of inulin have been supposed to be related to the higher moisture content in relation to the protein content and to the increase in filler volume [39].

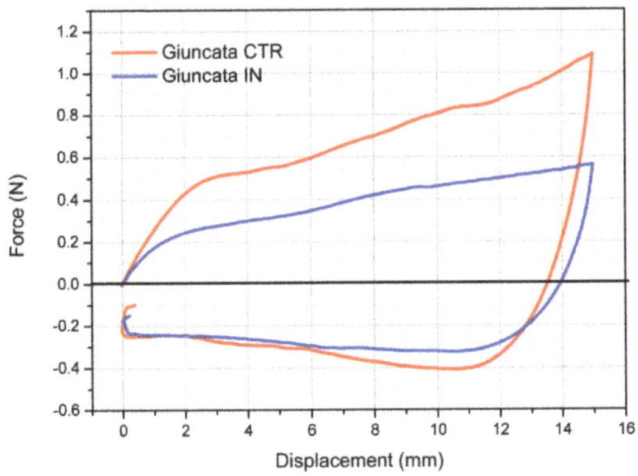

Figure 5. Displacement (mm) vs. force (N) for control giuncata cheese (Giuncata CTR) and inulin-fortified giuncata (Giuncata IN) registered during the penetration of 20 mm flat probe at room temperature.

The effect of inulin fortification emerges also by considering the calculated texture parameters as compressive force (N), stiffness (N mm) and adhesiveness, which are reported in Table 3.

Table 3. Textural parameter obtained from compression analysis of control giuncata cheese (Giuncata CTR) and inulin-fortified giuncata (Giuncata IN): Positive F_{max} "Compressive Force" (N), Negative Fmax (N), Positive area "Stiffness" (N mm), and Negative area "Adhesiveness" (N mm).

	Positive F_{max} "Compressive Force" (N)	Negative F_{max} (N)	Positive Area "Stiffness" (N mm)	Negative Area "Adhesiveness" (N mm)
Giuncata CTR	1.055 ± 0.291 [a]	0.384 ± 0.083 [a]	10.03 ± 3.24 [a]	3.88 ± 1.43 [a]
Giuncata IN	0.611 ± 0.127 [b]	0.296 ± 0.041 [b]	6.57 ± 1.10 [b]	3.39 ± 0.46 [a]

[a,b] Values with different uppercase letters are significantly different ($p < 0.05$).

All these parameters showed a decrease in their value after the incorporation of inulin, for which the statistical comparison was significant for the compressive force ($p = 0.0065$) and stiffness ($p = 0.0327$) but not for adhesiveness ($p = 0.4428$). Compression force, also referred as firmness, is the force required to obtain a certain deformation and it can be partially related to cheese chewiness. The obtained results suggest that the incorporation of inulin at the investigated concentrations can affect the chewability of giuncata cheese, despite the fact that this parameter is much more relevant for "hard cheese" with respect to "soft cheese" as giuncata. The positive area or stiffness is related to the area under the force vs. displacement plots, and it is a measure of the resistance to the penetration of the probe inside the cheese, having a similar meaning in terms of texture properties as the compression force. Adhesiveness, instead, is related to the work necessary to detach the material, such as cheese, from the testing probe. Being the values among the two tested giuncata, it can be supposed that the two cheeses have a similar sliceability.

3.3.4. Water Content and Thermal Analysis

The water content was firstly determined by using a thermobalance and turned out to be $63.98 \pm 1.82\%$ for Giuncata CTR and $69.16 \pm 0.39\%$ for Giuncata IN, underlining that the presence of inulin increases the amount of water in the cheese. These results were confirmed by TGA analysis, showing a decrease in the sample weight for both samples between 64 and 72% (specifically, $64.76 \pm 0.37\%$ for Giuncata CTR and $71.61 \pm 0.99\%$ for Giuncata IN), related to the loss of water before 150 °C. The TGA technique was also able to discriminate between bound water and free water in the cheese since two or three slopes in the weight (%) trace can be recognized (Figure 6), one of them below 100 °C related to free water and the second and third one between 100 °C and 150 °C related to bound water [40]. This is also confirmed by the derivative of the weight signal, showing two or three minima at the temperatures at which the loss of free and bounded water is faster. The slight differences observed between the control and the samples could be attributed to a different kinetics of the water release from the cheese matrix [41]. This analysis also underlines that, as for the other soft cheese, the amount of free water in giuncata is much higher than the amount of bound water, being in the range of 68–72% with respect to the total water content. The endothermic transitions, evidenced from the heat flow signals (DTA analysis), having maxima corresponding to the minima in the derivate weight signals, confirm the release and evaporation of water from giuncata cheese occurring at temperatures at which the sample weight decreases. The degradation of the residual components of both giuncata cheese occurs at temperatures above 300 °C, at which the weight (%) decreases down to 10%.

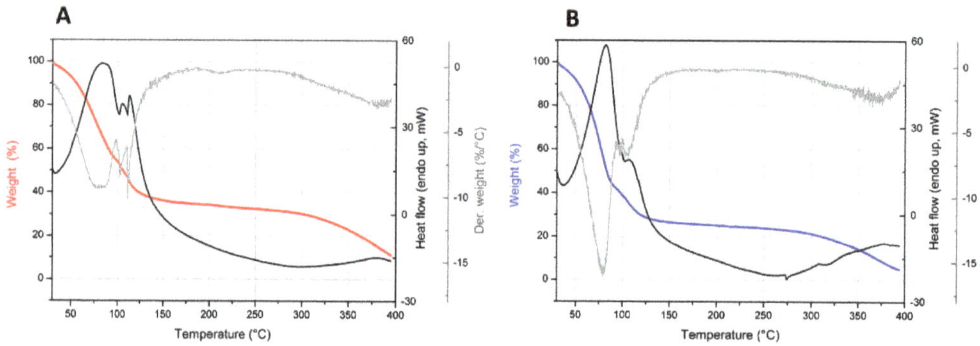

Figure 6. Weight (%), its derivative and heat flow (mW) vs. temperature plots for control giuncata cheese (Giuncata CTR) (**A**) and inulin-fortified giuncata (Giuncata IN) (**B**) as obtained from thermogravimetric analysis (TGA).

4. Conclusions

This study led to the development of a new functional Italian "giuncata" cheese enriched with inulin. The fortification process achieved an inulin concentration of ~4% *w*/*w* in the fresh cheese, thereby claiming this dairy food product as a "source of dietary fiber" (inulin > 3 g/100 g according to European regulations on food). The addition of inulin determined some changes in the physical-chemical and mechanical properties, without altering the general appearance of the cheese. Only a little effect on the color, which becomes lighter, and a slight fat reduction have been observed in the inulin-fortified cheese with the respect to the control. A slightly "softening effect" was exerted by the incorporation of inulin in giuncata, as observed from rheological and mechanical analyses on cheese. Overall, these results demonstrate the feasibility of producing giuncata cheese as a functional dairy product, having possible benefits on human health such as the modulation of the composition of colonic microbiota, thanks to the prebiotic effect of inulin. Therefore, the results of this study can be supportive for an industrial scale-up in the production of functional cheese.

Supplementary Materials: The following supporting information can be downloaded at: https://www.mdpi.com/article/10.3390/foods12183499/s1, Figure S1: (A) Sample of enriched cheese and Carrez reagents before clarification. (B) Extraction of inulin after clarification with Carrez reagents; Figure S2: Inulin extraction procedure from giuncata cheese; Table S1: Repeatability of the analytical method.

Author Contributions: Conceptualization, D.R.P. and A.S.; methodology, D.R.P. and A.S.; software, F.M.; validation, G.S., F.M. and G.C.; formal analysis, D.R.P. and A.S.; investigation, D.R.P. and A.S.; resources, G.S.; data curation, G.C.; writing—original draft preparation D.R.P. and A.S.; writing—review and editing, G.C. and F.M.; visualization, G.B.; supervision, G.S.; project administration, G.C.; funding acquisition, G.S. and S.V. All authors have read and agreed to the published version of the manuscript.

Funding: This research was partially funded by the Sabelli Group (Ascoli Piceno, Italy).

Data Availability Statement: The data used to support the findings of this study can be made available by the corresponding author upon request.

Acknowledgments: The authors are grateful to Sabelli Group (Ascoli Piceno, Italy) for the technical assistance and the partially economic support for this research project.

Conflicts of Interest: The authors declare no conflict of interest.

References

1. Apolinário, A.C.; de Lima Damasceno, B.P.G.; de Macêdo Beltrão, N.E.; Pessoa, A.; Converti, A.; da Silva, J.A. Inulin-type fructans: A review on different aspects of biochemical and pharmaceutical technology. *Carbohydr. Polym.* **2014**, *101*, 368–378. [CrossRef]
2. Judprasong, K.; Tanjor, S.; Puwastien, P.; Sungpuag, P. Investigation of Thai plants for potential sources of inulin-type fructans. *J. Food Compos. Anal.* **2011**, *24*, 642–649. [CrossRef]
3. Roberfroid, M.B.; Van Loo, J.A.; Gibson, G.R. The bifidogenic nature of chicory inulin and its hydrolysis products. *J. Nutr.* **1998**, *128*, 11–19. [CrossRef]
4. Jackson PP, J.; Wijeyesekera, A.; Theis, S.; van Harsselaar, J.; Rastall, R.A. Food for thought! Inulin-type fructans: Does the food matrix matter? *J. Funct. Foods* **2022**, *90*, 104987. [CrossRef]
5. Vandeputte, D.; Falony, G.; Vieira-Silva, S.; Wang, J.; Sailer, M.; Theis, S.; Verbeke, K.; Raes, J. Prebiotic inulin-type fructans induce specific changes in the human gut microbiota. *Gut* **2017**, *66*, 1968–1974. [CrossRef]
6. Hughes, R.L.; Alvarado, D.A.; Swanson, K.S.; Holscher, H.D. The prebiotic potential of inulin-type fructans: A systematic review. *Adv. Nutr.* **2022**, *13*, 492–529. [CrossRef]
7. Krupa-Kozak, U.; Drabińska, N.; Rosell, C.M.; Piłat, B.; Starowicz, M.; Jeliński, T.; Szmatowicz, B. High-Quality Gluten-Free Sponge Cakes without Sucrose: Inulin-Type Fructans as Sugar Alternatives. *Foods* **2020**, *9*, 1735. [CrossRef]
8. Borges, J.V.; de Souza, J.A.; Fagnani, R.; Costa, G.N.; Dos Santos, J.S. Reduced fat Frescal sheep milk cheese with inulin: A first report about technological aspects and sensory evaluation. *J. Dairy Res.* **2019**, *86*, 368–373. [CrossRef]
9. Popoola-Akinola, O.O.; Raji, T.J.; Olawoye, B. Lignocellulose, dietary fibre, inulin and their potential application in food. *Heliyon* **2022**, *8*, e10459. [CrossRef]
10. Mudannayake, D.C.; Jayasena, D.D.; Wimalasiri, K.M.; Ranadheera, C.S.; Ajlouni, S. Inulin fructans-food applications and alternative plant sources: A review. *Int. J. Food Sci. Technol.* **2022**, *57*, 5764–5780. [CrossRef]
11. Lourencetti, R.E.; Benossi, L.; Marques, D.R.; Joia, B.M.; Monteiro, A.R.G. Development of biscuit type cookie with partial replacement of fat by inulin. *Int. J. Nutr. Food Sci.* **2013**, *2*, 261–265.
12. Mansouripour, S.; Mizani, M.; Rasouli, S.; Gerami, A.; Sharifan, A. Effect of inulin and galactooligosaccharides on particle size distribution and rheological properties of prebiotic ketchup. *Int. J. Food Prop.* **2017**, *20*, 157–170. [CrossRef]
13. Ren, Y.; Song, K.Y.; Kim, Y. Physicochemical and retrogradation properties of low-fat muffins with inulin and hydroxypropyl methylcellulose as fat replacers. *J. Food Process. Preserv.* **2020**, *44*, e14816. [CrossRef]
14. Cardarelli, H.R.; Buriti, F.C.; Castro, I.A.; Saad, S.M. Inulin and oligofructose improve sensory quality and increase the probiotic viable count in potentially synbiotic petit-suisse cheese. *LWT-Food Sci. Technol.* **2008**, *41*, 1037–1046. [CrossRef]
15. Hennelly, P.J.; Dunne, P.G.; O'sullivan, M.; O'riordan, E.D. Textural, rheological and microstructural properties of imitation cheese containing inulin. *J. Food Eng.* **2006**, *75*, 388–395. [CrossRef]
16. Keenan, D.F.; Resconi, V.C.; Kerry, J.P.; Hamill, R.M. Modelling the influence of inulin as a fat substitute in comminuted meat products on their physico-chemical characteristics and eating quality using a mixture design approach. *Meat Sci.* **2014**, *96*, 1384–1394.
17. Koca, N.; Metin, M. Textural, melting and sensory properties of low-fat fresh kashar cheeses produced by using fat replacers. *Int. Dairy J.* **2004**, *14*, 365–373. [CrossRef]
18. Pagliarini, E.; Beatrice, N. Sensory and Rheological properties of low-fat filled 'pasta filata'cheese. *J. Dairy Res.* **1994**, *61*, 299–304. [CrossRef]
19. Eur-lex.europa. Available online: https://eur-lex.europa.eu/legal-content/IT/TXT/PDF/?uri=CELEX:02006R1924-20141213&from=IT (accessed on 11 May 2023).
20. Eur-lex.europa. Available online: https://eur-lex.europa.eu/legal-content/IT/TXT/?uri=celex%3A02006R1924-20141213 (accessed on 11 May 2023).
21. Wang, R.; Wan, J.; Liu, C.; Xia, X.; Ding, Y. Pasting, thermal, and rheological properties of rice starch were partially replaced by inulin with different degrees of polymerization. *Food Hydrocoll.* **2019**, *92*, 228–232. [CrossRef]

22. Li, Y.; Ma, X.; Liu, X. Physicochemical and rheological properties of cross-linked inulin with different degree of polymerization. *Food Hydrocoll.* **2019**, *95*, 318–325. [CrossRef]
23. Teferra, T.F. Possible actions of inulin as prebiotic polysaccharide: A review. *Food Front.* **2021**, *2*, 407–416. [CrossRef]
24. Meyer, D.; Bayarri, S.; Tárrega, A.; Costell, E. Inulin as a texture modifier in dairy products. *Food Hydrocoll.* **2011**, *25*, 1881–1890. [CrossRef]
25. Barclay, T.; Ginic-Markovic, M.; Johnston, M.R.; Cooper, P.D.; Petrovsky, N. Analysis of the hydrolysis of inulin using real time 1H NMR spectroscopy. *Carbohydr. Res.* **2012**, *352*, 117–125. [CrossRef]
26. Petkova, N.I.; Todorova, M.; Vlaseva, R.; Denev, P. Spectrophotometric Method for Determination of Inulin and Fructooligosaccharides in Lactic Acid Fermented Dairy Products. *Conf. Pap.* **2013**. Available online: https://www.researchgate.net/publication/258519310 (accessed on 15 September 2023).
27. Besir, A.; Yazici, F.; Mortas, M.; Gul, O. A novel spectrophotometric method based on Seliwanoff test to determine 5-(Hydroxymethyl) furfural (HMF) in honey: Development, in house validation and application. *LWT* **2021**, *139*, 110602. [CrossRef]
28. Santanatoglia, A.; Nzekoue, F.K.; Sagratini, G.; Ricciutelli, M.; Vittori, S.; Caprioli, G. Development and application of a novel analytical method for the determination of 8 plant sterols/stanols in 22 legumes samples. *J. Food Compos. Anal.* **2023**, *118*, 105195. [CrossRef]
29. Manirakiza, P.; Covaci, A.; Schepens, P. Comparative study on total lipid determination using Soxhlet, Roese-Gottlieb, Bligh & Dyer, and modified Bligh Dyer extraction methods. *J. Food Compos. Anal.* **2001**, *14*, 93–100.
30. Caprioli, G.; Kamgang Nzekoue, F.; Fiorini, D.; Scocco, P.; Trabalza-Marinucci, M.; Acuti, G.; Tardella, F.M.; Sagratini, G.; Catorci, A. The effects of feeding supplementation on the nutritional quality of milk and cheese from sheep grazing on dry pasture. *Int. J. Food Sci. Nutr.* **2020**, *71*, 50–62. [CrossRef]
31. Seuvre, A.M.; Turci, C.; Voilley, A. Effect of the temperature on the release of aroma compounds and on the rheological behaviour of model dairy custard. *Food Chem.* **2008**, *108*, 1176–1182. [CrossRef]
32. Santanatoglia, A.; Nzekoue, F.K.; Alesi, A.; Ricciutelli, M.; Sagratini, G.; Suo, X.; Torregiani, E.; Vittori, S.; Caprioli, G. Development of Innovative Vitamin D Enrichment Designs for Two Typical Italian Fresh Cheeses: Burrata and Giuncata. *Molecules* **2023**, *28*, 1049. [CrossRef]
33. Salvatore, E.; Pes, M.; Mazzarello, V.; Pirisi, A. Replacement of fat with long-chain inulin in a fresh cheese made from caprine milk. *Int. Dairy J.* **2014**, *34*, 1–5. [CrossRef]
34. Foegeding, E.A.; Drake, M.A. Invited review: Sensory and mechanical properties of cheese texture. *J. Dairy Sci.* **2007**, *90*, 1611–1624. [CrossRef]
35. Brown, J.A.; Foegeding, E.A.; Daubert, C.R.; Drake, M.A.; Gumpertz, M. Relationships among rheological and sensorial properties of young cheeses. *J. Dairy Sci.* **2003**, *86*, 3054–3067.
36. Diaz-Bustamante, M.L.; Reyes, L.H.; Solano, O.A.A. Application of a Multiscale Approach in the Substitution and Reduction of NaCl in Costeño-Type Artisan Cheese. *Appl. Sci.* **2020**, *10*, 9008. [CrossRef]
37. Dimitreli, G.; Thomareis, A.S. Instrumental textural and viscoelastic properties of processed cheese as affected by emulsifying salts and in relation to its apparent viscosity. *Int. J. Food Prop.* **2009**, *12*, 261–275. [CrossRef]
38. Sołowiej, B.; Glibowski, P.; Muszyński, S.; Wydrych, J.; Gawron, A.; Jeliński, T. The effect of fat replacement by inulin on the physicochemical properties and microstructure of acid casein processed cheese analogues with added whey protein polymers. *Food Hydrocoll.* **2015**, *44*, 1–11. [CrossRef]
39. Karimi, R.; Azizi, M.H.; Ghasemlou, M.; Vaziri, M. Application of inulin in cheese as prebiotic, fat replacer and texturizer: A review. *Carbohydr. Polym.* **2015**, *119*, 85–100. [CrossRef]
40. Saldo, J.; Sendra, E.; Guamis, B. Changes in water binding in high-pressure treated cheese, measured by TGA (thermogravimetrical analysis). *Innov. Food Sci. Emerg. Technol.* **2002**, *3*, 2013–2207.
41. Chen, Y.; MacNaughtan, W.; Jones, P.; Yang, Q.; Foster, T. The state of water and fat during the maturation of Cheddar cheese. *Food Chem.* **2020**, *303*, 125390. [CrossRef]

foods

MDPI

Article

Augmenting Functional and Sensorial Quality Attributes of Kefir through Fortification with Encapsulated Blackberry Juice

Vanja Travičić [1,*], Olja Šovljanski [1], Ana Tomić [1], Milica Perović [1], Maja Milošević [1], Nenad Ćetković [2] and Mirjana Antov [1]

[1] Faculty of Technology Novi Sad, University of Novi Sad, Bulevar Cara Lazara 1, 21000 Novi Sad, Serbia; oljasovljanski@uns.ac.rs (O.Š.); anav@uns.ac.rs (A.T.); perovicmilica@uns.ac.rs (M.P.); majamilosevicpuhacin@eurofins.se (M.M.); mantov@uns.ac.rs (M.A.)

[2] Medical Faculty, University of Novi Sad, Hajduk Veljkova 3, 21000 Novi Sad, Serbia; nenad.cetkovic@mf.uns.ac.rs

* Correspondence: vanjaseregelj@tf.uns.ac.rs

Abstract: Kefir is a fermented dairy product claimed to confer many health-promoting effects, but its acidic taste is not appealing to some consumers. Therefore, the aim of this study was to enhance the functional and sensorial quality attributes of kefir through fortification with encapsulated blackberry juice (EBJ). The blackberry juice was successfully encapsulated via freeze-drying using lentil protein isolate (LPI) as the carrier. The encapsulated blackberry juice showed good physicochemical, functional, and morphological properties, as well as microbiological safety for use as a food additive. The kefir was fortified with EBJ in concentrations of 1, 2.5, 5, and 7.5% (*w*/*w*), stored for up to 28 days under refrigeration, and periodically evaluated. Parameters such as the viscosity, titrable acidity, and pH indicate that the kefir fortification did not affect its stability during storage. EBJ significantly increased the antioxidant properties of the kefir, depending on the fortification level. Additionally, all the fortified samples provided more anthocyanins than the daily recommended intake. Microbiological profiling demonstrated that good laboratory practice and hygiene were implemented during the experiments. Finally, the panelists showed that higher EBJ concentrations in the kefir resulted in greater overall acceptability, indicating that this encapsulate has the potential to be a substitute synthetic color additive in the dairy industry.

Keywords: blackberry; lentil protein isolate; encapsulation; freeze-drying; kefir fortification; storage stability

1. Introduction

In developing countries, with rapid urbanization, less availability of time, and more concerns about a healthier diet, consumers are relying on adopting new instant food habits. The new lifestyles have driven the food industry to develop new functional foods and beverages with added health benefits. Antioxidants in fruit juices have become increasingly popular because of their potential health benefits and ability to combat oxidative stress and free radicals, as advertised by their manufacturers [1]. Nowadays, juices made from berries and red fruits represent frequent consumer choices due to their high purported nutritive value and high contents of bioactive compounds [2].

The application of bioactive compounds in the creation of functional food has some challenges, as bioactivity depends on preserving the stability and bioavailability of the active ingredients. Color and flavor are also very important quality parameters that are driving forces of consumers' acceptance of food. In the present study, blackberries were selected as a source of polyphenolic antioxidants, mostly anthocyanins, due to their diverse health benefits, appealing color, and pleasant flavor. During the processing and storage periods, natural polyphenolics can be negatively affected, as they are easily degradable during processing and storage because of their sensitivity to heat, light, oxygen, and pH [3,4]. Nowadays, one of the emerging trends is their incorporation into food products in encapsulated form. For this purpose, the selection of carriers/wall materials for bioactive compounds is of major importance [5].

Proteins are unique carriers that can efficiently encapsulate bioactive compounds and enable their protection against harsh storage, processing, and gastrointestinal tract (GIT) conditions. Among different plant-based proteins that can be used as encapsulation agents, lentil protein is an excellent choice because of its superior emulsifying properties, low cost, good nutritional value, and low allergen risk [6]. Numerous studies report the encapsulation of different bioactive substances with the usage of lentil proteins. Chang et al. [7] developed lentil protein isolate (LPI)-based microcapsules to encapsulate canola oil with an entrapment efficiency of 88%. Native and heated lentil proteins were effective encapsulation agents for the encapsulation of flaxseed oil [8]. Quiroz et al. [9] demonstrated the ability of lentil proteins to encapsulate annatto extract with encapsulation efficiencies of 69%, where the protein structure and amino acid content were the relevant factors to obtain a high encapsulation capability.

Due to the customer desire for food products that promote good health, kefir consumption has continuously grown. Kefir is a dairy product made from kefir grains that contains a specific combination of bacteria and yeast. It is made by mixing milk with a culture known as "kefir grains," which results in a creamy consistency, sour flavor, and minimal effervescence [10]. In general, the microbial composition of kefir varies depending on the type and composition of the milk, the culture medium, the temperature, and the fermentation period, as well as the storage conditions. The most common bacteria in kefir are *Lactobacillus*, *Lactococcus*, *Streptococcus*, and *Leuconostoc*, while the most common yeasts are *Saccharomyces*, *Kluyveromyces*, and *Candida* [11]. To achieve high-quality kefir, it is essential to maintain a balanced and diverse community of microorganisms, as they work in symbiosis to ferment milk and produce the desired characteristics. Any deviation from this process or contamination by unwanted pathogenic microorganisms can lead to off-flavors, spoilage, or health risks. Therefore, maintaining aseptic conditions during the fermentation process is crucial to prevent contamination by undesirable bacteria, fungi, and/or yeasts. Furthermore, keeping the right balance between the LAB and yeast populations is challenging, as the overgrowth of one group can alter the flavor and texture of the final product. It is important to emphasize that kefir is a perishable product, and maintaining its quality during storage can be challenging. Probiotic microorganisms may lose viability over time. Moreover, monitoring and maintaining the proper pH throughout the fermentation and storage processes are essential to ensure the microbiological quality of the kefir. To ensure good microbiological quality in kefir production and storage, it is important to use proper sanitation practices, control the fermentation conditions, and consider packaging methods that minimize exposure to oxygen and contaminants [12].

Kefir consumption has been linked with an increase in the number of gut health-beneficial bacteria and yeasts; thus, it can prevent digestion problems. In addition, kefir's anti-bacterial, anti-fungal, and anti-cancer properties have been approved in some research [13]. Despite all these health advantages, kefir's acidic flavor does not win over many people, especially young people. To increase kefir consumption among youths, kefir is made with a variety of fruit sauces (strawberry, forest fruit, honey–banana, and mixed fruit) and flavors [14]. To provide attractive color, various anthocyanin-rich fruit and vegetable juices (black mulberry [15], pomegranate [16], strawberry [17], and black carrot [18]) are added to kefir at various concentrations.

Studies on fortified kefirs with encapsulated phytochemicals and natural colorants are limited. Therefore, the main purpose of this research was to develop fortified kefir with encapsulated blackberry juice (EBJ). Four different levels of EBJ (1, 2.5, 5, and 7.5% (*w*/*w*) were added to the kefir formulation after fermentation. During the storage period of 28 days at 4 °C, analyses of the quality parameters, such as the color, rheology, titrable acidity, pH value, total polyphenol content, total anthocyanin content, microbiological profiling, and sensorial properties, were performed to confirm that a quality product was obtained. The current study was designed to (i) advance the scientific knowledge of the use of LPI for the encapsulation of phytochemicals such as polyphenols and anthocyanins from blackberry juice, (ii) evaluate the effect of fortification on the kefir physicochemical and sensorial properties, and (iii) establish the level of added ingredients based on consumer acceptance. This approach may be useful to dairy producers in the strategy of widening their product range considering that the interest in and consumption of functional food is increasing.

2. Material and Methods

2.1. Chemicals

Folin–Ciocalteu reagent, 2,2-diphenyl-1-picrylhydrazyl radical (DPPH$^{\bullet}$), 2,2′-azinobis 3-ethylbenzothiazoline-6-sulphonic acid (ABTS$^{\bullet +}$), Trolox, and trichloroacetic acid were purchased from Sigma Chemical Co. (St. Louis, MO, USA). Methanol, ferric chloride was obtained from J.T. Baker (Deventer, Holland). Methanol and acetic acid were obtained from JT Baker (Deventer, Holland). Manganese (IV) oxide was obtained from Alfa Aesar (Karlsruhe, Germany). Other chemicals and solvents used were of the highest analytical grade. Ultrapure water was produced using an Elix_3 Millipore water purification system coupled to a Milli-Q module (model Advantage10) (Millipore, Molsheim, France).

2.2. Plant Material

Berries were washed and the juice was extracted with a laboratory juicer (model JC400E, Gorenje, Velenje, Slovenia). Green lentils (*Lens culinaris*) were purchased locally, ground in an electric mill, and sieved through an 18-mesh sieve. The lentil flour was stored in a sealed plastic bag at 4 °C until use.

2.3. Protein Extraction

Protein isolates were obtained from green lentil flour via alkaline extraction followed by isoelectric precipitation and freeze-drying. Lentil flour was mixed with distilled water in a 1:10 (w/v) mixing ratio, and the pH was adjusted to 9 with 0.1 M NaOH solution. The alkaline extraction lasted 1 h at 25 °C with constant stirring (400 rpm) on a magnetic stirrer (RH basic, IKA, Staufen, Germany). After extraction, solids were removed via centrifugation (9000× g for 10 min at 4 °C, Sorvall RC-5B, DuPont Instruments, Boston, MA, USA). The supernatant was subjected to isoelectric precipitation by adjusting the pH to 4.2 using 0.1 M HCl. Centrifugation was repeated under the same conditions in order to collect the precipitated protein. Precipitate was redissolved in distilled water and freeze-dried at −40 °C for 24 h (Alpha 2–4 LSCbasic, Martin Christ, Osterode am Harz, Germany).

2.4. Blackberry Juice Encapsulation with Lentil Proteins

The freeze-drying encapsulation process, which is described by Čakaravеć et al. [19], implied the following steps: The carrier LPI was dispersed in blackberry juice in a core–carrier ratio of 1:20 w/v and homogenized with an Ultra-Turrax homogenizer at 11,000 rpm for 5 min, followed by shaking on a laboratory shaker at 400 rpm for 15 min. The prepared formulation was kept overnight at −20 °C in a freezer, and then freeze-dried (Christ Alpha 2–4 LSC, Martin Christ, Osterode am Harz, Germany) at −45 °C for 48 h to ensure complete drying. The freeze-dried encapsulate was stored at −20 °C pending further use.

2.5. Characterization of Lentil Protein Isolate (LPI) and Encapsulated Blackberry Juice (EBJ)

2.5.1. Physicochemical and Color Properties

The water activity, or a_w value, was determined via a LabSwift-aw meter (Novasina, Lachen, Switzerland) at 25 °C. The moisture content was determined using the air oven method at 105 °C, until a constant weight was obtained. For hygroscopic moisture analysis, about 2 g of encapsulates were placed at 25 °C in an airtight plastic container filled with NaCl-saturated solution (75.29% RH). The hygroscopic moisture was measured after one week, and the result is expressed as grams of moisture per 100 g of dry solids (g/100 g). Solubilities of the LPI and EBJ were determined in distilled water at pH 9 according to Perović et al. [20]. The water-holding capacity (WHC) and oil-holding capacity (OHC) of the LPI and EBJ were determined according to Tan et al. [21]. The WHC and OHC are presented as grams of water or oil held by 1 g of sample. To define the encapsulate flow properties, the bulk density (Db), tapped density (Dt), compressibility ratio (CR), and Hausner ratio (HR) were determined according to Šeregelj et al. [22]. Further, the classification of the encapsulate flowability and cohesiveness was made as outlined by Jinapong et al. [23]. The measurements of the color properties were taken with a Minolta reflectance colorimeter (Minolta ChromaMeter CR-300, Minolta, Osaka, Japan) considering the CIELab color system (L*, a*, and b*). Chroma (C*) was calculated manually via the equation $\sqrt{(a^2 + b^2)}$.

2.5.2. Morphological Properties

The morphological properties of the lentil protein isolate and encapsulate were assessed via a HITACHI TM 3030 scanning electron microscope (Hitachi, Chiyoda, Tokyo) at a custom magnification.

2.5.3. Phytochemical Content and Antioxidant Properties

For the determination of the total polyphenol content (TPC), 100 mg of LPI and EBJ were mixed with 2 mL ethanol, acetic acid, and water (50:8:42) on vortex for one minute, and then centrifuged at 4000 rpm for three minutes. The liquid part was collected and filtered through a 0.45 µm filter. The TPC was determined spectrophotometrically via the Folin–Ciocalteu method adopted to microscale [24]. Results are expressed as milligrams of gallic acid equivalents per 100 g (mg GAE/100 g).

For the determination of the total anthocyanin content (TAC,) the same extraction procedure as for the total polyphenol content was used. The TAC was determined spectrophotometrically according to the pH differential method described by Lee et al. [25]. Results are expressed as milligrams of cyanidin-3-glycoside per 100 g (mg CyGE/100 g).

The encapsulation efficiencies of the polyphenols and anthocyanins (EE TPC and EE TAC, respectively) were determined according to the Tumbas Šaponjac et al. [26] protocol. The quantification of the polyphenols and anthocyanins was performed following the previously mentioned spectrophotometric procedure. The EE TPC (EE TAC) was calculated by using the following equation:

$$\text{EE TPC} = [(\text{CPC} - \text{SPC})/\text{CPC}] \times 100$$

$$\text{EE TAC} = [(\text{CAC} - \text{SAC})/\text{CAC}] \times 100$$

where CPC (CAC) is the polyphenols (anthocyanins) inside the core of the encapsulate, while SPC (SAC) is the surface polyphenol (anthocyanin) content. The control sample was the carrier without the blackberry juice.

The analysis of the antioxidant activity, expressed as micromole Trolox equivalent (TE) per 100 g of LPI or EBJ, was performed via three methods adapted to microscale: the 2,2-diphenyl-1-picrylhydrazyl (DPPH) method, the 2,2-azino-bis-3-ethylbenzothiazoline-6 acid (ABTS) method by Tumbas Šaponjac et al. [27], and the reducing power (RP) method by Oyaizu [28].

2.5.4. Microbiological Profiles

The microbiological profiles of the LPI and EBJ were determined via ISO standardized analyses: aerobic mesophilic bacteria (ISO 4833-1:2013) [29], yeast and mold (ISO 21527-1:2008) [30], Enterobacteriaceae (ISO 21528-2:2017) [31], and *Clostridium perfringens*

(ISO 7937:2004) [32]. For each analysis, the samples were mixed with Peptone Water solution (HiMedia, Mumbai, India) in a ratio of 1:10 using an EasyMIX blender (BioMerieux, Marcy-l'Étoile, France). Subsequently, a series of dilutions were prepared, and aliquots were streaked into a (non)selective medium following the mentioned standards. After all incubation steps, colonies were counted, and the obtained results were presented as log CFU/g. The complete procedures and all recommendations from the standardized methodology were followed in detail. All incubations were carried out in temperature-controlled incubators (Medline Scientific, Oxfordshire, UK), and all measurements were performed in triplicate.

2.6. Preparation of Kefir Fortified with Encapsulated Blackberry Juice (EBJ)

Kefir was made in the Laboratory of Microbiology at the Faculty of Technology Novi Sad, the University of Novi Sad. Briefly, 6 g of commercial lyophilized kefir grains (containing *Lactococcus lactis* (ssp. *cremoris*, ssp. *lactis*, and ssp. *lactis* biovar diacetylactis), *S. thermophilus*, *Lactobacillus bulgaricus*, *L. kefir*, *L. parakefir*, and *Saccharomyces cerevisiae*) was activated in pasteurized cow milk ("Imlek", Belgrade, Serbia) in a ratio of 1:8 (w/v) at 25 °C for 24 h to develop the inoculant. The obtained inoculants, without kefir grains, were added to the pasteurized milk and kefir fermentation was performed at 25 °C for 24 h. The obtained kefir was divided into five equal parts. The control was intact (without the addition of any compounds), while the other four parts were fortified with different concentrations of the prepared encapsulate: 1, 2.5, 5, and 7.5% (w/w). The fortification was performed on the same day that the kefir was produced in sterile conditions at the Laboratory of Microbiology. In order to avoid contamination, the process of fortification was performed using sterile bottles, which were quickly and carefully filled and immediately closed. The homogenization was performed with the shaker for 1 min, and the obtained final kefir drinks were refrigerated at 4 °C until further experimental steps.

2.7. Monitoring of Kefir Quality Parameters during Storage

During the storage period of 28 days at 4 °C, the control and fortified samples were monitored at the initial point (0 days) and every seven days (the total number of sampling points was 5). In order to determine the quality of the obtained kefirs, the following analyses were monitored during the storage period: color, rheological, titrable acidity, pH value, total polyphenol content, total anthocyanin content, antioxidant activity, and microbiological profiling analyses.

2.7.1. Color Analysis and Influence of Encapsulated Blackberry Juice (EBJ) Fortification on Viscosity, pH, and Total Solids

Color analysis was conducted via the Minolta reflectance colorimeter described above. Color quantifications were measured at five points for each sample with a minimum of 10 readings, and the results are presented as average values using the CIE L*a*b* system.

Using a rotational viscometer (DV1 Digital Viscometer, Brookfield, WI, USA), the viscosity was determined (at 25 °C, cylindrical spindle, and a shear rate of 51 s^{-1}). The equation of the power-law model ($\delta = K(\gamma)^n$) [33] for the calculation of the viscosity was used, where δ and γ demonstrate the shear stress (Pa) and the shear rate (s^{-1}), respectively.

The titrable acidity and pH values of the control and fortified samples were monitored using the AOAC method (2000) and pH meter HI99161 (Hanna Instruments, Vunsakat, USA). The titrable acidity is expressed as g/100 mL in anhydrous lactic acid.

The extraction for the TPC and TAC estimation in the kefir samples was carried out according to Kabakci et al. [34]. Kefir samples (10 g) were exhaustively extracted using acidified 30 mL methanol (0.1% HCl, v/v), followed by centrifugation at 4000 rpm (Centrifuge, model EBA 21, from Hettich Zentrifugen, Tuttlingen, Germany) and the collection of supernatants. Methanol in the supernatants was evaporated at 40 °C in a rotary evaporator (Büchi Labortechnik, Flawil, Switzerland). Acidified ultrapure water

(0.01% HCl, v/v) was used for dissolving the residue. The resulting extracts were used for the TPC, TAC, and antioxidant activity determination using the above-described protocols.

2.7.2. Microbiological Profiling of Encapsulated Blackberry Juice (EBJ)-Fortified Kefir

In order to monitor the kefir culture profile, as well as the selected hygienic and microbiological safety parameters, a set of microbiological analyses were conducted during the complete storage period. For this purpose, 25 g of sample was suspended in 225 mL of double-concentrated Buffered Peptone Water solution (HiMedia, Mumbai, India) and mixed in a homogenizator (AES Chemunex, Barcelona, Spain). The series of dilutions were prepared, and aliquots of selected dilutions were transferred into nutrient media according to the standardized procedures. Briefly, the following ISO standardized analyses were performed: lactic acid bacteria (ISO 7889:2003) [35]; aerobic mesophilic bacteria (ISO 4833-1:2013) [29]; yeast and mold (ISO 21527-1:2008) [30]; *Escherichia coli* (ISO 7251:2005) [36]; *Listeria monocytogenes* (ISO 11290-2:2017) [37]; *Staphylococcus aureus* (ISO 6888-1:1999/AMD2:2018) [38]; *Salmonella* spp. (ISO 6579-1:2017) [39]; Enterobacteriaceae (ISO 21528-2:2017) [31]; and *Clostridium perfringens* (ISO 7937:2004) [32]. The complete procedures and all recommendations from the standardized methodology were followed in detail. All incubations were carried out in temperature-controlled incubators (Medline Scientific, Oxfordshire, UK), and all measurements were performed in triplicate.

2.8. Sensory Evaluation

Sensory characteristics of kefir samples were evaluated by a panel of 24 trained assessors (12 female and 12 male), with a mean age of 25 years. Kefir samples were presented in a random order in plastic cups. A nine-point hedonic scale was applied to determine the sensory acceptance of the kefir samples [40]; values from 1 to 9 represented from "dislike extremely" to "like extremely". The evaluated attributes were taste, appearance, aroma, color, texture, and overall acceptance.

2.9. Statistical Analyses

All analyses were performed in triplicate and the results are presented as mean values ± standard deviations. One-way analysis of variance (ANOVA) with a Tukey's HSD test was used to determine significant differences among the data at the significance level $p < 0.05$. Statistical analyses were carried out using the Origin 8.0 SRO software package and Microsoft Office Excel 2010.

3. Results and Discussion

3.1. Lentil Protein Isolate (LPI) and Encapsulated Blackberry Juice (EBJ)

3.1.1. Physicochemical and Color Properties

A mixture of the blackberry juice and lentil protein isolate with no further ingredients was prepared and processed into juice encapsulates via freeze-drying. The general appearances of the LPI and EBJ are shown in Figure 1, and the physicochemical properties of these samples are presented in Table 1.

(a) (b)

Figure 1. Images of (**a**) lentil protein isolate and (**b**) encapsulated blackberry juice.

Table 1. Physicochemical and color properties of lentil protein isolate (LPI) and encapsulated blackberry juice (EBJ).

Properties	Lentil Protein Isolate (LPI)	Encapsulated Blackberry Juice (EBJ)
Water activity (a_w)	0.09 ± 0.00	0.24 ± 0.00
Moisture content (g/100 g)	1.64 ± 0.00	2.17 ± 0.01
Hygroscopic moisture (g/100 g)	2.71 ± 0.01	2.98 ± 0.02
Solubility (%)	56.50 ± 0.26	29.30 ± 0.26
Water-holding capacity (WHC) (g/g)	1.72 ± 0.06	1.55 ± 0.03
Oil-holding capacity (OHC) (g/g)	1.48 ± 0.02	3.66 ± 0.36
Bulk density (Db) (g/mL)	0.91 ± 0.00	0.87 ± 0.00
Tapped density (Dt) (g/mL)	1.79 ± 0.00	1.10 ± 0.00
Compressibility ratio (CR)	49.16 ± 0.00	20.91 ± 0.00
Hausner ratio (HR)	1.97 ± 0.00	1.26
Flowability	Very bad	Fair
Cohesiveness	High	Intermediate
L*	17.90 ± 0.02	16.30 ± 0.04
a*	17.75 ± 0.02	35.80 ± 0.04
b*	54.05 ± 0.03	13.90 ± 0.03
C*	56.89 ± 0.02	38.40 ± 0.04

The water activity and moisture content are valuable quality parameters, as they determine the stability, flowability, drying efficiency, stickiness, oxidation of bioactive molecules, and potential of microbial growth on the final encapsulates. For both samples, the a_w values (LPI: 0.09; EBJ: 0.24) and moisture content (LPI: 1.64 g/100 g; EBJ: 2.17 g/100 g) were within the maximum a_w and moisture specification for dried encapsulates in the food industry, which are 0.6 and 3–4 g/100 g, respectively [41]. The obtained results suggest that the final encapsulates could be stable during storage.

Hygroscopicity is an important parameter for food encapsulation technology and presents the ability of encapsulates to absorb moisture from a relatively humid environment. Therefore, defining this parameter offers the best possibility of maintaining the best storage conditions. The hygroscopicity of encapsulates can be classified as follows: non-hygroscopic (<10%), slightly hygroscopic (10–15%), and hygroscopic (15–20%) [42]. According to this classification, LPI and EBJ could be considered non-hygroscopic and stable during storage.

Among the different physicochemical properties, the solubility and ability of proteins to bind water or oil are the most important for their usage in food applications. The obtained results of these characteristics showed that the LPI had higher solubility (56%) compared to the EBJ (29%). These results are in good agreement with the obtained results for the water-holding capacity. The WHC results showed the superior water-binding ability of the LPI compared to the EBJ. Moreover, the EBJ showed a better ability to retain oil, resulting in a higher oil-holding capacity (3.66 g/g). The higher oil-holding capacity of proteins greatly influences the shelf life and textural and other quality properties of food products [43].

Bulk density (Db) is also a significant characteristic of encapsulates. To ensure consistency during packaging, the encapsulates must meet Db criteria. Via the Db and Dt parameters, it is possible to present two parameters: the CR, which is an indicator of the flowability, and the HR, which is a measurement of the cohesiveness. According to these parameters, the obtained EBJ had fair flowability and intermediate cohesiveness. The free-flow properties of encapsulates are highly dependent on the carrier characteristics, as well as the technique applied for encapsulation [44].

Because color influences the sensory attractiveness of foods, this quality indicator is of importance for consumers, particularly for enriched products in which encapsulates are used as a colorant. The EBJ was characterized by lower lightness ($L^* = 16.30$) and yellowness ($b^* = 13.90$), as well as greater redness ($a^* = 35.80$), compared to the LPI ($L^* = 17.90$; $a^* = 17.75$; $b^* = 54.05$). The darker color and superior redness of the EBJ are explained by the encapsulated anthocyanins, which are the main features of blackberries. Jiménez-Aguilar et al. [45] also reported that the coordinate a* is attributed to the anthocyanin content of the fruit. In addition, color changes in food products enriched with encapsulated bioactive compounds can occur due to the formation of polymeric anthocyanins [46]. According to Weiss et al. [47], the relative amounts of flavylium cations (red or orange), quinoidal forms (blue), carbinol pseudobase (colorless), and chalcones (colorless or slightly yellowish) in equilibrium conditions might vary depending upon the pH and the anthocyanin structure.

3.1.2. Morphological Properties

The morphologies of the carrier (LPI) and encapsulate (EBJ) are shown in Figure 2. The carrier was similar to broken glass and had a square–rectangular structure with smooth, straight edges (Figure 2a). The reason for this structure may be in the obtaining procedure, namely in the low temperature and the lack of forces that alter the surface topology during the evaporation process [24,48]. Figure 2b shows the structure of the freeze-dried encapsulate in which the base is the described carrier with particles different in diameter, irregular shapes, pores, and wrinkles. According to Fernandes et al. [49], this structure originates from encapsulated phytochemicals and indicates the encapsulating agent's suitability.

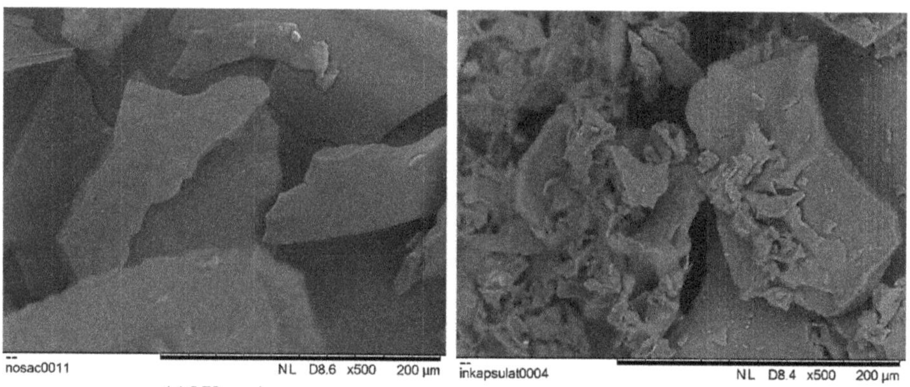

(a) LPI carrier

(b) EBJ encapsulate

Figure 2. SEM imaging of surface morphology of (**a**) lentil protein isolate (LPI) and (**b**) encapsulated blackberry juice (EBJ) at a magnification of $\times 500$.

3.1.3. Phytochemical Content and Antioxidant Properties

The encapsulation technique must ensure a high retention rate of the bioactive ingredients to effectively reach the site of action in high quantities. The encapsulation efficiency (EE) is defined as the amount of the bioactive ingredient that is effectively encapsulated in the carrier material. The EE is mainly influenced by the properties of the carrier and core materials, as well as the drying parameters. Table 2 summarizes the results of the phytochemical content and antioxidant properties of the LPI and EBJ.

Table 2. Functional properties of lentil protein isolate (LPI) and encapsulated blackberry juice (EBJ). nd—not detected.

Properties	Lentil Protein Isolate (LPI)	Encapsulated Blackberry Juice (EBJ)
Total polyphenol content (TPC) (mg GAE/100 g)	124.08 ± 7.76	411.06 ± 9.87
Total anthocyanin content (TAC) (mg CyGE/100 g)	nd	323.81 ± 14.65
EE TPC (%)	-	89.68 ± 2.14
EE TAC (%)	-	91.12 ± 4.55
DPPH (μmol TE/100 g)	116.41 ± 3.68	1024.09 ± 14.11
RP (μmol TE/100 g)	75.99 ± 4.56	547.12 ± 17.22
ABTS (μmol TE/100 g)	214.03 ± 13.57	1452.71 ± 28.30

According to the EE results (EE TPC = 89.68%; EE TAC = 91.12%), the combination of the freeze-drying technique and LPI as a carrier for the encapsulation of polyphenols/anthocyanins from blackberry juice was very successful. Robert et al. (2010) obtained lower values for the encapsulation of polyphenols/anthocyanins from pomegranate juice using soy proteins (76.2% and 58.5%, respectively). Dumitrascu et al. [50] also reported a lower EE of anthocyanins from cornelian cherry juice (68%) when using soy protein concentrate as the coating material and freeze-drying. Further, Diaz et al. [51] investigated the effect of wall materials on the physicochemical and antioxidant properties of blackberry juice; the authors reported that guma arabica had the highest percentage of encapsulation (71.62%), followed by maltodextrin (66.45%) and whey protein concentrate (23.33%). In this study, high encapsulating efficiencies of anthocyanins by individual coating agents and the freeze-drying technique were obtained, which is not in agreement with the assertion reported by Tao et al. [52], who reported that mixtures of coating agents in different combinations are more efficient than the individual compounds for the encapsulation of anthocyanins.

Bioactive compounds present in blackberry juice are considered to exert health-beneficial properties mainly through antioxidant activities. Currently, there is no single antioxidant assay that can accurately evaluate the contribution of bioactive compounds to the total antioxidant activity of plants or food. The amount and type of antioxidants present in biological samples, as well as the free radicals or the oxidants used in the tests, play a significant role in measuring the antioxidant activity. To evaluate the antioxidant activity in fruit juices, it is important to use different antioxidant assays, instead of relying on a single assay. In addition, results that are not seen when testing individual ingredients may vary as a result of synergistic effects and the concentration [53]. The encapsulated juice presented good antioxidant activity with a DPPH value of 1024.09 μmol TE/100 g, an RP of 547.12 μmol TE/100 g, and an ABTS of 1452.71 μmol TE/100 g.

3.1.4. Microbiological Profiles

To proceed to the next steps of the experimental part, it was necessary to determine the microbiological statuses of the carrier and encapsulate used in this study. According to the values of all the selected microbiological parameters presented in Table 3, it can be concluded that there was no contamination observed in the EBJ. However, relatively low numbers of aerobic mesophilic bacteria (1.2 log CFU/g) and yeast and mold (1.4 log CFU/g) were observed in the case of the tested carrier, which was expected considering its origin (lens protein). Taking all into account, it can be concluded that the tested LPI and EBJ are acceptable for further use in the creation of fortified kefir from the microbiological point of view.

Table 3. Microbiological profiles of lentil protein isolate (LPI) and encapsulated blackberry juice (EBJ).

Microbiological Parameter Log CFU/g	Lentil Protein Isolate (LPI)	Encapsulated Blackberry Juice (EBJ)
Aerobic mesophilic bacteria	1.20 ± 0.54	<1
Yeast and mold	1.40 ± 0.00	<1
Enterobacteriaceae	<1	<1
Clostridium perfringens	<1	<1

3.2. Kefir Fortified with Encapsulated Blackberry Juice (EBJ)

The appearances of the kefirs after fortification with EBJ at concentration levels of 1, 2.5, 5, and 7.5% (*w*/*w*) are shown in Figure 3. With the increasing encapsulate concentrations, the produced color became more intense. The introduced encapsulate was not dispersed homogeneously within the product; the appearance of stains or pigmented grains can be observed. Because the solubility of EBJ is low (29.30%, Table 1), this appearance of the kefir samples was expected.

Figure 3. Appearances of the kefir control sample and samples with EBJ fortification at different concentration levels: 1, 2.5, 5, and 7% (*w*/*w*).

3.2.1. Influence of Encapsulated Blackberry Juice (EBJ) Kefir Fortification on Color, Titratable Acidity, and pH during Storage

Table 4 summarizes the color parameters L*, a*, b*, and C*. The L* parameter reveals that the EBJ addition significantly decreased the kefir luminosity with progressively lower values with the increasing encapsulate concentration. This observation is in agreement with the sample appearance shown in Figure 3, as well as with the anthocyanin content (as the anthocyanins gave color to the samples). The L* varied for all concentration levels during the storage period; even though these variations were statistically significant, the behavior was minimal for all samples; thus, from a technological standpoint, it should not have an impact on the visual perception.

Parameter a* was very sensitive to the EBJ addition; even at the lowest concentration (1%, *w*/*w*), the value changed from negative to positive, indicating the predominance of the red color. With the EBJ addition, the b* color parameter changed from positive to negative, indicating a predominantly blue color. Although blackberry juice has a dark-red color, at acidic pH values (pH~4.0), anthocyanins present a light-purple color. The parameter C* values, or Chroma, also followed the increase in the EBJ concentrations; the kefirs with greater anthocyanin concentrations presented higher C* values. Du and Myracle [54] also reported positive a* and negative b* values for aronia kefir; these values indicated that the product presented a bluish-red color. The authors expected the bluish-red color in the products due to the acidic environment of kefir.

Table 4. Color properties during storage.

Kefir Sample	Parameter	Storage Period (Days)				
		0	7	14	21	28
Control	L*	97.20 ± 1.54 [a]	98.20 ± 1.11 [a]	95.14 ± 1.20 [a]	96.20 ± 0.24 [a]	97.10 ± 1.65 [b]
	a*	−1.60 ± 0.20 [a]	−1.91 ± 0.01 [b]	−2.00 ± 0.01 [b]	−2.31 ± 0.00 [c]	−2.40 ± 0.01 [c]
	b*	3.31 ± 0.04 [a]	3.45 ± 0.01 [a]	3.41 ± 0.0 [a]	3.36 ± 0.00 [a]	3.29 ± 0.04 [a]
	Cr*	3.68 ± 0.25 [a]	3.94 ± 0.02 [b]	3.95 ± 0.01 [b]	4.08 ± 0.00 [b]	4.07 ± 0.05 [b]
1%	L*	74.80 ± 1.31 [a]	74.51 ± 1.49 [a]	76.50 ± 1.27 [a]	75.11 ± 1.14 [a]	74.20 ± 1.61 [a]
	a*	10.72 ± 0.33 [b]	9.85 ± 0.24 [a]	11.31 ± 0.21 [c]	10.62 ± 0.17 [b]	11.11 ± 0.08 [c]
	b*	−4.51 ± 0.02 [a]	−4.42 ± 0.00 [a]	−4.91 ± 0.17 [b]	−4.42 ± 0.02 [a]	−4.64 ± 0.13 [a]
	Cr*	11.61 ± 0.55 [b]	10.80 ± 0.25 [a]	12.33 ± 0.40 [c]	11.50 ± 0.20 [b]	12.04 ± 0.34 [c]
2.5%	L*	62.53 ± 1.32 [b]	59.11 ± 1.07 [a]	62.50 ± 0.99 [b]	62.52 ± 1.00 [b]	60.31 ± 0.56 [a,b]
	a*	21.90 ± 1.60 [a]	22.32 ± 2.04 [a]	22.31 ± 0.74 [a]	22.40 ± 1.08 [a]	22.32 ± 0.27 [a]
	b*	−11.20 ± 0.63 [a,b]	−12.11 ± 1.65 [c]	−12.10 ± 0.00 [c]	−10.79 ± 0.19 [a]	−11.62 ± 0.68 [b,c]
	Cr*	24.60 ± 2.25 [a]	25.39 ± 3.70 [a]	25.38 ± 0.75 [a]	24.86 ± 1.30 [a]	25.16 ± 1.00 [a]
5%	L*	43.72 ± 1.84 [b]	40.50 ± 0.53 [a]	43.75 ± 0.00 [b]	41.71 ± 1.51 [a,b]	41.00 ± 0.02 [a]
	a*	33.86 ± 0.64 [a]	34.11 ± 0.35 [a]	33.89 ± 0.65 [a]	33.25 ± 0.21 [a]	33.41 ± 0.18 [a]
	b*	−11.52 ± 0.00 [a]	−11.51 ± 0.07 [a]	−11.42 ± 0.18 [a]	−11.44 ± 0.06 [a]	−10.99 ± 0.02 [a]
	Cr*	35.77 ± 0.65 [a]	34.55 ± 0.42 [a]	35.77 ± 0.85 [a]	35.16 ± 0.27 [a]	35.17 ± 0.20 [a]
7.5%	L*	34.30 ± 1.84 [a]	36.21 ± 0.00 [b]	37.92 ± 0.91 [b]	33.45 ± 0.21 [a]	33.72 ± 1.17 [a]
	a*	45.00 ± 0.01 [b,c]	43.54 ± 1.11 [a,b]	41.78 ± 0.97 [a]	47.31 ± 1.61 [c]	44.12 ± 0.40 [b]
	b*	−10.42 ± 0.65 [a]	−11.11 ± 0.07 [a]	−10.82 ± 1.96 [a]	−10.36 ± 0.75 [a]	−10.47 ± 0.07 [a]
	Cr*	46.19 ± 0.66 [b,c]	44.94 ± 1.19 [a]	43.16 ± 2.02 [a]	48.43 ± 2.40 [b,c]	45.35 ± 0.05 [a,b]

L* represents lightness (L* = 0 (black) and L* = 100 (white)), a* represents red and green color coordinates (−a* = greenness and +a* = redness), while b* represents yellow and blue color coordinates (−b* = blueness and +b* = yellowness); Cr*—Chroma. Different letters in the same row indicate significant differences among samples at $p \leq 0.05$.

Table 5 provides data on various parameters, namely the viscosity, titrable acidity, and pH values, for the kefir control sample and EBJ-fortified kefir samples over a 28-day storage period. These parameters impact the sensory attributes, safety, and product consistency. Monitoring and controlling these parameters throughout the storage process are essential for delivering a high-quality kefir product that meets consumer expectations and safety standards. The viscosity is one of the most important parameters in kefir quality assessment. It measures the thickness and flow properties of this dairy product, which are influenced by the interaction of proteins, microbes, and other components in the matrix. Viscosity impacts the mouthfeel and texture of kefir, making it a critical attribute for consumer acceptance. For quality control, monitoring the viscosity is essential to ensure consistency and meet consumer expectations. In the control sample, the viscosity increased slightly from 1.89 at day 0 to 1.93 at day 28. The 1% EBJ (w/w) sample showed a more significant increase, with the viscosity ranging from 1.91 to 2.17 Pa·s. The 2.5% EBJ (w/w) sample exhibited a similar trend, with the viscosity increasing from 1.94 to 2.14 Pa·s. Moreover, the 5% EBJ (w/w) sample displayed a consistent increase in viscosity, ranging from 1.95 to 2.23 Pa·s, while the 7.5% EBJ (w/w) sample showed a steady increase from 2.03 to 2.26 Pa·s. The typical viscosity values for kefir can range from 1.5 to 2.5 Pa·s, although these values can vary depending on factors like the fat content, fermentation time, and specific microbial strains used.

Table 5. Viscosity (Pa·s), titrable acidity (g/100 g of lactic acid), and pH of kefirs during storage.

Kefir Sample	Parameters	Storage Period (Days)				
		0	7	14	21	28
Control	Viscosity	1.89 ± 0.01 ^{a}	1.91 ± 0.04 ^{a}	1.87 ± 0.00 ^{a}	1.91 ± 0.52 ^{a}	1.93 ± 0.11 ^{a}
	Titrable acidity	71.00 ± 1.5 ^{c}	69.00 ± 0.56 ^{c}	69.00 ± 0.00 ^{c}	64.50 ± 0.56 ^{b}	61.00 ± 0.00 ^{a}
	pH	4.21 ± 0.02 ^{a}	4.27 ± 0.00 ^{a}	4.31 ± 0.56 ^{a,b}	4.47 ± 0.07 ^{b,c}	4.67 ± 0.05 ^{c}
1%	Viscosity	1.91 ± 0.06 ^{a}	2.04 ± 0.04 ^{b}	2.03 ± 0.10 ^{b}	2.07 ± 0.16 ^{b,c}	2.17 ± 0.56 ^{c}
	Titrable acidity	71.00 ± 1.00 ^{b}	70.00 ± 0.00 ^{b}	69.00 ± 0.00 ^{b}	69.00 ± 1.0 ^{b}	64.00 ± 0.56 ^{a}
	pH	4.12 ± 0.05 ^{a}	4.17 ± 0.00 ^{a}	4.23 ± 0.56 ^{a}	4.27 ± 0.14 ^{a}	4.21 ± 0.04 ^{a}
2.5%	Viscosity	1.94 ± 0.00 ^{a}	2.10 ± 0.24 ^{b}	2.12 ± 0.17 ^{b}	2.12 ± 0.56 ^{b}	2.14 ± 0.41 ^{b}
	Titrable acidity	74.00 ± 2.00 ^{a}	74.00 ± 1.00 ^{a}	73.00 ± 0.00 ^{a}	72.00 ± 1.00 ^{a}	72.00 ± 0.00 ^{a}
	pH	4.01 ± 0.03 ^{a}	4.08 ± 0.07 ^{a}	4.17 ± 0.01 ^{a}	4.20 ± 0.00 ^{a}	4.19 ± 0.34 ^{a}
5%	Viscosity	1.95 ± 0.02 ^{a}	2.15 ± 0.01 ^{b}	2.21 ± 0.17 ^{b}	2.23 ± 0.23 ^{b}	2.23 ± 0.00 ^{b}
	Titrable acidity	81.00 ± 1.00 ^{a}	82.00 ± 0.00 ^{a}	82.00 ± 0.00 ^{a}	83.00 ± 0.00 ^{a}	82.00 ± 0.00 ^{a}
	pH	3.88 ± 0.10 ^{a}	3.79 ± 0.09 ^{a}	3.81 ± 0.03 ^{a}	3.84 ± 0.51 ^{a}	3.82 ± 0.07 ^{a}
7.5%	Viscosity	2.03 ± 0.19 ^{a}	2.21 ± 0.11 ^{b}	2.19 ± 0.25 ^{b}	2.25 ± 0.41 ^{b}	2.26 ± 0.09 ^{b}
	Titrable acidity	83.00 ± 0.50 ^{b,c}	84.00 ± 2.00 ^{c}	81.00 ± 0.00 ^{b,c}	79.00 ± 0.00 ^{b}	67.00 ± 1.00 ^{a}
	pH	3.81 ± 0.00 ^{a}	3.64 ± 0.52 ^{a}	3.74 ± 0.14 ^{a}	4.18 ± 0.21 ^{b}	4.34 ± 0.05 ^{b}

Different letters in the same row indicate significant differences among samples at $p \leq 0.05$.

As for other fermented dairy products, acidity plays a critical role in kefir quality control. It is primarily measured as the titrable acidity, which quantifies the concentration of acids in the kefir. The specific tangy flavor and preservative properties of kefir are closely associated with its acidity. In quality control, monitoring and maintaining the titrable acidity within a specified range are crucial. Typical values for the titrable acidity in kefir usually fall within the range of 60–120 g/100 g of lactic acid, although this can vary based on regional and consumer preferences. In the case of the control samples, this value decreased from 71 to 61 g/100 g of lactic acid in 28 storage days, which reflects a significant decrease. The 1% EBJ (*w*/*w*) sample displayed relatively stable titrable acidity levels within the range of 64–71 g/100 g of lactic acid. The 2.5% EBJ (*w*/*w*) sample showed a slight decrease, ranging from 74 to 72 g/100 g of lactic acid. The 5% EBJ (*w*/*w*) sample maintained relatively stable titrable acidity levels within the range of 81–83 g/100 g of lactic acid. The 7.5% EBJ (*w*/*w*) sample exhibited a decrease in titrable acidity from 83 to 67 g/100 g of lactic acid, with a significant decrease observed on day 28.

The last but not least important parameter for kefir quality monitoring is the pH value. The pH value is a key indicator of the microbial activity in the kefir and is inversely related to the acidity, meaning that as the pH decreases, the acidity increases. Monitoring the pH value during storage helps in assessing the kefir's overall stability and the progression of the fermentation. Typical pH values for kefir fall within the range of 4.0–4.6. In the control sample, the pH value increased from 4.21 at day 0 to 4.67 at day 28, indicating a progressive pH increase over time. The 1% EBJ (*w*/*w*) sample displayed an increase in pH from 4.12 to 4.21, with some fluctuations. The 2.5% EBJ (*w*/*w*) sample showed a gradual increase in pH, ranging from 4.01 to 4.19. The 5% EBJ (*w*/*w*) sample maintained a relatively stable pH within the range of 3.79–3.88. In contrast, the 7.5% EBJ (*w*/*w*) sample exhibited a decrease in pH from 3.81 to 4.34, with significant fluctuations and a sharp increase observed on day 28. It must be noted that monitoring the pH is important not only for the flavor but also for safety, as extreme pH values can indicate spoilage or the growth of undesirable microorganisms.

3.2.2. Influence of Encapsulated Blackberry Juice (EBJ) Kefir Fortification on Phytochemical Content and Antioxidant Activity during Storage

Table 6 shows the phytochemical content and antioxidant properties of the kefir samples during storage. Increases in the TPC, TAC, and antioxidant activity are the main aims of kefir fortification. Initially, the TPC content varied from 98.64 to 181.12 mg GAE/100 g kefir, while the TAC varied from 18.25 to 44.56 mg CyGE/100 g kefir, according to the increase in the EBJ concentration added. The addition of EBJ, for all concentrations, showed a similar behavior during storage. A slight decrease in the phytochemical content with time was observed. A temporary decrease in the TPC in the kefir samples could be explained by the decomposition of polymeric phenolics in the presence of lactic acid bacteria during refrigerated storage [55]. The antioxidant activity followed the same trend. Increasing the fortification level of the EBJ led to an increase in the antioxidant activity of the kefir against DPPH and ABTS, as well as reducing power. The high increase in the antioxidant activity of the fortified kefirs was due to the presence of phenolic compounds with prominent antioxidant activity, such as anthocyanins (probably cyanidin-3-O-glucoside and cyanidin-3-O-dioxalylglucoside) and certain phenolic acids (probably ellagic acid, chlorogenic acid, caffeic acid, and gallic acid) [56]. Barretto et al. [57] investigated the stability of encapsulated and non-encapsulated anthocyanins in yogurt produced with natural dye obtained from Solanum melongena L. bark. The authors reported the greater stability of the anthocyanins during storage, as well as antioxidant activity, when added in encapsulated form in yogurt. Kabacki et al. [34] also reported a gradual degradation of anthocyanins during storage; the authors showed a negative correlation between glucose and anthocyanins during the storage of kefir fortified with anthocyanin-rich juices. Thus, among the previously mentioned factors, such as heat, light, oxygen, and pH, the differences in the anthocyanin stability may also result from variations in the sugar composition in the fermented matrices. Esatbeyoglu et al. [58] recently reported that the total anthocyanin content of water kefir made of aronia pomace and juice decreased with the fermentation (72 h). The authors reported that the kefir with aronia pomace had higher TACs (17.8 and 13.3 mg/100 mL kefir for 0 h and 72 h, respectively) compared to those with aronia juice (7.30 and 5.34 mg/100 mL kefir at 0 h and 72 h, respectively). Overall, the addition of encapsulated anthocyanins in the step after fermentation is a promising strategy for maintaining the good stability of these compounds throughout storage. Based on the 240 mL serving size (1 cup), all the produced kefirs in this study could provide more anthocyanins than the daily recommended intake, which is 12.5 mg/day/person [59].

Table 6. Functional properties of kefir fortified with encapsulated blackberry juice (EBJ).

Kefir Sample	Parameter	Storage Period (Days)				
		0	7	14	21	28
Control	TPC	nd	nd	nd	nd	nd
	TAC	nd	nd	nd	nd	nd
	DPPH	20.01 ± 0.06 [b]	20.65 ± 1.02 [b]	18.14 ± 0.26 [a]	18.11 ± 1.35 [a]	17.41 ± 0.23 [a]
	RP	nd	nd	nd	nd	nd
	ABTS	44.11 ± 2.15 [b]	45.18 ± 2.13 [b]	41.56 ± 1.101 [a]	40.35 ± 0.89 [a]	40.15 ± 1.65 [a]
1%	TPC	98.64 ± 0.78 [a]	99.11 ± 2.14 [a]	98.96 ± 3.58 [a]	98.05 ± 2.14 [a]	98.71 ± 3.44 [a]
	TAC	18.25 ± 0.63 [a]	18.12 ± 2.26 [a]	19.21 ± 1.14 [b]	19.06 ± 2.17 [a,b]	18.11 ± 1.09 [a]
	DPPH	181.16 ± 9.81 [a]	180.99 ± 10.12 [a]	176.58 ± 13.69 [a]	179.36 ± 9.11 [a]	180.47 ± 21.07 [a]
	RP	10.13 ± 0.51 [a]	11.79 ± 0.65 [a]	10.34 ± 0.74 [a]	11.55 ± 0.98 [a]	11.10 ± 0.36 [a]
	ABTS	265.04 ± 13.44 [c]	249.11 ± 7.29 [b]	233.15 ± 10.41 [a]	251.29 ± 17.21 [b]	248.15 ± 16.54 [b]

Table 6. *Cont.*

Kefir Sample	Parameter	Storage Period (Days)				
		0	7	14	21	28
2.5%	TPC	126.42 ± 9.15 [a,b]	129.87 ± 5.13 [b]	125.30 ± 8.91 [a,b]	125.52 ± 11.60 [a,b]	123.10 ± 9.56 [a]
	TAC	21.62 ± 1.51 [b]	21.98 ± 3.22 [b]	20.87 ± 2.09 [a]	20.91 ± 2.86 [a,b]	20.44 ± 1.13 [a]
	DPPH	205.23 ± 14.63 [a]	211.58 ± 10.55 [a]	209.78 ± 13.17 [a]	206.34 ± 9.15 [a]	204.19 ± 5.68 [a]
	RP	12.78 ± 0.98 [b]	13.71 ± 1.63 [c]	11.25 ± 0.44 [a]	13.41 ± 0.97 [b,c]	13.55 ± 1.11 [c]
	ABTS	287.10 ± 22.65 [a,b]	292.51 ± 13.74 [b]	276.77 ± 13.81 [a]	296.86 ± 15.19 [b]	289.16 ± 5.00 [a,b]
5%	TPC	163.71 ± 9.08 [a]	160.12 ± 2.50 [a]	161.41 ± 3.82 [a]	160.99 ± 7.85 [a]	159.84 ± 5.06 [a]
	TAC	38.40 ± 1.25 [b]	34.13 ± 2.81 [a]	34.54 ± 3.96 [a]	33.41 ± 2.21 [a]	34.07 ± 4.20 [a]
	DPPH	256.08 ± 5.01 [a]	249.66 ± 11.40 [a]	247.35 ± 6.39 [a]	243.17 ± 16.58 [a]	244.96 ± 7.89 [a]
	RP	17.99 ± 3.47 [d]	15.81 ± 2.07 [b,c]	16.33 ± 2.50 [c]	15.36 ± 1.18 [b]	14.18 ± 2.00 [a]
	ABTS	328.10 ± 15.06 [a]	324.45 ± 11.43 [a]	313.06 ± 12.98 [a]	310.55 ± 17.62 [a]	312.69 ± 5.14 [a]
7.5%	TPC	181.12 ± 9.10 [a]	179.57 ± 8.97 [a]	175.13 ± 12.01 [a]	174.41 ± 6.33 [a]	172.08 ± 6.25 [a]
	TAC	44.56 ± 2.96 [b]	43.91 ± 0.81 [a,b]	44.63 ± 1.17 [b]	42.09 ± 2.74 [a]	43.36 ± 1.25 [a,b]
	DPPH	221.41 ± 11.52 [b]	215.31 ± 14.52 [a,b]	208.12 ± 17.65 [a]	208.41 ± 13.58 [a]	205.10 ± 6.28 [a]
	RP	17.97 ± 1.36 [d]	16.95 ± 2.83 [c]	14.25 ± 1.08 [a]	15.35 ± 2.97 [b]	14.51 ± 1.88 [a]
	ABTS	301.20 ± 28.15 [b]	295.42 ± 15.14 [b]	276.77 ± 13.81 [a]	280.77 ± 19.09 [a]	280.10 ± 14.53 [a]

Different letters in the same row indicate significant differences among samples at $p \leq 0.05$; nd—not detected.

3.2.3. Influence of Encapsulated Blackberry Juice (EBJ) Kefir Fortification on Microbiological Profile during Storage

The tracking of the microbiological statuses of the kefirs over the storage time was performed on the day of kefir fortification and then every seven days until the 28th day. According to the obtained results presented in Table 7, no *L. monocytogenes*, *Salmonella* spp., *S. aureus*, *E. coli*, Enterobacteriaceae, or *Cl. pefrigens* were found in any of the kefir samples at any time of the analysis. This result is an excellent indicator of the good laboratory practice and hygiene implemented during the experiments. Also, in all samples at the starting point, the numbers of LAB 1, LAB 2, and AMB were between 7 and 8 log CFU/mL, while the numbers of yeast and mold were approximately 5 log CFU/mL. The Fermented Milks Codex [60] lists 10^4 CFU/g as the minimum yeast content in kefir and 10^7 CFU/g as the minimum load of total microorganisms constituting the starter culture. Thus, the obtained initial kefirs met the requirements appointed by the Fermented Milks Codex. Slightly higher numbers of AMB and yeast and mold could be noticed in all the fortified kefirs in comparison to the control kefir sample. This result is a consequence of the addition of an encapsulate and suitable carrier, which was demonstrated to carry low numbers of AMB and yeast and mold (Table 7), to the kefir during the process of fortification, resulting in increases in the total numbers of these microorganisms at the starting point. A similar situation was earlier presented by Goncu et al. [61], whose work has demonstrated that the addition of dietary fibers in kefir leads to the stimulated growth of LAB. During the storage period, slightly decreasing numbers of LAB 1, LAB2, and AMB could be noticed in all the samples. After the 28th day, the initial numbers of the mentioned groups of microorganisms were lower than the initial numbers by approximately 1–2 log CFU/mL. Contrarily, the numbers of yeast and mold increased by approximately 1 log CFU/mL in all the tested samples over time. The most intensive changes in the number of tested groups of microorganisms were observed for the samples with the addition of 5% and 7.5% of EBJ, while the samples with less EBJ added did not significantly vary from the control kefir sample. The potential reasons for the observed behavior could be not only the type and concentration of bioactive substances and the initial concentrations of individual groups of bacteria and yeast and their mutual relationship, but also an antimicrobial effect of the microbial products, as well as stimulated storage conditions for only one group of tested microorganisms. Depending on the specific bioactive substances and their concentration, fortification can lead to changes in the microbial composition [62]. Briefly, the fortification of kefir with some types of additional nutrients can either support the growth of beneficial microorganisms or potentially compete with yeast and bacteria for resources [63]. Moreover, yeast and lactic acid bacteria are sensitive to pH changes, so changes in the pH value can influence their growth and survival [12]. On the one hand, bioactive substances can have an inhibitory effect on certain microorganisms, and their encapsulated forms can enable a prolonged influence of microbial populations in kefirs. On the other hand, some bioactive compounds may act as prebiotics, providing a source of nutrients for microorganisms and encouraging the growth of specific strains [64]. Additionally, the antioxidant potential can be correlated with changes in the kefir microbial community due to a strong influence on the oxidative stress and, consequently, on the microbial viability during the storage period [64]. Considering that the obtained results indicate that higher EBJ concentrations have a stronger influence on the microbial community, specific concentrations of these compounds can be a crucial factor and the first step in further determining the influences on kefir's microbial composition during storage.

Table 7. Microbiological profiling of kefirs during storage (results are represented as log CFU/mL for all quantification analyses, while results of detection of *L. monocytogenes* and *Salmonella* spp. are presented as detected/not detected).

Kefir Sample	Parameter	Storage Period (Days)				
		0	7	14	21	28
Control	LAB 1	8.02 ± 0.33 [c]	8.10 ± 0.74 [c]	7.80 ± 0.33 [b]	7.40 ± 0.64 [b]	6.90 ± 0.00 [a]
	LAB 2	7.10 ± 0.00 [d]	6.90 ± 0.56 [c]	6.80 ± 0.00 [c]	6.50 ± 0.56 [a]	6.10 ± 0.33 [a]
	AMB	7.20 ± 0.64 [d]	7.00 ± 0.00 [c]	7.00 ± 0.33 [c]	6.70 ± 0.37 [b]	6.50 ± 0.00 [a]
	Yeast and mold	4.80 ± 0.37 [a]	4.90 ± 0.33 [a]	4.90 ± 0.56 [a]	5.30 ± 0.00 [b]	5.70 ± 0.74 [c]
	L. monocytogenes	nd				
	Salmonella spp.					
	S. aureus	<1				
	E. coli					
	Enterobacteriaceae					
	Cl. perfringens					
1%	LAB 1	8.20 ± 0.00 [d]	8.00 ± 0.00 [c,d]	7.70 ± 0.74 [c]	7.10 ± 0.33 [b]	6.80 ± 0.27 [a]
	LAB 2	7.10 ± 0.33 [c]	6.80 ± 0.33 [b]	6.60 ± 0.27 [b]	6.20 ± 0.74 [a]	5.90 ± 0.00 [a]
	AMB	7.50 ± 0.56 [c]	7.50 ± 0.74 [c]	7.20 ± 0.00 [b,c]	6.90 ± 0.64 [a,b]	6.60 ± 0.33 [a]
	Yeast and mold	5.10 ± 0.00 [a]	5.20 ± 0.64 [a]	5.50 ± 0.33 [b]	5.70 ± 0.37 [b]	6.10 ± 0.56 [c]
	L. monocytogenes	nd				
	Salmonella spp.					
	S. aureus	<1				
	E. coli					
	Enterobacteriaceae					
	Cl. perfringens					

Table 7. *Cont.*

Kefir Sample	Parameter	Storage Period (Days)				
		0	7	14	21	28
2.5%	LAB 1	8.20 ± 0.00 [c]	8.10 ± 0.74 [c]	7.80 ± 0.33 [b,c]	7.40 ± 0.00 [a,b]	7.00 ± 0.64 [a]
	LAB 2	7.00 ± 0.00 [c]	6.80 ± 0.33 [b,c]	6.50 ± 1.32 [b]	6.00 ± 0.00 [a]	5.70 ± 0.33 [a]
	AMB	7.70 ± 0.64 [c]	7.60 ± 0.00 [c]	7.10 ± 0.37 [b]	6.80 ± 0.32 [a,b]	6.50 ± 0.33 [a]
	Yeast and mold	5.20 ± 0.56 [a]	5.30 ± 0.74 [a]	5.70 ± 0.64 [c]	6.00 ± 0.74 [c,d]	6.30 ± 0.00 [d]
	L. monocytogenes	nd				
	Salmonella spp.					
	S. aureus	<1				
	E. coli					
	Enterobacteriaceae					
	Cl. perfringens					
5%	LAB 1	8.10 ± 0.33 [d]	7.60 ± 0.27 [c]	7.10 ± 0.00 [b]	6.70 ± 0.00 [a]	6.50 ± 0.33 [a]
	LAB 2	7.00 ± 0.33 [c]	6.80 ± 0.00 [c]	6.40 ± 0.37 [b]	6.00 ± 0.74 [a]	5.80 ± 0.37 [a]
	AMB	7.90 ± 0.0 [d]	7.40 ± 0.64 [c]	6.90 ± 0.33 [b]	6.70 ± 0.64 [a,b]	6.50 ± 0.74 [a]
	Yeast and mold	5.50 ± 0.00 [a]	5.80 ± 0.74 [a,b]	6.10 ± 0.00 [b]	6.10 ± 0.33 [b]	6.50 ± 1.32 [c]
	L. monocytogenes	nd				
	Salmonella spp.					
	S. aureus	<1				
	E. coli					
	Enterobacteriaceae					
	Cl. perfringens					
7.5%	LAB 1	8.10 ± 0.33 [d]	7.40 ± 0.00 [c]	7.10 ± 0.64 [c]	6.60 ± 0.00 [b]	6.20 ± 0.33 [a]
	LAB 2	7.00 ± 0.00 [d]	6.50 ± 0.56 [c]	6.20 ± 0.00 [b,c]	6.00 ± 0.74 [a,b]	5.80 ± 1.32 [a]
	AMB	8.10 ± 0.37 [e]	7.60 ± 0.37 [d]	7.10 ± 0.00 [c]	6.50 ± 0.74 [b]	6.10 ± 0.00 [a]
	Yeast and mold	5.90 ± 0.00 [a]	6.30 ± 0.74 [b]	6.60 ± 0.33 [b,c]	6.80 ± 0.33 [c]	6.90 ± 0.64 [c]
	L. monocytogenes	nd				
	Salmonella spp.					
	S. aureus	<1				
	E. coli					
	Enterobacteriaceae					
	Cl. perfringens					

LAB 1—*Lactobacillus* species; LAB 2—*Streptococcus* and *Lactococcus* species; AMB—aerobic mesophilic bacteria; nd—not detected. Different letters in the same row indicate significant differences among samples at $p \leq 0.05$.

The exact influence of kefir fortification with bioactive substances on the yeast and bacteria concentrations will vary depending on the specific ingredients used and the microbial community in the kefir. To better understand these effects, it is essential to conduct controlled experiments with different concentrations of bioactive substances and monitor changes in the microbial population over time.

3.2.4. Influence of Encapsulated Blackberry Juice (EBJ) Kefir Fortification on Sensory Attributes

The sensory evaluation showed that different percentages of EBJ strongly modified the sensory attributes of the kefir (Figure 4). The panelists demonstrated a clear preference for the kefir fortified with EBJ at greater concentrations. The most preferred samples by the panelists were the kefir containing 7.5% EBJ (6.50 overall acceptability score) GAP, followed by the kefir containing 5% EBJ (6.42 overall acceptability score). The least preferred concentration among all the kefirs was 1% (5.34 score for overall acceptability). A similar trend was reported by Šeregelj et al. [65] when evaluating the sensory acceptance of yogurt fortified with red pepper waste encapsulate. In addition, similar results of acceptance were reported by Rubio et al. [66] for colored yogurts using colorants extracted from grape pomace and jabuticaba by-products and encapsulated via spray-drying using brewer's spent yeast as the carrier. In the study by Domingos Pinto et al. [67], the panelists showed good acceptance for all the tested attributes of yogurt colored with anthocyanin-rich pigment prepared from jabuticaba skin. The authors noted that natural pigments have the potential to be a substitute synthetic color additive to produce mixed-berry yogurts.

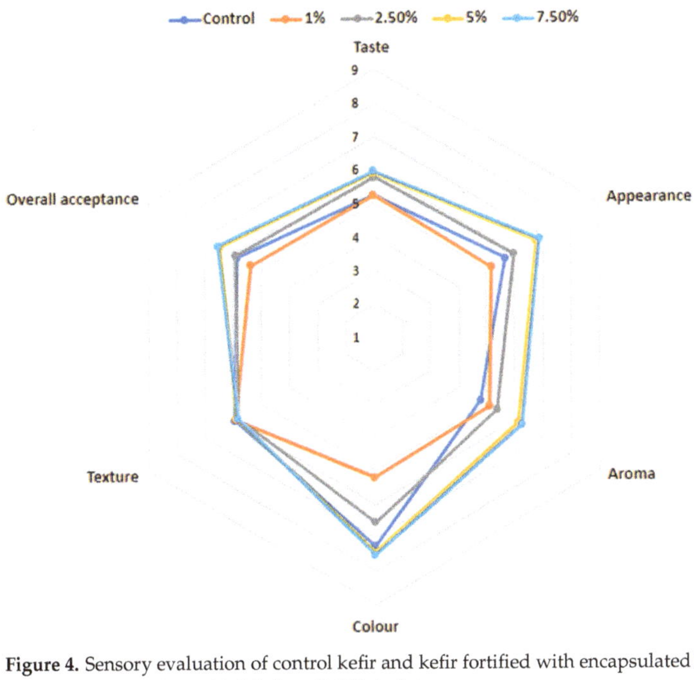

Figure 4. Sensory evaluation of control kefir and kefir fortified with encapsulated blackberry juice (EBJ) in concentrations of 1, 2.5, 5, and 7.5% (w/w).

4. Conclusions

This study specifies that bioactive compounds (polyphenols and anthocyanins) from blackberry juice were successfully encapsulated with LPI using a freeze-drying technique. Kefir fortified with EBJ in different concentrations displayed enhanced antioxidant properties, making it a promising functional food ingredient. The microbiological analysis confirmed the safety of the carrier and encapsulate, indicating their suitability for use in kefir fortification. Moreover, the stability of the fortified products was confirmed by monitoring the viscosity, titrable acidity, and pH values. The sensory evaluation revealed that higher EBJ concentrations in kefir were preferred by the panelists, demonstrating the potential for product diversification.

As the main milestones of this study, the high encapsulation efficiency, safety assurance, and enhanced antioxidant properties and sensory attributes of the final products can be emphasized. The study's sensory evaluation showed that the addition of EBJ at higher concentrations led to increased overall acceptability, indicating the potential for creating appealing kefir variants. Furthermore, the encapsulated kefir with improved antioxidant activity offers the potential for the development of functional food products with health benefits. Future research should focus on enhancing the stability of encapsulated bioactive compounds during storage, potentially through the use of modified encapsulation techniques. Further studies can explore consumer preferences and demands for encapsulated kefir products, leading to tailored offerings for different market segments. Moreover, investigating the specific health benefits associated with the enhanced antioxidant properties of encapsulated kefir could open doors to functional food products that target particular health concerns. Future research should also address the environmental aspects of encapsulation materials (i.e., the potential use of encapsulation materials isolated from different plant by-products). In summary, this study presents a promising approach to enhancing the quality and functionality of kefir through encapsulation. While challenges related to the stability and appearance should be addressed, the potential for creating appealing and health-beneficial kefir products is evident, offering opportunities for future research and product development.

Author Contributions: Conceptualization, V.T. and O.Š.; methodology, A.T. and M.P.; software, M.M.; validation, V.T. and O.Š.; formal analysis, A.T., M.P. and M.M.; investigation, V.T.; resources, O.Š.; data curation, M.P. and M.M.; writing—original draft preparation, V.T.; writing—review and editing, M.A.; visualization, N.Ć.; supervision, M.A. and N.Ć.; project administration, V.T. All authors have read and agreed to the published version of the manuscript.

Funding: This research study was funded by the Ministry of Education, Science, and Technological Development of the Republic of Serbia (grant no. 451-03-47/2023-01/200134).

Institutional Review Board Statement: The study was approved by the Ethics Committee of the Institute of Food Technology in Novi Sad, the University of Novi Sad, Serbia (Ref. No. 175/I/40-3).

Informed Consent Statement: Informed consent was obtained from all subjects involved in the study.

Data Availability Statement: Data are contained within the article.

Conflicts of Interest: The authors declare no conflict of interest.

References

1. Roy, B.; Singh, J.; Dewan, T. A comparative study of antioxidant activity and total phenolic content of fresh juices of some common Indian fruits with their commercial counterparts. *Curr. Sci.* **2020**, *118*, 300–304. [CrossRef]
2. Fidelis, M.; Santos, J.S.; Coelho, A.L.K.; Rodionova, O.Y.; Pomerantsev, A.; Granato, D. uthentication of juices from antioxidant and chemical perspectives: A feasibility quality control study using chemometrics. *Food Cont.* **2017**, *73*, 796–805. [CrossRef]
3. Rosales, T.K.O.; Fabi, J.P. Valorization of polyphenolic compounds from food industry by-products for application in polysaccharide-based nanoparticles. *Front. Nutr.* **2023**, *10*, 1144677. [CrossRef]
4. Rathee, P.; Sehrawat, R.; Rathee, P.; Khatkar, A.; Akkol, E.K.; Khatkar, S.; Redhu, N.; Türkcanoğlu, G.; Sobarzo-Sánchez, E. Polyphenols: Natural Preservatives with Promising Applications in Food, Cosmetics and Pharma Industries; Problems and Toxicity Associated with Synthetic Preservatives; Impact of Misleading Advertisements; Recent Trends in Preservation and Legislation. *Materials* **2023**, *16*, 4793. [CrossRef]
5. Costa, M.; Sezgin-Bayindir, Z.; Losada-Barreiro, S.; Paiva-Martins, F.; Saso, L.; Bravo-Díaz, C. Polyphenols as Antioxidants for Extending Food Shelf-Life and in the Prevention of Health Diseases: Encapsulation and Interfacial Phenomena. *Biomedicines* **2021**, *9*, 1909. [CrossRef]
6. Boye, J.I.; Aksay, S.; Roufik, S.; Ribéreau, S.; Mondor, M.; Farnworth, E.; Rajamohamed, S.H. Comparison of the Functional Properties of Pea, Chickpea and Lentil Protein Concentrates Processed Using Ultrafiltration and Isoelectric Precipitation Techniques. *Food Res. Int.* **2010**, *43*, 537–546. [CrossRef]
7. Chang, C.; Varankovich, N.; Nickerson, M.T. Microencapsulation of Canola Oil by Lentil Protein Isolate-Based Wall Materials. *Food Chem.* **2016**, *212*, 264–273. [CrossRef] [PubMed]
8. Avramenko, N.A.; Chang, C.; Low, N.H.; Nickerson, M.T. Encapsulation of Flaxseed Oil within Native and Modified Lentil Protein-Based Microcapsules. *Food Res. Int.* **2016**, *81*, 17–24. [CrossRef]

9. Quiroz, J.Q.; Velazquez, V.; Corrales-Garcia, L.L.; Torres, J.D.; Delgado, E.; Ciro, G.; Rojas, J. Use of Plant Proteins as Microencapsulating Agents of Bioactive Compounds Extracted from Annatto Seeds (*Bixa orellana* L.). *Antioxidants* **2020**, *9*, 310. [CrossRef]
10. Dimidi, E.; Cox, S.R.; Rossi, M.; Whelan, K. Fermented Foods: Definitions and Characteristics, Impact on the Gut Microbiota and Effects on Gastrointestinal Health and Disease. *Nutrients* **2019**, *11*, 1806. [CrossRef]
11. Bellilci-Koyu, E.; Sarer-Yurkeli, B.P.; Akyon, Y.; Aydin-Kose, F.; Karagozlu, C.; Ozgen, A.G.; Brinkmann, A.; Nitsche, A.; Ergunay, K.; Yilmaz, E.; et al. Effects of Regular Kefir Consumption on Gut Microbiota in Patients with Metabolic Syndrome: A Parallel-Group, Randomized, Controlled Study. *Nutrients* **2019**, *11*, 2089. [CrossRef] [PubMed]
12. de Oliveira Leite, A.M.; Miguel, M.A.; Peixoto, R.S.; Rosado, A.S.; Silva, J.T.; Paschoalin, V.M. Microbiological, technological and therapeutic properties of kefir: A natural probiotic beverage. *Braz. J. Microbiol.* **2013**, *44*, 341–349. [CrossRef] [PubMed]
13. Ershadi, A.; Parastouei, K.; Khaneghah, A.M.; Hadian, Z.; Lorenzo, J.M. Encapsulation of Curcumin in Persian Gum Nanoparticles: An Assessment of Physicochemical, Sensory, and Nutritional Properties. *Coatings* **2021**, *11*, 841. [CrossRef]
14. Melo, A.; Silva, M.A. Development of fermented and flavoured kefir milk. *BMC Proc.* **2014**, *8*, 15. [CrossRef]
15. Tomas, M.; Toydemir, G.; Boyacıoğlu, D.; Hall, R.; Beekwilder, J.; Çapanoğlu, E. The effects of juice processing on black mulberry antioxidants. *Food Chem.* **2015**, *186*, 277–284. [CrossRef] [PubMed]
16. Türkyılmaz, M.; Tağı, Ş.; Dereli, U.; Özkan, M. Effects of various pressing programs and yields on the antioxidant activity, antimicrobial activity, phenolic content and colour of pomegranate juices. *Food Chem.* **2013**, *138*, 1810–1818. [CrossRef]
17. Oszmiánski, J.; Wojdyło, A. Comparative study of phenolic content and antioxidant activity of strawberry puree, clear, and cloudy juices. *Eur. Food Res. Technol.* **2009**, *228*, 623–631. [CrossRef]
18. Türkyılmaz, M.; Yemiş, O.; Özkan, M. Clarification and pasteurization effects on monomeric anthocyanins and percent polimeric colour of black carrot (*Daucus carota* L.) juice. *Food Chem.* **2012**, *134*, 1052–1058. [CrossRef]
19. Čakarević, J.; Šeregelj, V.; Tumbas Šaponjac, V.; Ćetković, G.; Čanadanović Brunet, J.; Popović, S.; Kostić, M.H.; Popović, L. Encapsulation of beetroot juice: A study on the application of pumpkin oil cake protein as new carrier agent. *J. Microencapsul.* **2020**, *37*, 121–133. [CrossRef]
20. Perović, M.N.; Pajin, B.S.; Antov, M.G. The Effect of Enzymatic Pretreatment of Chickpea on Functional Properties and Antioxidant Activity of Alkaline Protein Isolate. *Food Chem.* **2022**, *374*, 131809. [CrossRef]
21. Tan, E.S.; Ying-Yuan, N.; Gan, C.Y. A Comparative Study of Physicochemical Characteristics and Functionalities of Pinto Bean Protein Isolate (PBPI) against the Soybean Protein Isolate (SPI) after the Extraction Optimisation. *Food Chem.* **2014**, *152*, 447–455. [CrossRef] [PubMed]
22. Šeregelj, V.; Cetković, G.; Čanadanović-Brunet, J.; Šaponjac, V.T.; Vulić, J.; Lević, S.; Nedović, V.; Brandolini, A.; Hidalgo, A. Encapsulation of Carrot Waste Extract by Freeze and Spray Drying Techniques: An Optimization Study. *LWT* **2021**, *138*, 110696. [CrossRef]
23. Jinapong, N.; Suphantharika, M.; Jamnong, P. Production of Instant Soymilk Powders by Ultrafiltration, Spray Drying and Fluidized Bed Agglomeration. *J. Food Eng.* **2008**, *84*, 194–205. [CrossRef]
24. Šeregelj, V.; Ćetković, G.; Čanadanović-Brunet, J.; Tumbas Šaponjac, V.; Vulić, J.; Stajčić, S. Encapsulation and Degradation Kinetics of Bioactive Compounds from Sweet Potato Peel During Storage. *Food Technol. Biotechnol.* **2020**, *58*, 314–324. [CrossRef]
25. Lee, J.; Durst, R.W.; Wrosltad, R.E. Determination of total monomeric anthocyanin pigment content of fruit juices, beverages, natural colorants, and wines by the pH differential method: Collaborative study. *J AOAC Int.* **2005**, *88*, 1269–1278. [CrossRef] [PubMed]
26. Šaponjac, V.T.; Ćetković, G.; Čanadanović-Brunet, J.; Pajin, B.; Đilas, S.; Petrović, J.; Lončarević, I.; Stajčić, S.; Vulić, J. Sour cherry pomace extract encapsulated in whey and soy proteins: Incorporation in cookies. *Food Chem.* **2016**, *207*, 27–33. [CrossRef]
27. Šaponjac, V.T.; Gironés-Vilaplana, A.; Djilas, S.; Mena, P.; Cetkovi'c, G.; Moreno, D.A.; Kruni´c, M. Anthocyanin profiles and biological properties of caneberry (*Rubuss* pp.) press residues. *JSFA* **2014**, *94*, 2393–2400. [CrossRef]
28. Oyaizu, M. Studies on products of browning reaction–antioxidant activities of products of browning reaction prepared from glucosamine. *Jpn. J. Nutr.* **1986**, *44*, 307–315. [CrossRef]
29. *ISO 4833-1:2013*; Microbiology of the Food Chain—Horizontal Method for the Enumeration of Microorganisms—Part 1: Colony Count at 30 °C by the Pour Plate Technique. ISO: Geneva, Switzerland, 2013.
30. *ISO 21527-1:2008*; Microbiology of Food and Animal Feeding Stuffs Horizontal Method for the Enumeration of Yeasts and Moulds—Part 1: Colony Count Technique in Products with Water Activity Greater than 0. ISO: Geneva, Switzerland, 2008.
31. *ISO 21528-2:2017*; Microbiology of the Food Chain—Horizontal Method for the Detection and Enumeration of *Enterobacteriaceae*—Part 2: Colony-Count Technique. ISO: Geneva, Switzerland, 2017.
32. *ISO 7937:2004*; Microbiology of Food and Animal Feeding Stuffs—Horizontal Method for the Enumeration of Clostridium Perfringens—Colony-Count Technique. ISO: Geneva, Switzerland, 2004.
33. Gürsoy, O.; Kocatürk, K.; Güler Dal, H.O.; Yakalı, H.N.; Yılmaz, Y. Physicochemical and rheological properties of commercial kefir drinks. *Akad. Gıda* **2020**, *18*, 375–381. [CrossRef]
34. Kabakcı, S.A.; Türkyılmaz, M.; Özkan, M. Changes in the Quality of Kefir Fortified with Anthocyanin-Rich Juices during Storage. *Food Chem.* **2020**, *326*, 126977. [CrossRef]
35. *ISO 7889:2003*; Yogurt—Enumeration of Characteristic Microorganisms—Colony-Count Technique at 37 °C. ISO: Geneva, Switzerland, 2003.

36. *ISO 7251:2005*; Microbiology of Food and Animal Feeding Stuffs—Horizontal Method for the Detection and Enumeration of Presumptive *Escherichia coli*—Most Probable Number Technique. ISO: Geneva, Switzerland, 2005.
37. *ISO 11290-2:2017*; Microbiology of the Food Chain—Horizontal Method for the Detection and Enumeration of *Listeria monocytogenes* and of *Listeria* spp.—Part 2. Switzerland: ISO. Enumeration Method. ISO: Geneva, Switzerland, 2017.
38. *ISO 6888-1:1999/AMD 2:2018*; Microbiology of Food and Animal Feeding Stuffs—Horizontal Method for the Enumeration of Coagulase-Positive Staphylococci (*Staphylococcus aureus* and Other Species)—Part 1: Technique Using Baird-Parker Agar Medium—Amendment 2: Inclusion of an Alternative Confirmation Test Using RPFA Stab Method. ISO: Geneva, Switzerland, 2018.
39. *ISO 6579-1:2017*; Microbiology of the Food Chain—Horizontal Method for the Detection, Enumeration and Serotyping of *Salmonella*—Part 1: Detection of *Salmonella* spp. ISO: Geneva, Switzerland, 2017.
40. Carpenter, R.; Lyon, D.; Hasdell, T. *Guidelines for Sensory Analysis in Food Product Development and Quality Control*; Aspen Publishers, Inc.: Gaithersburg, MD, USA, 2000; pp. 71–91. [CrossRef]
41. Nthimole, T.; Kaseke, T.; Fawole, O.A. Micro-Encapsulation and Characterization of Anthocyanin-Rich Raspberry Juice Powder for Potential Applications in the Food Industry. *Processes* **2022**, *10*, 1038. [CrossRef]
42. GEA Niro. Hygroscopicity—Method A14a. GEA Niro Analytical Methods of Dry Milk Products, GEA Niro, Soeborg, Denmark. 2005, pp. 1–3. Available online: https://efps.gr/products/dryers-particle-processing-plants/spray-dryers/food-dairy-products/analytical-methods-dry-milk-products/ (accessed on 22 September 2023).
43. Zayas, J. *Functionality of Proteins in Food*; Springer: Berlin/Heidelberg, Germany, 1997; ISBN 9783642624759.
44. Vulić, J.; Šeregelj, V.; Tumbas Šaponjac, V.; Karadžić Banjac, M.; Kovačević, S.; Šovljanski, O.; Ćetković, G.; Čanadanović-Brunet, J.; Jevrić, L.; Podunavac-Kuzmanović, S. From Sweet Corn By-Products to Carotenoid-Rich Encapsulates for Food Applications. *Processes* **2022**, *10*, 1616. [CrossRef]
45. Jiménez-Aguilar, D.M.; Ortega-Regules, A.E.; Lozada-Ramírez, J.D.; PérezPérez, M.C.I.; Vernon-Carter, E.J.; Welti-Chanes, J. Color and chemical stability of spray-dried blueberry extract using mesquite gum as wall material. *J. Food Comp. Anal.* **2011**, *24*, 889–894. [CrossRef]
46. Ahmed, M.; Akter, M.S.; Lee, J.-C.; Eun, J.-B. Encapsulation by spray drying of bioactive components, physicochemical and morphological properties from purple sweet potato. *LWT-Food Sci. Technol.* **2010**, *43*, 1307–1312. [CrossRef]
47. Weiss, V.; Okun, Z.; Shpigelman, A. Utilization of hydrocolloids for the stabilization of pigments from natural sources. *COCIS* **2023**, *68*, 101756. [CrossRef]
48. Chen, C.; Chi, Y.-J.; Xu, W. Comparisons on the Functional Properties and Antioxidant Activity of Spray-Dried and Freeze-Dried Egg White Protein Hydrolysate. *Food Bioprocess Technol.* **2011**, *5*, 2342–2352. [CrossRef]
49. Fernandes, L.P.; Candido, R.C.; Oliveira, W.P. Spray drying microencapsulation of *Lippia sidoides* extracts in carbohydrate blends. *Food Bioprod. Process.* **2012**, *90*, 425–432. [CrossRef]
50. Dumitrașcu, L.; Stănciuc, N.; Aprodu, I. Encapsulation of Anthocyanins from Cornelian Cherry Fruits Using Heated or Non-Heated Soy Proteins. *Foods* **2021**, *10*, 1342. [CrossRef]
51. Díaz, D.I.; Beristain, C.I.; Azuara, E.; Luna, G.; Jimenez, M. Effect of wall material on the antioxidant activity and physicochemical properties of *Rubus fruticosus* juice microcapsules. *J. Microencapsul.* **2015**, *32*, 247–254. [CrossRef]
52. Tao, Y.; Wang, P.; Wang, J.; Wu, Y.; Han, Y.; Zhou, J. Combining various wall materials for encapsulation of blueberry anthocyanin extracts: Optimization by artificial neural network and genetic algorithm and a comprehensive analysis of anthocyanin powder properties. *Powder Technol.* **2017**, *311*, 77–87. [CrossRef]
53. Kaur, C.; Kapoor, H.C. Antioxidants in Fruits and Vegetables—The Millennium's Health. *Int. J. Food Sci. Technol.* **2001**, *36*, 703–725. [CrossRef]
54. Du, X.; Myracle, A.D. Development and evaluation of kefir products made with aronia or elderberry juice: Sensory and phytochemical characteristics. *Int. Food Res. J.* **2018**, *25*, 4. Available online: http://www.ifrj.upm.edu.my/25%20(04)%202018/(7).pdf (accessed on 22 September 2023).
55. Barrett, A.J. The classes of proteolytic enzymes. In *Plant Proteolytic Enzymes*; Dalling, M.J., Ed.; CRC Press: Boca Raton, FL, USA, 1986; Volume 1, pp. 1–16.
56. Kopjar, M.; Ćorković, I.; Buljeta, I.; Šimunović, J.; Pichler, A. Fortification of Pectin/Blackberry Hydrogels with Apple Fibers: Effect on Phenolics, Antioxidant Activity and Inhibition of α-Glucosidase. *Antioxidants* **2022**, *11*, 1459. [CrossRef] [PubMed]
57. Barretto, F.J.d.F.P.; Clemente, H.A.; Santana, A.L.B.D.; Vasconcelo, M.A.d.S. Stability of Encapsulated and Non-Encapsulated Anthocyanin in Yogurt Produced with Natural Dye Obtained from *Solanum melongena* L. Bark. *Rev. Bras. Frutic.* **2020**, *42*, e137. [CrossRef]
58. Esatbeyoglu, T.; Fischer, A.; Legler, A.D.S.; Oner, M.E.; Wolken, H.F.; Kopsel, M.; Ozogul, Y.; Ozyurt, G.; Biase, D.; Ozogul, F. Physical, chemical, and sensory properties of water kefir produced from *Aronia melanocarpa* juice and pomace. *Food Chem. X* **2023**, *18*, 100683. [CrossRef]
59. Wu, X.; Beecher, G.R.; Holden, J.M.; Haytowitz, D.B.; Gebhardt, S.E.; Prior, R.L. Concentrations of Anthocyanins in Common Foods in the United States and Estimation of Normal Consumption. *J. Agric. Food Chem.* **2006**, *54*, 4069–4075. [CrossRef]
60. Codex Alimentarius Commision, International Food Standards (2003). Standards For Fermented Milks CXS 243-2003, Food and Agriculture Organization of the United Nations, WHO. Available online: https://www.fao.org/fao-who-codexalimentarius/sh-

proxy/en/?lnk=1&url=https%253A%252F%252Fworkspace.fao.org%252Fsites%252Fcodex%252FStandards%252FCXS%2B243-2003%252FCXS_243e.pdf (accessed on 22 September 2023).
61. Goncu, B.; Celikel, A.; Guler-Akin, M.B.; Serdar Akin, M. Some properties of kefir enriched with apple and lemon fiber. *Mljekarstvo* **2017**, *67*, 208–216. [CrossRef]
62. Aiello, F.; Restuccia, D.; Spizzirri, U.G.; Carullo, G.; Leporini, M.; Loizzo, M.R. Improving Kefir Bioactive Properties by Functional Enrichment with Plant and Agro-Food Waste Extracts. *Fermentation* **2020**, *6*, 83. [CrossRef]
63. Bourrie, B.C.T.; Willing, B.P.; Cotter, P.D. The microbiota and health promoting characteristics of the fermented beverage kefir. *Front. Microbiol.* **2016**, *7*, 647. [CrossRef]
64. Vieira, C.P.; Rosario, A.; Lelis, C.A.; Rekowsky, B.S.S.; Carvalho, A.P.A.; Rosario, D.K.A.; Elias, T.A.; Costa, M.P.; Foguel, D.; Conte-Junior, C.A. Bioactive Compounds from Kefir and Their Potential Benefits on Health: A Systematic Review and Meta-Analysis. *Oxidative Med. Cell. Longev.* **2021**, *2021*, 9081738. [CrossRef]
65. Šeregelj, V.; Tumbas Šaponjac, V.; Lević, S.; Kalušević, A.; Ćetković, G.; Čanadanović-Brunet, J.; Vidaković, A. Application of encapsulated natural bioactive compounds from red pepper waste in yogurt. *J. Microencapsul.* **2019**, *36*, 704–714. [CrossRef] [PubMed]
66. Rubio, F.T.V.; Haminiuk, C.W.I.; Santos, M.M.; Thomazini, M.; Moraes, I.C.F.; Martelli-Tosi, M.; Fávaro-Trindade, C.S. Development of Natural Pigments Microencapsulated in Waste Yeast *Saccharomyces cerevisiae* Using Spray Drying Technology and Their Application in Yogurt. *Food Funct.* **2021**, *12*, 8946–8959. [CrossRef] [PubMed]
67. Pinto, K.R.D.; Tulini, F.L.; Guimarães, J.d.G.L.; Moraes, I.C.F.; Ditchfield, C.; Lima, C.G.d.; Silva, V.L.d.S.; Favaro-Trindade, C.S. Production and Evaluation of Yogurt Colored with Anthocyanin-Rich Pigment Prepared from Jabuticaba (*Myrciaria cauliflora* Mart.) Skin. *Processes* **2023**, *11*, 526. [CrossRef]

foods

MDPI

Article

Identification of Protein–Phenol Adducts in Meat Proteins: A Molecular Probe Technology Study

Fenhong Yang [1,†], Yingying Zhu [2,†], Xiaohan Li [1], Fengtao Xiang [1], Moru Deng [1], Wei Zhang [1], Wei Song [1], Hao Sun [1] and Changbo Tang [1,*]

[1] State Key Laboratory of Meat Quality Control and Cultured Meat Development, Key Laboratory of Meat Processing, Ministry of Agriculture, Key Lab of Meat Processing and Quality Control, Ministry of Education, Jiangsu Collaborative Innovation Center of Meat Production and Processing, College of Food Science and Technology, Nanjing Agricultural University, Nanjing 210095, China; 2020108080@stu.njau.edu.cn (F.Y.); lixiaohanxh@stu.njau.edu.cn (X.L.); xiangfengtao@stu.njau.edu.cn (F.X.); d1670677025@163.com (M.D.); lww99706@163.com (W.Z.); wsong@njau.edu.cn (W.S.); haosun@njau.edu.cn (H.S.)

[2] Engineering Research Center of Magnetic Resonance Analysis Technology, Department of Food Nutrition and Test, Suzhou Vocational University, Suzhou 210005, China; yingy_zhu@163.com

* Correspondence: changbotang@hotmail.com

† These authors contributed equally to this work.

Abstract: Plant polyphenols with a catechol structure can form covalent adducts with meat proteins, which affects the quality and processing of meat products. However, there is a lack of fast and effective methods of characterizing these adducts and understanding their mechanisms. This study aimed to investigate the covalent interaction between myofibrillar protein (MP) and caffeic acid (CA), a plant polyphenol with a catechol structure, using molecular probe technology. The CA-MP adducts were separated via sodium dodecyl sulfate–polyacrylamide gel electrophoresis (SDS-PAGE) and detected via Western blot and LC-MS/MS analyses. The Western blot analysis revealed that various specific adducts were successfully enriched and identified as bands around 220 kDa, 45 kDa, and two distinct bands between 95 and 130 kDa. Combined with the LC-MS/MS analysis, a total of 51 peptides were identified to be CA-adducted, corresponding to 31 proteins. More than 80% of the adducted peptides carried one adducted site, and the rest carried two adducted sites. The adducted sites were located on cysteine (C/Cys), histidine (H/His), arginine (R/Arg), lysine (K/Lys), proline (P/Pro), and N-terminal (N-Term) residues. Results showed that the covalent interaction of CA and MP was highly selective for the R side chain of amino acids. Moreover, the adducts were more likely to form via C-N bonding than C-S bonding. This study provides new insights into the covalent interaction of plant polyphenols and meat proteins, which has important implications for the rational use of plant polyphenols in the meat processing industry.

Keywords: myofibrillar protein; caffeic acid; molecular probe technology; phenol–protein adducts

1. Introduction

Plant polyphenols are natural antioxidants that can protect meat products from oxidative deterioration by scavenging free radicals and inhibiting peroxidation [1,2]. However, plant polyphenols with a catechol structure, such as caffeic acid (CA), can also undergo oxidation to form quinones, which can react covalently with muscle proteins and form phenol–protein adducts which can affect the structure and functional properties of meat proteins [3,4], as well as the quality, nutritional, and digestive properties of meat products [5,6]. Therefore, it is important to understand the mechanism and characteristics of the covalent interaction between plant polyphenols and meat proteins and its impact on meat quality and processing.

Previous studies have shown that CA adduction can affect the structure and gel properties of myofibrillar protein (MP) [2,4]. However, due to the complexity of meat protein systems, there is still a lack of analytical methods of accurately characterizing phenol–protein adducts and their formation mechanisms. Most of the existing methods are based on indirect measurements, such as colorimetric assays, spectrophotometric assays, or gel electrophoresis, which have low sensitivity, specificity, and resolution [7,8]. Moreover, these methods cannot provide information on the identity, location, and number of the adducted sites on the protein molecules, which are crucial for understanding the molecular basis of the phenol–protein interaction [2,9]. Furthermore, these methods cannot quantify the extent of protein modification by plant polyphenols, which is essential for evaluating their effect on meat quality and processing [2,4]. Therefore, there is a need for a fast and effective method of characterizing phenol–protein adducts and their formation mechanisms at the molecular level.

Molecular probe technology is a powerful method for identifying the modification of proteins by natural active compounds (e.g., plant polyphenols) in complex systems. This method involves labeling the active compound with a molecular probe that can bind strongly to the target proteins and carry a detectable tag, such as fluorophores, isotopes, biotin, or bio-orthogonal groups [10,11]. The molecular probe–protein complexes can then be separated via intracellular imaging or SDS-PAGE in-gel fluorescence scanning for probes with isotopes or fluorophores [12] or via streptavidin magnetic beads for probes with biotin tags [13], which undergo a series of non-specific washing and specific elution operations to obtain the purified proteins or proteomes modified by the active compound [14]. The purified proteins or proteomes modified by the active compound can then be analyzed using methods such as Western blot and mass spectrometry to identify the peptide sequences and the adducted sites on the protein molecules [15] so that the mechanisms of interaction between the active compounds and proteins can be deeply explored. This method can provide information on the identity, location, and number of the adducted sites on the protein molecules, which are crucial for understanding the molecular basis of the active compound–protein interaction. Moreover, this method can also quantify the extent of protein modification by plant polyphenols by comparing the relative abundances of modified and unmodified peptides or proteins. In this strategy, multiple controls are required to exclude non-specific interacting proteins.

In this study, we synthesized a biotinylated derivative of caffeic acid (BioC) that can act as a molecular probe to identify protein targets responsible for caffeic acid's covalent modification. Based on the molecular probe technology, we aimed to identify protein–phenol adducts in meat proteins. A method to enrich, isolate, and identify adducts was firstly established and its feasibility was validated; it was then applied to identify and preliminarily quantify specific phenol–protein adducts in the MP system. Combined with LC-MS/MS, the adducted peptides and amino acid sites were also comprehensively identified and quantified. Finally, the mechanism of the adduction reaction between CA and MP was explored based on the adducted sites and their relative abundance. This study provides new insights into the covalent interactions of plant polyphenols and meat proteins and their implications for meat quality and processing.

2. Materials and Methods

2.1. Materials

Caffeic acid and streptavidin magnetic beads were both obtained from Cell Signaling Technology (#5947; Beverly, MA, USA). Bovine serum albumin, biotin, trypsin, and polyvinylidene fluoride membranes were purchased from Sigma-Aldrich (Taufkirchen, Germany). An anti-biotin antibody (abs153474) was obtained from Absin Bioscience Inc. (Shanghai, China). Goat anti-rabbit IgG was purchased from Beijing Cowin Biotech Co., Ltd. (Beijing, China). Skimmed milk powder was obtained from Sinopharm Chemical Reagent Co., Ltd. (Shanghai, China). Other reagents were purchased from Genscript Biotech Co., Ltd. (Nanjing, China) unless otherwise specified.

2.2. Sample Preparation

The synthesis of the BioC was carried out according to a published experimental protocol [16]. Briefly, CA and N-Boc-ethylenediamine were dissolved in N,N-dimethylformamide, which was stirred at room temperature for 15 h in the presence of 1-(3-dimethylaminopropyl)-3-ethylcarbodiimide hydrochloride and 4-dimethylaminopyridine. When the reaction was then carried out with methyl tert-butyl ether, a yellow solid precipitated (intermediate 1). It was purified using a silica column (methylene chloride/methanol = 20:1, v/v) and recrystallized with methylene chloride/petroleum ether. Next, intermediate 1 was dissolved in dichloromethane, the treatment of which with methanolic hydrochloric acid solution for 15 h in an ice bath and subsequent concentration generated the desired intermediate 2. Finally, the amidation of intermediate 2 with D-(+)-biotin in a similar way as above afforded the BioC (Figure S1).

The extraction of the MP and the BioC solution's oxidation treatment were carried out following the methods of Park [17] and Yang [16]. The MP was stirred and dissolved in phosphate buffer (PBS, 10 mM; 0.6 M NaCl; pH 6.2). For a concentration-dependent experiment, the BioC was dispersed into MP suspensions (1 mg/mL) to reach final concentrations of 0, 10, 50, 250, and 1250 μmol/g MP which were then incubated on a shaker (QB-210; Haimen Kylin-Bell Lab Instruments Co., Ltd., Nantong, China) at 4 °C for 12 h. For competition inhibition experiments, different concentrations of CA (0×, 1×, 4×, 16× of 250 μmol), the competitive inhibitor of the BioC, were firstly added and incubated at 4 °C for 12 h. The BioC was then added to obtain a final concentration of 250 μM with a final MP concentration of 1 mg/mL, and the samples were incubated for another 12 h at 4 °C.

2.3. Identification Method for Phenol–Protein Adducts

The separation of the MP-BioC adducts was performed according to Yang [17] with slight modifications. Streptavidin magnetic beads were gently mixed and moved (10 μL for each sample) into 1.5 mL microcentrifugation tubes, which were then placed on the magnetic rack so that they could be magnetically separated. The collected beads were washed with 1 mL of PBST buffer three times. The beads then could be mixed and incubated with 100 μL of the MP-BioC samples (1 mg/mL) at 4 °C for 40 min so that the BioC-derivatized proteins fully bound to the beads. After incubation, the beads were collected, and the supernatant was discarded. The beads were then subjected to non-specific protein washing with 1 mL of a washing buffer at least five times, with as much supernatant as possible removed and retained at the last moment. To release the MP-BioC adducts from the beads, the beads were resuspended in PBS buffer and mixed with 5 × Sample Buffer (125 mM Tris, 10% SDS, 0.25% BPB). The samples were boiled at 95 °C for 5 min, followed by cooling and centrifugation at 14,000 rpm for 1 min to collect the adducts in the supernatant, which were then subjected to SDS-PAGE or a Western blot analysis (see Section 2.6) in order to accomplish the identification and initial quantification of the polyphenol–protein adducts.

2.4. Validation of Method

Bovine serum albumin (BSA), as a model protein, has been widely studied in various fields due to the diversity of its physiological functions [18]. Consequently, to verify the validity of the above-mentioned method of identifying phenol–protein adducts, a BSA-BioC reaction system was established by using BSA as the substrate protein. First, the BioC was added to a BSA solution to obtain final concentrations of 50 and 250 μM with a final BSA concentration of 1 mg/mL. Meanwhile, the BSA solution was supplemented with the same concentration (50, 250 μM) of biotin or an equal volume of the solvent DMSO (1%, v/v) as a control. All the samples were incubated at 4 °C for 12 h. Then, the BSA-BioC adducts in the system were isolated and characterized using the method described in Section 2.3

2.5. Gel Electrophoresis

Samples were loaded onto SDS-PAGE equipment according to the method of Liu [19] with a 4–20% gradient gel. Small amounts of protein samples from Section 2.2 or Section 2.4 were first taken for SDS-PAGE to illustrate that there was no problem with the original protein samples prior to incubation with the streptavidin magnetic beads. Then, two 20 μL samples in Section 2.3 (i.e., enriched phenol–protein adducts) were subjected to SDS-PAGE. After staining with Coomassie brilliant blue, one gel was used for a Western blot analysis for the phenol–protein adducts' identification and the other for a mass spectrometry analysis after in-gel digestion.

2.6. Western Blot Analysis

A Western blot analysis was used to detect bands of the BioC-adducted proteins according to Pi [20]. Briefly, after SDS-PAGE, an eBlot L1 fast wet protein transfer system (GenScript, Nanjing, China) was used to transfer the proteins to polyvinylidene fluoride (PVDF) membranes (0.45 mm) which were then blocked with 5% skimmed milk powder in Tris-buffered saline with Tween-20 (TBST, pH 7.4) for 1.5 h. After blocking, the membranes were washed with TBST three times (10 min/wash) and incubated with the anti-biotin antibody diluted 1:1000 in TBST (20 mM Tris-base, 137 mM NaCl and 5 mM KCl, 0.05% Tween-20) overnight at 4 °C, followed by incubation with goat anti-rabbit IgG (diluted 1:2000) for 2 h. A Chemiluminescence Imager (ImageQuant LAS 4000; General Electric Company, Boston, MA, USA) was used to capture the membranes.

2.7. LC-MS/MS Analysis

Samples were prepared using a filter-aided sample preparation (FASP) workflow, as described by Wiśniewski et al. [21]. The peptides digested by enzymes were subsequently cleaned, concentrated, and enriched using the stop-and-go-extraction tips (StageTips) method, as described by Rappsilber et al. [22]. The peptides were dissolved in a buffer containing 2% acetonitrile and 0.1% trifluoroacetic acid (TFA) for MS analysis. The elute was dried completely in a speedvac centrifuge. The peptides were suspended in a buffer containing 2% acetonitrile and 0.1% TFA ready for MS analysis. The peptides were separated via LC–MS/MS, using an Easy-nLC1200 (Thermo Fisher Scientific, Altham, GA, USA) coupled with an Orbitrap Exploris 480 Mass Spectrometer (Thermo Fisher Scientific, USA). A trap column (100 μm × 2 cm) and an analytical column (75 μm × 25 cm) packed with Reprosil-Pur C18 particles (Dr Maisch GmbH) were used to separate the peptides with mobile phase A (0.1% formic acid (FA) in water) and mobile phase B (0.1% FA in acetonitrile) in a 60 min gradient: 6 to 23% B in 38 min, 23 to 32% B in 12 min, and 32 to 80% B in 5 min, and then B was kept at 80% for 5 min. The flow rate was set at 300 nL/min. The Orbitrap Exploris 480 Mass Spectrometer was operated in a data-dependent acquisition mode with a spray voltage of 2 kV and a heated capillary temperature of 320 °C. MS1 data were collected at a high resolution of 60,000 with a mass range of 350 to 1500 m/z; the precursor intensity threshold was set at 5.0e^4, and a maximum injection time of 20 ms was used. For each full MS scan, the top 10 most abundant precursor ions were selected for MS2 with an isolation window of 1.6 m/z and a higher-energy collision dissociation with a normalized collision energy of 30. MS2 spectra were collected at a resolution of 15,000. The target value was 5.0e^3 with a maximum fill time of 20 ms and a dynamic exclusion time of 30 s.

2.8. Data Search

Raw data were processed and analyzed using Proteome Discoverer (PD) 2.4 SP1 (Thermo Fisher Scientific, USA) with default settings. PD was set up to search the database from NCBI. Human keratin and trypsin sequences were used as the contaminated database for proteomic searching. Trypsin was used as the digestion enzyme. Carbamidomethyl (C) was specified as the fixed modification. Oxidation (M), acetyl (Protein N-term), the BioC ($C_{21}H_{28}N_4O_5S$) (C, K, R, H, P, N-term + 446.16239 Da) and the possible degradation product of the BioC ($C_9H_8O_4$) (C, K, R, H, P, N-term + 162.03169 Da) were specified as variable modifications. A q-value (FDR) cutoff for the precursor and protein level was applied at 1%. The processed data output was then exported to Excel for further usage.

2.9. Statistical Analysis

All the experiments were carried out in triplicate. Differences were evaluated using a one-way ANOVA, and means were compared using Duncan's multiple comparison under the SAS system (version 8.2); $p < 0.05$ was considered statistically significant.

3. Results

3.1. Experimental Strategy

We developed a novel method for the enrichment and separation of phenol–protein adducts by using a BioC rather than CA in the reaction system to label proteins that can adduct with CA. The BioC is a biotinylated derivative of CA which can be recognized and captured by streptavidin magnetic beads with high specificity and affinity. However, to avoid the possible adsorption of other molecules on the surface of the streptavidin magnetic beads, non-specific washing steps are often required [23]. Therefore, four different washing buffers were tested to evaluate their washing effects on MP-BioC (0 μM, control) samples, including reagent 1, PBS buffer; reagent 2, PBS buffer (1% SDS); reagent 3, 25 mM Tris, 0.15 M NaCl, pH = 7.2; and reagent 4, 20 mM Tris-HCl, 2% SDS, pH 7.4. The non-specific washing step was repeated five times, and the supernatant was retained after magnetic separation (the fifth washing supernatant), which was then subjected to SDS-PAGE sample preparation together with proteins eluted from the streptavidin magnetic beads (the fifth elution solution).

As shown in Figure 1a, there was no residual protein in the supernatant after five non-specific washings regardless of which kind of washing buffer was used, indicating that five non-specific washing steps were sufficient. At the same time, proteins remaining in the fifth eluent were observed. It was also found that there were almost no protein bands in the lane corresponding to reagent 4, indicating that it effectively removed a large number of non-specific proteins adsorbed on the surface of the streptavidin magnetic beads after washing. Therefore, the reagent 4, which had the best effect, was selected as the washing buffer. In the follow-up tests, it was found that only a few specific phenol–protein adducts were identified in the experimental group compared with the control group (as shown in Figure 1b). We speculated that reagent 4, containing 2% SDS, may cause the loss of specific adducts while washing away non-specific proteins. To achieve a comprehensive identification of the formation mechanism of phenol–protein adducts, reagent 3 (25 mM Tris, 0.15 M NaCl, pH = 7.2) was finally selected to carry out the next experiment, and the identification of specific phenol–protein adducts was achieved based on both SDS-PAGE and a Western blot analysis.

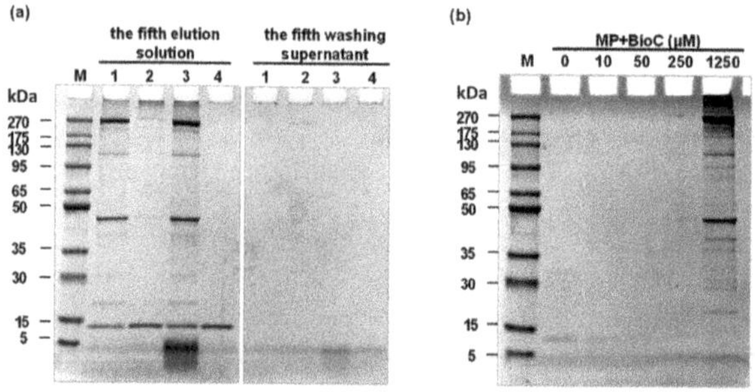

Figure 1. Non-specific washing using 4 kinds of washing buffers, reagent 1, 2, 3, and 4 (**a**); SDS-PAGE of MP-BioC adducts eluted from streptavidin magnetic beads after the fifth washing using washing reagent 4 (**b**).

3.2. Efficacy of the Identification Method of Phenol–Protein Adducts

Figure 2a shows a clear protein band around 66 kDa for BSA, which is consistent with its molecular mass [24]. Protein bands in all lanes are well defined, indicating that the protein samples were not degraded before incubation with streptavidin magnetic beads. Compared to the control group (BSA, BSA-DMSO, and BSA-Biotin), the band intensity at 66 kDa decreases and the bands near 95 kDa and above the separating gel increase as the BioC concentration increases. It is speculated that the BSA-BioC adducts were formed by a covalent reaction between the BioC and BSA which was proportional to the concentration. A series of control groups in the experiment could eliminate the interference of natural biotin-derived proteins in the samples. Previous studies have reported that phenolic compounds such as chlorogenic acid, catechin, and epicatechin can form covalent conjugates with BSA [25,26].

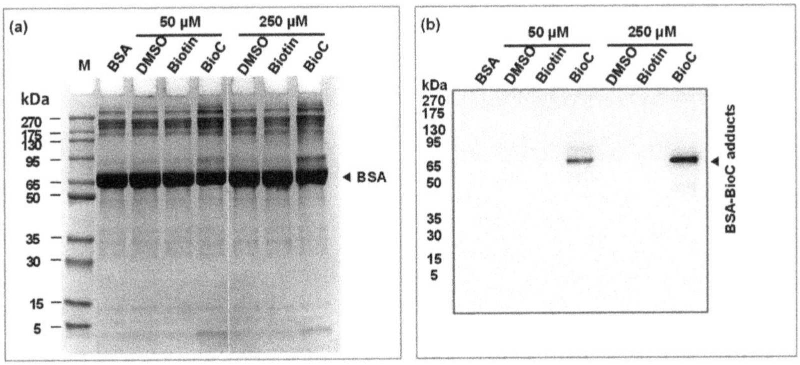

Figure 2. SDS-PAGE analysis of samples (**a**); Western blot analysis of BSA-BioC adducts (**b**).

Figure 2b shows the immunoblotting result of BSA-BioC adducts enriched and isolated by the biotin–streptavidin system. No bands were labeled in any of the control groups, indicating that the streptavidin magnetic beads did not bind any non-specific proteins after washing. In contrast, the experimental groups showed a specific band between 66 and 95 kDa, and its intensity increased significantly with an increase in the BioC concentration. This confirmed that the BSA-BioC adducts were successfully enriched and isolated via the method described in Section 2.3, which proved to be effective and reliable. It is worth noting that increases in the protein band intensity of two bands were observed in the BSA-BioC samples in Figure 2a, but only one band appears in the Western blot result, which could be attributed to the S-S-mediated cross-linking of BSA rather than the covalent interaction between the BioC and BSA [27].

3.3. Identification of BioC-MP Adducts

3.3.1. Concentration-Dependent Experiment

Figure 3a shows a typical electropherogram of the MP lane in which the main proteins are myosin heavy chain (MHC) around 220 kDa and Actin around 45 kDa [28,29]. When the BioC concentration increased, the MHC and Actin bands became weaker and moved to a higher molecular mass. As the concentration of the BioC increased, the MHC and Actin bands gradually weakened and shifted to a higher molecular mass. This suggests that MP-BioC adducts were generated in the system. Tang et al. [30] also found more than 20 proteins modified by rosmarinic acid in MP family proteins, including myosin heavy chain protein isoforms (MyHC1, MyHC2a, MyHC2x, and MyHC2b) and Actin. Since the feasibility of the method of identifying phenol–protein adducts was verified in Section 3.2, we carried out an enrichment of the adducts in the MP-BioC system following the same method and identified the MP-BioC adducts subsequently via a Western blot.

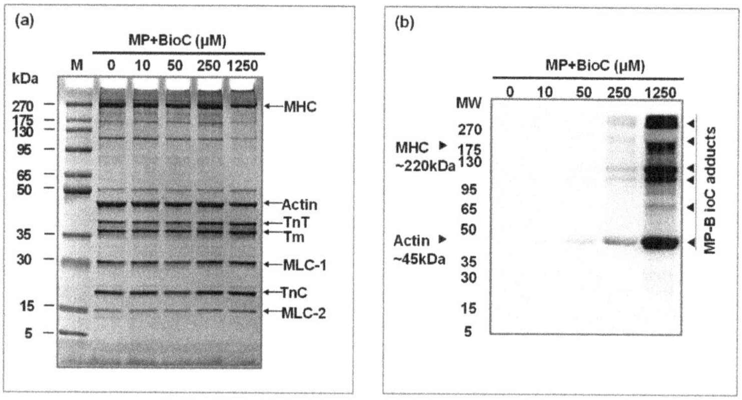

Figure 3. SDS-PAGE analysis of samples (**a**) and Western blot analysis of MP-BioC adducts (**b**) in the concentration-dependent experiment.

As shown in Figure 3b, the molecular probe, the BioC, marked out proteins at 220 kDa, 45 kDa, and 95–130 kDa as MP-BioC adducts, which depended on the concentration. Considering the low molecular mass of the BioC (MW = 448.185 Da), the position of the protein bands could hardly have shifted after the BioC modifications. Therefore, it was presumed that the protein bands around 220 kDa and 45 kDa were MHC-BioC adducts and Actin-BioC adducts respectively, and other the labeled bands were BioC covalent conjugates of the corresponding proteins.

In Figure 4, a preliminary quantitative analysis of MP-BioC adduct bands at 220 kDa, 95–130 kDa, and 45 kDa shows that the relative intensity of these adduct bands at a concentration of 1250 μM was significantly higher than that at other concentrations ($p < 0.05$). Specifically, the Actin-BioC and band-2-BioC adducts were firstly observed at a low BioC concentration (50 μM). MHC-BioC and band-1-BioC were observed at a BioC concentration of 250 μM. With an increasing BioC concentration, the intensity of all of adducts was significantly increased ($p < 0.05$). The results illustrate that the formation of MP-BioC adducts has concentration-dependent characteristics.

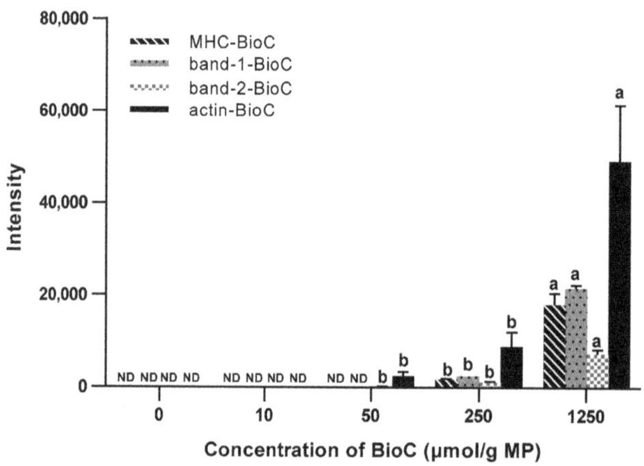

Figure 4. Types and intensity of MP-BioC adducts in the presence of different concentrations of BioC. Letters (a,b) indicate a significant difference ($p < 0.05$) in the relative intensity of MP-BioC adducts in groups with different concentrations of BioC; Band-1/2-BioC—the upper/lower one of two adducts between 95 and 130 kDa; ND—not detected.

3.3.2. Competition Suppression Experiment

The characteristic protein bands identified in the concentration-dependent experiments decreased gradually when we added different concentrations of highly active ligands. This was an effective way to confirm specific protein bands. In our study, we further tested the covalent interaction between molecule probes and highly active ligands that were added simultaneously to establish a competition control which could prevent the molecule probes from binding to target proteins via competition from highly active ligands. The addition of different concentrations of highly active ligands may lead to a gradual fading of the characteristic protein bands identified in concentration-dependent experiments, which was an effective method of determining specific protein bands. In this study, to further validate the covalent interaction between the BioC and MP, we allowed the BioC or CA to compete for binding to MP with or without an excess of the competitive inhibitor CA, and the BioC concentration was kept at 250 μM. Figure 5a shows that the MP-BioC samples in the competitive suppression experiment were normal. Figures 5b and 6 show that the MP-BioC adducts decreased significantly ($p < 0.05$) when the CA concentration increased. When the CA concentration was 16 times higher than the BioC concentration (4×10^3 μmol/g MP), the binding of the BioC to MP was almost completely blocked.

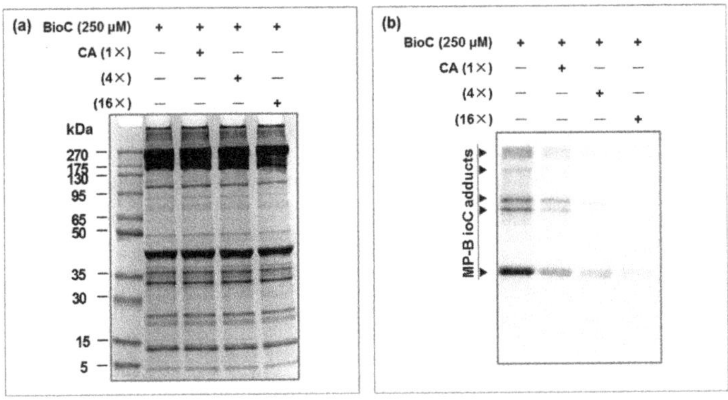

Figure 5. SDS-PAGE analysis of samples (**a**), and Western blot analysis of MP-BioC adducts (**b**) in the competition suppression experiment.

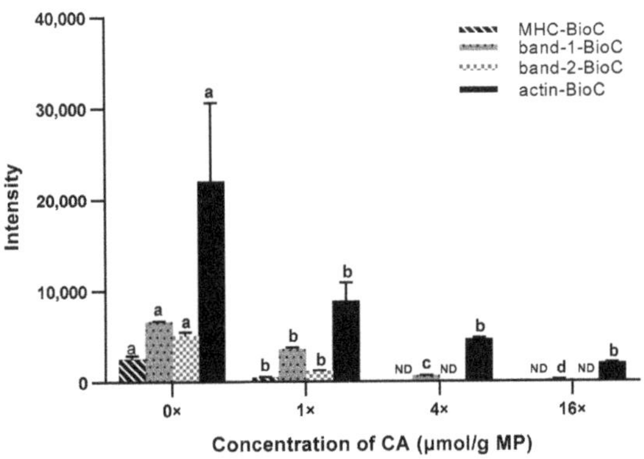

Figure 6. Types and intensity of MP-BioC adducts in the presence of different concentrations of CA. Letters (a–d) indicate a significant difference ($p < 0.05$) in the relative intensity of MP-BioC adducts with different concentrations of the competitive inhibitor CA (0×, 1×, 4×, 16× of 250 μmol); ND indicates not detected.

The protein bands at 220 kDa and 45 kDa, as well as two strong protein bands between 95 and 130 kDa, depended on the concentration and could be stopped by the competitive inhibitor CA. This indicated that all these protein bands were specific MP-BioC adducts. More importantly, the results showed that the BioC and MP interacted in the same way as the CA and MP. Since the BioC was modified from CA and had the same biological activity as CA (Figure S2), it was considered that adducts labeled by BioC such as MHC-BioC and Actin-BioC labeled were equivalent to adducts generated by the covalent interaction of CA and MP.

3.4. LC-MS/MS of BioC-MP Adducts

3.4.1. Identification of Adducted Proteins and Peptides

To further explore the formation mechanism of MP-BioC adducts, the BioC-adducted proteins in the MP sample (1250 μM) were identified using an Orbitrap Exploris 480 mass spectrometer. After filtering data with a score > 10, a total of 31 BioC-adducted proteins were found, including myosin, actin, troponin, and tropomyosin (Table 1). Six BioC-modified myosin isoforms were identified near 220 kDa, which included myosin-4 isoform X1 (223.1 kDa), myosin heavy chain 2x (223 kDa), myosin-2 isoform X1 (223 kDa), myosin-8 isoform X1 (222.9 kDa), embryonic skeletal myosin heavy chain 3 partial (222.8 kDa), and myosin-7B isoform X5 (223.8 kDa). Among them, the first five proteins had a BioC modification at the R site and the common adducted peptide was FIR*IHFGTTGK (* denotes the adducted site). BioC modifications at the N-Term site were also identified. At the same time, skeletal myosin heavy chain 3 had BioC modifications at the H and N-Term sites (IEDMAMLTH*LNEPAVLYNLK) and myosin 7b isomer X5 at the K site (LLGSL-DIDHSQYQFGHTK*). At around 45 kDa, skeletal alpha actin (42 kDa), actin aortic smooth muscle (42 kDa), cardiac α actin 1 (pdb 5NOL B Chain B Cardiac muscle alpha actin 1, 40.7 kDa), and actin-partial (27.4 kDa) were modified with BioC at the R-site (EIVR*DIK) and N-Term site (MQKEITALAPSTMK). Three BioC-adducted peptides were identified in cytoskeletal beta actin partial (44.8 kDa), two of which were the same as above, and the other was DLYANTVLSGGTTMYPGIADR, with the N-Term as the adducted site.

Table 1. BioC-adducted proteins in MP.

Gene	Protein	Coverage (%)	MW (kD)	Calc. pI	Score
XP 020931560.1	LOW QUALITY PROTEIN titin	59	3816	6.27	17,031.96
XP 020921695.1	myosin-4 isoform X1	75	223.1	5.76	7827.33
BAA82146.1	myosin heavy chain 2x	73	223	5.77	6990.96
XP 020921876.1	myosin-2 isoform X1	72	223	5.81	5401.78
XP 020921928.1	myosin-8 isoform X1	50	222.9	5.8	4251.75
XP 020931125.1	LOW QUALITY PROTEIN nebulin	67	856.7	9.07	4104.53
AAC48692.1	skeletal alpha actin	76	42	5.39	2977.1
/	pdb 5NOL B Chain B Cardiac muscle alpha actin 1	77	40.7	5.57	2436.99
NP 001158122.1	actin aortic smooth muscle	76	42	5.39	2145.94
ART85712.1	embryonic skeletal myosin heavy chain 3 partial	28	222.8	5.86	1905.26
AAS55927.1	cytoskeletal beta actin partial	51	44.8	5.83	1595.71
XP 003122540.1	alpha-actinin-3 isoform X2	69	103.2	5.45	1384.5
ACD13863.1	actin partial	81	27.4	5.05	1298.08
XP 020946026.1	plectin isoform X4	34	532.3	6.09	799.32
JAA53758.1	Plectin partial	37	402.7	5.68	673.79
XP 013840744.1	myosin-7B isoform X5	6	223.8	6	505.58
XP 003122636.2	glycogen phosphorylase muscle form	58	97.2	7.11	370.28
ABF81978.1	muscle glycogen phosphorylase partial	63	84	6.52	370.18
NP 001123421.1	creatine kinase M-type	68	43	7.09	251
ACE80199.1	troponin I	47	21.3	8.91	200.06
XP 001928934.3	C- U-editing enzyme APOBEC-2	57	25.8	4.81	89.24
XP 020949729.1	FH1/FH2 domain-containing protein 1 isoform X1	13	129.2	6.32	35.59
XP 003356007.1	EH domain-containing protein 2	20	61.2	6.37	35.33
NP 001008689.1	GTP-binding protein SAR1b	34	22.4	6.11	32.56

Table 1. Cont.

Gene	Protein	Coverage (%)	MW (kD)	Calc. pI	Score
BAX02569.1	alpha2 chain of type I collagen	9	129.1	9.07	31.72
NP 001231230.1	importin-5	12	123.6	4.94	29.83
XP 020922659.1	keratin type I cytoskeletal 14	7	51.6	5.16	27.27
NP 001258644.1	laminin subunit gamma-1 precursor	8	177.5	5.15	22.28
NP 001231936.1	eIF5-mimic protein 1	13	48.1	6.68	18.44
XP 013849669.2	LOW QUALITY PROTEIN neuroblast differentiation-associated protein AHNAK	12	654.3	6.05	13.91
XP 020939127.1	regulator of nonsense transcripts 1 isoform X1	6	125.5	6.7	13.08

In the molecular mass range of 95–130 kDa, the BioC modification of alpha-actinin-3 isoform X2 (103.2 kDa), glycogen phosphorylase muscle form (97.2 kDa), and muscle glycogen phosphorylase partial (84 kDa, adducted peptide FSAYLER) occurred at the N-Term site (DGLALCALIHR, FSAYLER, and FSAYLER, respectively), and that of FH1/FH2 domain-containing protein 1 isoform X1 (129.2 kDa) was located at two P sites (SPFPPPP-PAAP*LP*PSAPDGLALPTK). In addition, the alpha2 chain of type I collagen (129.1 kDa) had two adducted sites P and R (GEVGPAGPNGFAGPAGAAGQP*GAKGER*GTK). Furthermore, lower-molecular-mass adducts were identified, such as C-U-editing enzyme APOBEC-2 (25.8 kDa), GTP-binding protein SAR1b (22.4 kDa), and troponin I (21.3 kDa). BioC modifications of them occurred at the C or N-Term sites, and the adducted peptides involved were TFLC*YVIEAQSK, ELNARPLEVFMC*SVLK, and MSADAMLK, respectively.

Myosin, an important component of MP, has a molecular structure rich in charged amino acids and cysteines [31], which are susceptible to be attacked by the reactive electrophilic intermediate o-benzoquinone, leading to a Michael addition reaction. Tang et al. [32] investigated the interaction of RosA and meat proteins under the Fenton oxidizing system and showed that a diversity of myosins were involved in RosA modifications at K, R, and C sites. The difference was that the CA in our study mainly interacted with myosin at the alkaline amino acid (R, K, H) and N-Term sites of myosin but did not involve the C site. Instead, modifications at the C site were identified in low-molecular-mass proteins such as C-U-editing enzyme APOBEC-2 and GTP-binding protein SAR1b. Nikolantonaki et al. [33] noted that there was an adducted formation between caffeic acid and 3-sulfanylhexan-1-ol. It was speculated that oxidation conditions affect the covalent interaction of phenols and proteins to some extent. B-actin is a major component of cytoskeleton [34], and polyphenol modifications to it can have an effect on gel properties during meat processing [28]. Meanwhile, muscle glycogen phosphorylase is a key enzyme for glycogen degradation and affects the differentiation of skeletal muscle cells. Therefore, it is hypothesized that BioC modification may regulate energy metabolism.

3.4.2. Identification of Adducted Amino Acid Sites

Altogether, 51 BioC-adducted peptides were identified among 31 MP-BioC-adducted proteins (Table 2). BioC adduction occurred at six different amino acid sites, including the C, H, R, K, P, and N-Term sites. Secondary mass spectra of peptides with the BioC adducted on the C, H, R, K, and P sites, respectively, are shown in Figure 7a–e.

Distribution of adducted sites: Figure 8a shows that 59 BioC-adducted sites were identified in the BioC-adducted peptides, including 2 sites at C, 2 sites at H, 12 sites at R, 3 sites at K, 8 sites at P, and 32 sites at the N-term. Among them, alkaline amino acids (K, R, and H) accounted for 28.81% of the adducted sites. The proportions of C, P, and N-Term were 3.39%, 13.56%, and 54.24%, respectively. The result indicates that the BioC mainly modified the N-Term, followed by the alkaline amino acids P and C. In Figure 8b, a total of 43 BioC-adducted peptides (84.31%) contained one adducted site, and the remaining 8 (15.69%) carried two adducted sites, which suggests that the BioC preferred to react with MP at a single site. This result was similar to that of Tang et al. [2], who reported that RosA had the highest frequency of single-site addition reactions, followed by two-site addition reactions, and a low frequency of three or more site addition reactions ($p < 0.05$).

Table 2. BioC-adducted peptides and amino acid sites.

Gene	Protein	Adducted Peptide	Adducted Amino Acid Site
XP 020931560.1	LOW QUALITY PROTEIN titin	[K].NAAGVISK.[G] [K].VDGTPEIR.[I]	[N-Term] [P5]
XP 020921695.1	myosin-4 isoform X1	[K].FIRIHFGTTGK.[L] [K].QAFTQQIEELKR.[Q] [R].VIQYFATIAVTGEK.[K] [R].IEAQNKPFDAK.[T]	[R3] [N-Term] [N-Term] [N-Term]
BAA82146.1	myosin heavy chain 2x	[K].FIRIHFGTTGK.[L] [K].QAFTQQIEELKR.[Q] [R].VIQYFATIAVTGEK.[K] [R].IEAQNKPFDAK.[T]	[R3] [N-Term] [N-Term] [N-Term]
XP 020921876.1	myosin-2 isoform X1	[K].FIRIHFGTTGK.[L] [R].VIQYFATIAVTGEK.[K]	[R3] [N-Term]
XP 020921928.1	myosin-8 isoform X1	[K].FIRIHFGTTGK.[L] [R].VIQYFATIAVTGEK.[K] [R].IEAQNKPFDAK.[T]	[R3] [N-Term] [N-Term]
ART85712.1	embryonic skeletal myosin heavy chain 3 partial	[K].FIRIHFGTTGK.[L] [K].QAFTQQIEELKR.[Q] [R].IEDMAMLTHLNEPAVLYNLK.[D]	[R3] [N-Term] [H9]; [N-Term]
XP 003122540.1 XP 003122636.2 ABF81978.1 NP 001123421.1 ACE80199.1 XP 001928934.3 XP 003356007.1 NP 001008689.1 NP 001231936.1	alpha-actinin-3 isoform X2 glycogen phosphorylase muscle form muscle glycogen phosphorylase partial creatine kinase M-type troponin I C- U-editing enzyme APOBEC-2 EH domain-containing protein 2 GTP-binding protein SAR1b eIF5-mimic protein 1	[K].DGLALCALIHR.[H] [K].FSAYLER.[E] [K].FSAYLER.[E] [R].SIKGYTLPPHCSR.[G] [R].MSADAMLK.[A] [K].TFLCYVIEAQSK.[G] [R].TVTSALK.[E] [K].ELNARPLEVFMCSVLK.[R] [R].VQQSLGTR.[K]	[N-Term] [N-Term] [N-Term] [K3]; [N-Term] [N-Term] [C4]; [N-Term] [N-Term] 1xBioC [C12] [N-Term]
XP 020931125.1	LOW QUALITY PROTEIN nebulin	[R].KQLGHHVGAR.[N] [K].VHIMPDIPQIILAK.[A]	[H5] [P5]
AAC48692.1	skeletal alpha actin	[R].EIVRDIK.[E] [R].MQKEITALAPSTMK.[I]	[R4] [N-Term]
NP 001158122.1	actin aortic smooth muscle	[R].EIVRDIK.[E] [R].MQKEITALAPSTMK.[I]	[R4] [N-Term]
/	pdb 5NOL B Chain B Cardiac muscle alpha actin 1	[R].EIVRDIK.[E] [R].MQKEITALAPSTMK.[I]	[R4] [N-Term]
ACD13863.1	actin partial	[R].EIVRDIK.[E] [R].MQKEITALAPSTMK.[I]	[R4] [N-Term]
AAS55927.1	cytoskeletal beta actin partial	[R].EIVRDIK.[E] [R].MQKEITALAPSTMK.[I] [K].DLYANTVLSGGTTMYPGIADR.[M]	[R4] [N-Term] [N-Term]
XP 020946026.1 JAA53758.1 XP 013840744.1	plectin isoform X4 plectin partial myosin-7B isoform X5	[R].SMVEEGTGLR.[L] [R].SMVEEGTGLR.[L] [K].LLGSLDIDHSQYQFGHTK.[V]	[N-Term] [N-Term] [K]
XP 020949729.1	FH1/FH2 domain-containing protein 1 isoform X1	[K].SPFPPPPPPAAPLPPSAPDGLALPTK.[R]	[P12]; [P14]
BAX02569.1	alpha2 chain of type I collagen	[R].GEVGPAGPNGFAGPAGAAGQPGAKGER GTK.[G]	[P21]; [R27]
NP 001231230.1 XP 020922659.1	importin-5 keratin type I cytoskeletal 14	[K].FLFDSVSSQNMGLR.[E] [R].TKYETELNLR.[M]	[N-Term] [K2]; [N-Term]
NP 001258644.1	laminin subunit gamma-1 precursor	[R].KTLPSGCFNTPSIEKP.[-]	[P11]
XP 013849669.2	LOW QUALITY PROTEIN neuroblast differentiation-associated protein AHNAK	[K].AEGGGAEVQLPSLEGGLSMPDVK.[L] [K].GPGIDVKAPK.[M]	[P11]; [N-Term] [P2]; [N-Term]
XP 020939127.1	regulator of nonsense transcripts 1 isoform X1	[R].MHPALSAFPSNIFYEGSLQNGVTAADRVK.[K]	[R27]

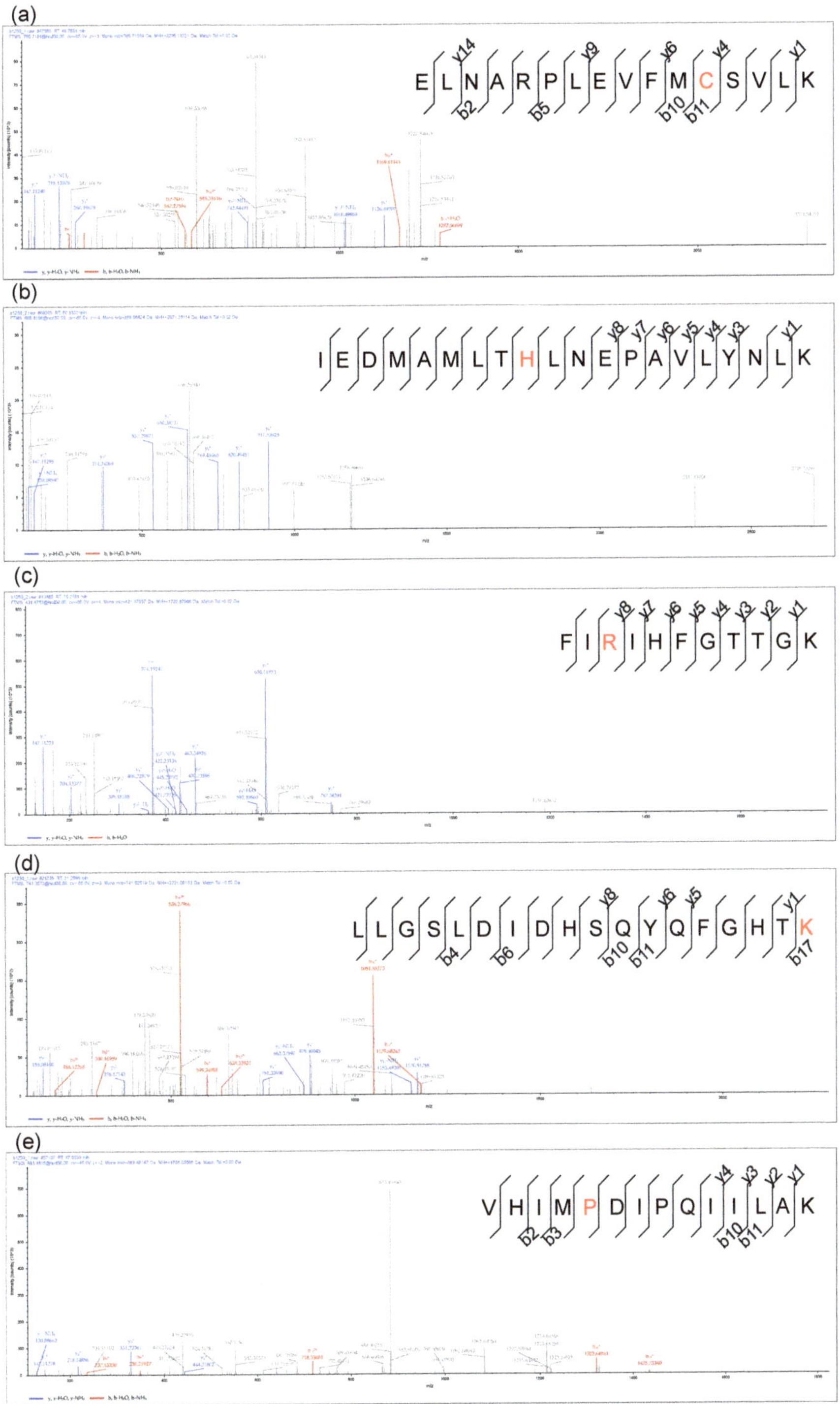

Figure 7. Secondary mass spectra of peptides with BioC adducted on different amino acid sites.

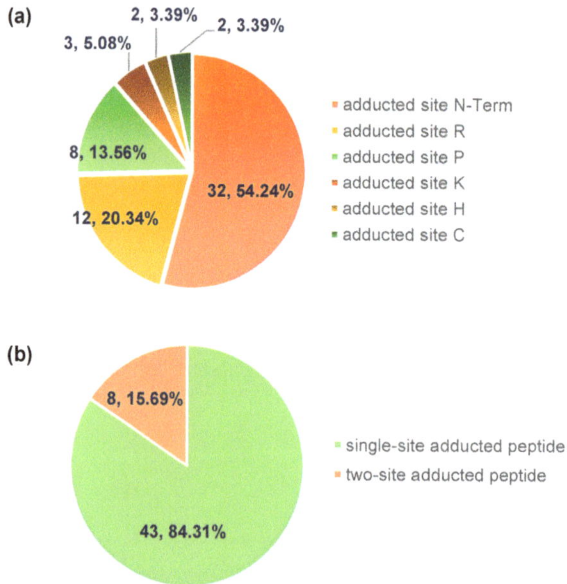

Figure 8. Distribution of different adducted sites (**a**); the number of adducted peptides carrying single or two sites (**b**).

In unoxidized phenolic compounds, hydroxyl groups are excellent hydrogen donors, which do not directly form irreversible covalent bonds with amino acids but allow for the formation of hydrogen bonds between phenols and carboxyl groups in proteins. Under alkaline conditions, hydroxyl groups are oxidized and converted into the active derivative quinone. The latter then undergoes irreversible covalent addition reactions with the sulfhydryl or amino groups of susceptible amino acid residues in proteins [35,36]. Some studies have found that the N-Term of protein is thought to be strongly related to the stability of protein structure [37]. Based on this, the high proportion of N-Term modifications might correspond to changes in protein structure. It has been reported that peptides containing alkaline amino acid are mostly located inside protein [2], so it could be inferred that the presence of steric hindrance made them less reactive with the BioC. It should also be noted that the oxidation level, type, and dose of phenolic compounds will also affect the extent of their interaction with proteins [38].

The sequence motif of adducted sites: The full-length sequence of the different adducted peptides was searched in the NCBI according to the gene number before selecting the adducted site and 10 amino acids on each side of it to form a polypeptide with 21 amino acids. The sequence logo (seqlogo) was drawn using Weblogo (Version 2.8.2) to further explore the sequence motif of the adducted sites. The height (%) of each letter in the seqlogo corresponds to the frequency of the occurrence of amino acid residues at that location, and that letters in each position are arranged in order of conservatism from the largest to smallest, which makes it easy to identify conserved sequences from the top letters. As shown in Figure 9, the most frequent amino acids at the BioC-adducted site (coordinate: 0) were R, P, K, C, and H, in decreasing order. About 85% of them were X-R/P/K-X motifs, and the rest were X-C/H-X motifs (X = any amino acid). This result suggests that R had the highest conservation. In addition, it was speculated that the sequence motif of the BioC-adducted peptide was DNSSAEGEFIRDIKGKLGKLA, which has implications for the future study of meat protein.

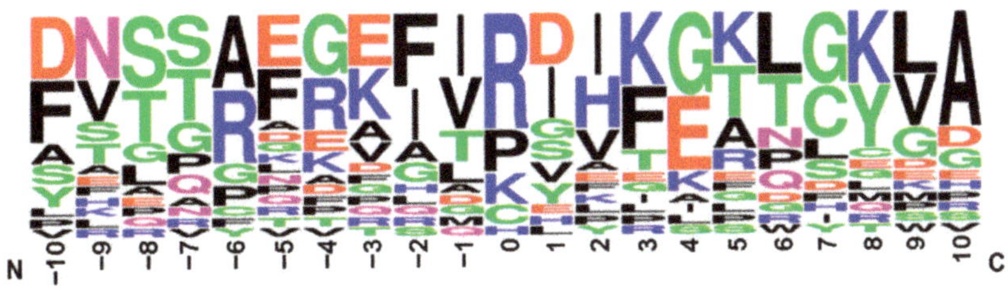

Figure 9. The seqlogo of the adducted peptides.

4. Conclusions

In this study, we developed and validated a novel method based on molecular probe technology to accurately identify phenol–protein adducts in meat protein systems, using BSA as a model protein. Our method involves the use of a BioC, a biotinylated derivative of CA which can covalently bind to MP in the same manner as CA. It was demonstrated that the BioC, a biotinylated derivative of CA, can interact with MP in the same way as CA, forming covalent bonds with specific amino acid residues. By using an LC-MS/MS analysis, 31 CA-adducted proteins, 51 CA-adducted peptides, and 6 kinds of CA-adducted sites were identified. Furthermore, it was revealed that CA prefers to react with the -NH_2 group (H, R, K, P, and N-term adducted sites) rather than the -SH group (C adducted site) of MP, resulting in various phenol–protein adducts. Our method provides a powerful tool for the isolation and identification of phenol–protein adducts in meat protein systems as well as the comprehensive analysis of the types and numbers of adducted sites, which reveal the mechanism of the covalent binding of phenols to proteins.

Although phenolic compounds have many beneficial effects on the quality, safety, and functionality of meat and meat products, there are some potential drawbacks and risks that need to be considered. One of the major challenges of using phenolic compounds in the meat industry is the oxidation of phenolic compounds with the formation of quinones, which could affect the quality, safety, and functionality of meat and meat products. The oxidation of phenolic compounds and the formation of quinones are influenced by many factors, such as the type, concentration, source of phenolic compounds, pH, temperature, oxygen, metal ions, enzymes, and other food components of the meat system, and the processing and storage conditions of the meat products. In this study, covalent interactions between plant polyphenols and meat proteins were investigated. However, only low-molecular-weight polyphenols were used, which may differ from high-molecular-weight polyphenols, such as proanthocyanidins and more complex tannins, in their interaction patterns and biological effects. High-molecular-weight polyphenols might form more stable covalent complexes with proteins. A future investigation is therefore required to take into account the molecular characteristics of polyphenols and the environmental factors that may influence the interactions between polyphenols and proteins and their effects on meat quality.

Supplementary Materials: The following supporting information can be downloaded at https://www.mdpi.com/article/10.3390/foods12234225/s1. Figure S1: The structure of BioC; Figure S2: The DPPH free radical scavenging ability of CA and BioC.

Author Contributions: Conceptualization, F.Y., C.T. and M.D.; data curation, F.Y., Y.Z., X.L., F.X., W.Z., W.S., H.S. and M.D.; formal analysis, F.Y. and X.L.; writing—original draft preparation, F.Y. and Y.Z.; writing—review and editing, Y.Z. and C.T.; methodology, Y.Z. and C.T.; visualization, Y.Z. and X.L.; investigation, F.X.; supervision, C.T.; funding acquisition, C.T. All authors have read and agreed to the published version of the manuscript.

Funding: This work was supported by the National Nature Science Foundation of China (32272345), the Fundamental Research Funds for the Central Universities (KYPT2023004, KYPT2023003), Key Technologies Research and Development Program (2021YFC2101400).

Data Availability Statement: The data used to support the findings of this study can be made available by the corresponding author upon request.

Conflicts of Interest: The authors declare no conflict of interest.

References

1. Fu, Q.; Liu, R.; Wang, H.; Hua, C.; Song, S.; Zhou, G.; Zhang, W. Effects of Oxidation in Vitro on Structures and Functions of Myofibrillar Protein from Beef Muscles. *J. Agric. Food Chem.* **2019**, *67*, 5866–5873. [CrossRef] [PubMed]
2. Tang, C.; Zhang, W.; Zou, Y.; Xing, L.; Zheng, H.; Xu, X.; Zhou, G. Influence of RosA-Protein Adducts Formation on Myofibrillar Protein Gelation Properties under Oxidative Stress. *Food Hydrocoll.* **2017**, *67*, 197–205. [CrossRef]
3. Xu, H.; Zhang, T.; Lu, Y.; Lin, X.; Hu, X.; Liu, L.; He, Z.; Wu, X. Effect of Chlorogenic Acid Covalent Conjugation on the Allergenicity, Digestibility and Functional Properties of Whey Protein. *Food Chem.* **2019**, *298*, 125024. [CrossRef] [PubMed]
4. Xue, F.; Li, C.; Adhikari, B. Physicochemical Properties of Soy Protein Isolates-Cyanidin-3-Galactoside Conjugates Produced Using Free Radicals Induced by Ultrasound. *Ultrason. Sonochem.* **2020**, *64*, 104990. [CrossRef] [PubMed]
5. Wang, S.; Li, X.; Zhu, J.; Liu, H.; Liu, T.; Yu, G.; Shao, M. Covalent Interaction between High Hydrostatic Pressure-Pretreated Rice Bran Protein Hydrolysates and Ferulic Acid: Focus on Antioxidant Activities and Emulsifying Properties. *J. Agric. Food Chem.* **2021**, *69*, 7777–7785. [CrossRef] [PubMed]
6. Zhang, L.; Zhou, C.; Xing, S.; Chen, Y.; Su, W.; Wang, H.; Tan, M. Sea Bass Protein-Polyphenol Complex Stabilized High Internal Phase of Algal Oil Pickering Emulsions to Stabilize Astaxanthin for 3D Food Printing. *Food Chem.* **2023**, *417*, 135824. [CrossRef] [PubMed]
7. Abdollahi, K.; Condict, L.; Hung, A.; Kasapis, S. Examination of β-Lactoglobulin-Ferulic Acid Complexation at Elevated Temperature Using Biochemical Spectroscopy, Proteomics and Molecular Dynamics. *Food Hydrocoll.* **2023**, *134*, 108053. [CrossRef]
8. Gallo, M.; Vinci, G.; Graziani, G.; De Simone, C.; Ferranti, P. The Interaction of Cocoa Polyphenols with Milk Proteins Studied by Proteomic Techniques. *Food Res. Int.* **2013**, *54*, 406–415. [CrossRef]
9. Condict, L.; Cavallo, J.; Hung, A.; Ashton, J.; Kasapis, S. MALDI-TOF-MS and in-Depth Dynamic Simulations on the Molecular Forces Determining the Stability of the 4-Hydroxybenzoic Acid—β-Casein Complex Following UHT-like Treatment. *Food Chem.* **2023**, *400*, 134047. [CrossRef]
10. Guerra, M.; Frey, D.; Hagner, M.; Dittrich, S.; Paulsen, M.; Mall, M.A.; Schultz, C. Cathepsin G Activity as a New Marker for Detecting Airway Inflammation by Microscopy and Flow Cytometry. *ACS Cent. Sci.* **2019**, *5*, 539–548. [CrossRef]
11. Kawatani, M.; Osada, H. Affinity-Based Target Identification for Bioactive Small Molecules. *MedChemComm* **2014**, *5*, 277–287. [CrossRef]
12. Wang, P.; Yu, L.; Gong, J.; Xiong, J.; Zi, S.; Xie, H.; Zhang, F.; Mao, Z.; Liu, Z.; Kim, J.S. An Activity-Based Fluorescent Probe for Imaging Fluctuations of Peroxynitrite (ONOO−) in the Alzheimer's Disease Brain. *Angew. Chem. Int. Ed.* **2022**, *61*, e202206894. [CrossRef] [PubMed]
13. Benns, H.J.; Wincott, C.J.; Tate, E.W.; Child, M.A. Activity- and Reactivity-Based Proteomics: Recent Technological Advances and Applications in Drug Discovery. *Curr. Opin. Chem. Biol.* **2021**, *60*, 20–29. [CrossRef] [PubMed]
14. Saghatelian, A.; Jessani, N.; Joseph, A.; Humphrey, M.; Cravatt, B.F. Activity-Based Probes for the Proteomic Profiling of Metalloproteases. *Proc. Natl. Acad. Sci. USA* **2004**, *101*, 10000–10005. [CrossRef]
15. Konziase, B. Biotinylated Probes of Artemisinin with Labeling Affinity toward Trypanosoma Brucei Brucei Target Proteins. *Anal. Biochem.* **2015**, *482*, 25–31. [CrossRef] [PubMed]
16. Park, D.; Xiong, Y.L.; Alderton, A.L. Concentration Effects of Hydroxyl Radical Oxidizing Systems on Biochemical Properties of Porcine Muscle Myofibrillar Protein. *Food Chem.* **2007**, *101*, 1239–1246. [CrossRef]
17. Yang, F.; Jin, S.; Li, X.; Shen, J.; Zeng, X.; Wang, Y.; Zhou, G.; Tang, C. Biotinylated Caffeic Acid Covalent Binding with Myofibrillar Proteins in Alkaline Conditions: Identification of Protein-Phenol Adducts and Alterations in Protein Properties. *Food Chem.* **2023**, *416*, 135818. [CrossRef] [PubMed]
18. Zhang, L.; Guan, Q.; Tang, L.; Jiang, J.; Sun, K.; Manirafasha, E.; Zhang, M. Effect of Cu^{2+} and Al^{3+} on the Interaction of Chlorogenic Acid and Caffeic Acid with Serum Albumin. *Food Chem.* **2023**, *410*, 135406. [CrossRef]
19. Liu, R.; Fu, Q.; Lonergan, S.; Huff-Lonergan, E.; Xing, L.; Zhang, L.; Bai, Y.; Zhou, G.; Zhang, W. Identification of S-Nitrosylated Proteins in Postmortem Pork Muscle Using Modified Biotin Switch Method Coupled with Isobaric Tags. *Meat Sci.* **2018**, *145*, 431–439. [CrossRef]
20. Pi, X.; Sun, Y.; Liu, J.; Wang, X.; Hong, W.; Cheng, J.; Guo, M. Characterization of the Improved Functionality in Soybean Protein-Proanthocyanidins Conjugates Prepared by the Alkali Treatment. *Food Hydrocoll.* **2023**, *134*, 108107. [CrossRef]
21. Wiśniewski, J.R.; Zougman, A.; Nagaraj, N.; Mann, M. Universal Sample Preparation Method for Proteome Analysis. *Nat. Methods* **2009**, *6*, 359–362. [CrossRef]
22. Rappsilber, J.; Mann, M.; Ishihama, Y. Protocol for Micro-Purification, Enrichment, Pre-Fractionation and Storage of Peptides for Proteomics Using StageTips. *Nat. Protoc.* **2007**, *2*, 1896–1906. [CrossRef] [PubMed]
23. Wang, R.E.; Hunt, C.R.; Chen, J.; Taylor, J.-S. Biotinylated Quercetin as an Intrinsic Photoaffinity Proteomics Probe for the Identification of Quercetin Target Proteins. *Bioorg. Med. Chem.* **2011**, *19*, 4710–4720. [CrossRef] [PubMed]

24. Benjakul, S.; Singh, A.; Chotphruethipong, L.; Mittal, A. Chapter Eight—Protein-Polyphenol Conjugates: Preparation, Functional Properties, Bioactivities and Applications in Foods and Nutraceuticals. In *Advances in Food and Nutrition Research*; Granato, D., Ed.; Application of Polyphenols in Foods and Food Models; Academic Press: Pittsburgh, PA, USA, 2021; Volume 98, pp. 281–320.
25. Prigent, S.V.; Voragen, A.G.; Visser, A.J.; van Koningsveld, G.A.; Gruppen, H. Covalent Interactions between Proteins and Oxidation Products of Caffeoylquinic Acid (Chlorogenic Acid). *J. Sci. Food Agric.* **2007**, *87*, 2502–2510. [CrossRef]
26. Roy, D.; Dutta, S.; Maity, S.S.; Ghosh, S.; Singha Roy, A.; Ghosh, K.S.; Dasgupta, S. Spectroscopic and Docking Studies of the Binding of Two Stereoisomeric Antioxidant Catechins to Serum Albumins. *J. Lumin.* **2012**, *132*, 1364–1375. [CrossRef]
27. Jansson, T.; Rauh, V.; Danielsen, B.P.; Poojary, M.M.; Waehrens, S.S.; Bredie, W.L.P.; Sørensen, J.; Petersen, M.A.; Ray, C.A.; Lund, M.N. Green Tea Polyphenols Decrease Strecker Aldehydes and Bind to Proteins in Lactose-Hydrolyzed UHT Milk. *J. Agric. Food Chem.* **2017**, *65*, 10550–10561. [CrossRef]
28. Chen, K.; Chen, X.; Liang, L.; Xu, X. Gallic Acid-Aided Cross-Linking of Myofibrillar Protein Fabricated Soluble Aggregates for Enhanced Thermal Stability and a Tunable Colloidal State. *J. Agric. Food Chem.* **2020**, *68*, 11535–11544. [CrossRef]
29. Liu, Z.; Xiong, Y.L. Oxidation Desensitizes Actomyosin to Magnesium Pyrophosphate-Induced Dissociation. *Food Chem.* **2013**, *141*, 662–668. [CrossRef]
30. Tang, C.; Zhang, W.; Wang, Y.; Xing, L.; Xu, X.; Zhou, G. Identification of Rosmarinic Acid-Adducted Sites in Meat Proteins in a Gel Model under Oxidative Stress by Triple TOF MS/MS. *J. Agric. Food Chem.* **2016**, *64*, 6466–6476. [CrossRef]
31. Levitsky, D.I. Actomyosin Systems of Biological Motility. *Biochemistry* **2004**, *69*, 1177–1189. [CrossRef]
32. Tang, C.; Zhang, W.; Dai, C.; Li, H.; Xu, X.; Zhou, G. Identification and Quantification of Adducts between Oxidized Rosmarinic Acid and Thiol Compounds by UHPLC-LTQ-Orbitrap and MALDI-TOF/TOF Tandem Mass Spectrometry. *J. Agric. Food Chem.* **2015**, *63*, 902–911. [CrossRef] [PubMed]
33. Nikolantonaki, M.; Jourdes, M.; Shinoda, K.; Teissedre, P.-L.; Quideau, S.; Darriet, P. Identification of Adducts between an Odoriferous Volatile Thiol and Oxidized Grape Phenolic Compounds: Kinetic Study of Adduct Formation under Chemical and Enzymatic Oxidation Conditions. *J. Agric. Food Chem.* **2012**, *60*, 2647–2656. [CrossRef] [PubMed]
34. Yates, L.D.; Greaser, M.L.; Huxley, H.E. Quatitative Determination of Myosin and Actin in Rabbit Skeletal Muscle. *J. Mol. Biol.* **1983**, *168*, 123–141. [CrossRef] [PubMed]
35. Rawel, H.M.; Czajka, D.; Rohn, S.; Kroll, J. Interactions of Different Phenolic Acids and Flavonoids with Soy Proteins. *Int. J. Biol. Macromol.* **2002**, *30*, 137–150. [CrossRef] [PubMed]
36. Estévez, M.; Heinonen, M. Effect of Phenolic Compounds on the Formation of α-Aminoadipic and γ-Glutamic Semialdehydes from Myofibrillar Proteins Oxidized by Copper, Iron, and Myoglobin. *J. Agric. Food Chem.* **2010**, *58*, 4448–4455. [CrossRef]
37. Gawron, D.; Ndah, E.; Gevaert, K.; Van Damme, P. Positional Proteomics Reveals Differences in N-Terminal Proteoform Stability. *Mol. Syst. Biol.* **2016**, *12*, 858. [CrossRef]
38. Zhao, X.; Xu, X.; Zhou, G. Covalent Chemical Modification of Myofibrillar Proteins to Improve Their Gelation Properties: A Systematic Review. *Compr. Rev. Food Sci. Food Saf.* **2021**, *20*, 924–959. [CrossRef]

foods

MDPI

Review

Current Innovations in the Development of Functional Gummy Candies

Mohammad Tarahi [1], Sima Tahmouzi [2], Mohammad Reza Kianiani [3], Shiva Ezzati [4], Sara Hedayati [5,*] and Mehrdad Niakousari [1]

1 Department of Food Science and Technology, School of Agriculture, Shiraz University, Shiraz 7144165186, Iran; tarahimohammad@yahoo.com (M.T.); niakosar@shirazu.ac.ir (M.N.)
2 Department of Food Science and Technology, School of Public Health, Shahid Sadoughi University of Medical Sciences, Yazd 8916978477, Iran; simaa.tahmouzi@gmail.com
3 Department of Food Science and Technology, Faculty of Agriculture, Ferdowsi University of Mashhad (FUM), Mashhad 9177948978, Iran; rezakianiani@gmail.com
4 Department of Food Science and Technology, Faculty of Agriculture, University of Tabriz, Tabriz 5166616471, Iran; sh_ezzati@tabrizu.ac.ir
5 Nutrition Research Center, School of Nutrition and Food Sciences, Shiraz University of Medical Sciences, Shiraz 7193635899, Iran
* Correspondence: s_hedayati@sums.ac.ir

Abstract: Nowadays, consumers are aware of the necessity of following a healthy diet and there is demand for natural and nutritious food products, especially for children. Consequently, new trends in the food industry are focused on the development of foods with low levels of sucrose and artificial additives (e.g., flavors and colorants), as well as high antioxidant, protein, and fiber content. On the other hand, some consumers demand vegan, halal, and kosher-certified food products. In this regard, conventional confectionary products such as gummy candies (GCs) are increasingly losing their popularity. Therefore, the development of plant-based and functional GCs has gained the attention of researchers and manufacturers. This review highlights recent innovations in the development of GCs with alternative gelling agents and sweeteners, natural flavors and colorants, and the incorporation of medicines, fiber, protein and antioxidants into GCs. Additionally, it summarizes their effects on the techno-functional, sensory, and nutritional properties of GCs.

Keywords: functional foods; novel formulations; confectionary industry; soft candy; jelly candy

1. Introduction

In recent decades, consumer demand in the field of food production has changed significantly, as they have become increasingly aware of food quality and the health benefits associated with different food products. Nowadays, foods are intended not only to satisfy hunger and provide necessary nutrients for humans, but also to prevent diseases related to nutrition and improve the physical and mental health of consumers. In this regard, it is necessary to develop novel functional foods to meet these demands [1]. The term "functional food" was first used in Japan in the 1980s for food products enriched with certain compounds that have beneficial physiological effects, such as pre- and probiotics, as well as cholesterol-lowering substances. These types of food products can be consumed in order to improve the general condition of the body, reduce the risk of some diseases, and can even be used to treat some illnesses [2]. In 2014, the Functional Food Center (FFC) defined functional food products as "natural or processed foods containing effective and non-toxic amounts of bioactive compounds, which provide a clinically proven and documented health benefit through the use of specific biomarkers to prevent, manage, or treat chronic diseases or their symptoms" [3]. However, despite these different definitions, scientists generally agree that functional foods contain bioactive substances that provide health benefits for the human body beyond the usual dietary supplements.

The confectionery industry has a wide range of products with approximately USD 298.23 billion global market value in 2021, which is projected to expand at a compound annual growth rate (CAGR) of 4.3% from 2022 to 2028 [4]. Among these products, soft confectionery products have attracted the attention of consumers in all age groups, especially children, due to their unique texture, appearance, and taste [5]. Gummy candy (GC) is one the most popular types of soft confectionery products, which is also known as pastille, gummy, jelly candy, gummy jelly, and so on, with a global market value of USD 21.40 billion in 2022 [6,7]. GCs are traditionally made from high amounts of sucrose and glucose syrup combined with gelling agents, commonly known as gelatin, along with artificial additives such as flavors and colors. The raw materials are first heated and then poured into the molds with different shapes and sizes. After molding, GCs are cooled and dried by air convection in chambers or tunnels to achieve a dry-sticky texture in the shortest possible time, with a relatively high-water content varying between 15 and 30% [8]. Although gelatin offers a variety of functional properties in GC preparation, such as firmness, consistency, and gelling properties, some consumers demand vegan, halal, and kosher-certified food products [9]. Besides, these popular foodstuffs are often of low nutritional value and high sugar content, which may increase the risk of several chronic diseases, such as obesity, tooth decay, hyperglycemia, and type 2 diabetes. In addition, the overconsumption of low-nutrition confections by children is gradually increasing, which has become a serious concern for their parents [10]. In this respect, the development of new formulations can be a promising step for GC lovers in order to provide desirable health benefits and pleasant sensory attributes to certain types of diets.

In recent years, the incorporation of alternative gelling agents, sugar substitutes, natural flavorings and colorants, antioxidants, and other value-added ingredients into GC formulation has revolutionized the confectionery industry, offering exciting new possibilities for taste, texture, and nutritional values to meet consumer demands. Therefore, this study aims to comprehensively investigate the preparation and development of GCs based on these innovative formulations.

2. Preparation of GCs Based on Novel Formulations

Novel formulations have opened doors for customization and personalization of food products, by introducing a variety of value-added ingredients, flavors, and other combinations. One significant aspect in the preparation of novel GCs is the incorporation of gelatin alternatives, such as pectin and agar gum, which allows manufacturers to create vegan gummies with unique textures and sensory properties. Furthermore, sugar-free formulations have become more prevalent, allowing individuals with dietary restrictions to indulge in these delightful treats without compromise [11]. Manufacturers have also begun to explore alternatives to artificial colors and flavors, opting for natural fruit extracts and plant-based dyes. This shift toward more natural ingredients not only enhances the appeal of GCs but also aligns with the growing demand for healthier confectionary foods (Figure 1).

2.1. Alternative Gelling Agents

As the demand for vegan and vegetarian-friendly options continues to grow, alternative gelling agents such as pectin, gums, and starches have gained more attention [12]. However, the incorporation of these components in the preparation of GCs requires careful consideration of several factors, including optimization of formulation, selection of appropriate gelling agent, and pH adjustment, as well as their effects on texture, flavor, and shelf life of the final product. Therefore, it is important to evaluate various functional and physicochemical aspects of the alternative gelling agents to adjust the desired formulation [13]. For example, pectin requires an acidic environment to activate its gelling properties, so it may need to be used with acidic ingredients, such as lemon juice. On the other hand, agar gum has high gelling strength and can create a firmer texture in GCs compared to gelatin, so it may be used less than other gelling agents [14].

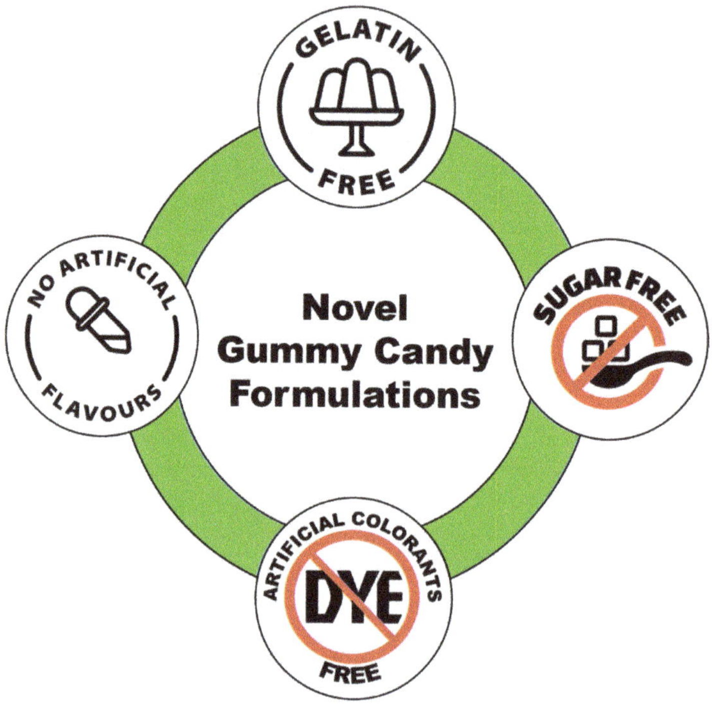

Figure 1. A schematic of GC preparation based on novel formulations.

Pectin is one of the most popular alternative gelling agents in the production of jams and jellies, which is commonly derived from different fruits. In order to use pectin as a gelling agent in GCs, it is first dissolved in water and then heated before being added to the candy mixture. Pectin has a different gelling profile compared to gelatin, and it can create a softer and more delicate texture for GCs. However, pectin may require more sugar than gelatin to achieve the desired sweetness and flavor. Besides, DeMars and Ziegler [14] showed that the gelatin-pectin combination at different levels (i.e., 3.0, 4.5, and 6.0% gelatin and 0.5, and 1.5% pectin) can significantly affect the texture of GCs. Here, the addition of pectin reduced the strain at the fracture of gelatin gels. In addition, Renaldi et al. [15] showed almost high sensory score ranges (6.9–7.5) for GCs formulated with 1.42:8.58% pectin-gelatin mixture. Guar gum is another natural polysaccharide derived from the seeds of the guar plant, which has been widely used in the food industry for its unique thickening, stabilizing, and emulsifying properties [16,17]. Guar gum can swell and disperse in both cold and hot water; while, other gums such as locust bean gum, disperse only in hot water [18]. Additionally, guar gum can improve the textural and color parameters of gummy candies. Thus, it can be considered as a promising gelling agent in candy production. In this regard, Dinesh Kumar et al. [19] found that the addition of guar gum in GC production resulted in novel products with better texture and brighter color, which were preferred by the panelists. Therefore, guar gum can be employed as an alternative gelling agent, to achieve the desired appearance, texture, and mouthfeel.

Agar gum and carrageenan are other important alternative gelling agents, which can be obtained from seaweeds. Similar to pectin, to use agar gum in GC production, it must first be dissolved in water and heated to activate its gelling properties [20]. However, too much agar gum can create a rubbery texture in GCs; hence, careful testing and experimentation are necessary to achieve the desired results. Ge et al. [21] found that incorporating different hydrocolloids, such as pectin and agar gum, into gelatin-based GCs had a synergistic effect on the gel network. These effects are mainly impacted by their nature and ratio to gelatin in the GC formulation. On the other hand, carrageenan has a gelling profile similar to gelatin and can create a delicate texture in GCs. However, it is more challenging to work with carrageenan compared to other alternative gelling agents and may require additional ingredients, such as calcium, to activate its gelling properties. More recently, Song et al. [22] evaluated the effects of carboxy methyl cellulose (CMC) on the texture properties and storage stability of kappa-carrageenan (KC) GCs. The researchers found that the addition of CMC to KC formed flexible GCs with low fragility and limited water syneresis during storage. This study also found that KC promoted side-by-side intermolecular aggregation of CMC helices through hydrogen bonds, which stabilized a denser network structure. These results provide preliminary evidence for utilizing hydrocolloids to adjust texture and control water migration in KC gels to promote vegan GCs. In another study, Hamka [23] investigated the physical and sensory properties of dragon fruit GCs with the addition of carrageenan at different concentrations. The results showed that the water and ash content of GCs fulfilled the required standards at all carrageenan concentrations. Also, the organoleptic test represented that the panelists preferred GCs formulated with 3% carrageenan.

Alginate is an important polysaccharide derived from seaweeds, which gained much attention in the food industry due to its unique properties, such as gel-forming ability, non-toxicity, and low cost. In GC production, alginate can be used as both a gelling agent and an encapsulation material. For instance, Ali et al. [24] showed the successful production of strawberry and red beetroot GCs incorporated with alginate with favorable nutritional and textural properties. Also, the possible application of cold-set gelation technique was evaluated for GC production by de Avelar and Efraim [25] using a mixture of alginate and pectin. In terms of textural properties, alginate/pectin-based GCs showed a higher hardness value (16.2 N) compared to pectin-based GCs (6.2 N). However, no significant difference ($p > 0.05$) was observed between alginate/pectin and pectin GCs for all sensory attributes, proving the cold-set gelation method as a sustainable alternative technique for the production of GCs.

Starches are also widely used in various sectors of the food industry as a thickening agent, texturizer, and binder to provide the desired texture and mouthfeel for final food products, such as GCs [26,27]. For instance, corn starch has a high water-holding capacity, which can help to improve the moisture content and overall texture of candies. Marfil et al. [28] investigated the microstructure and texture of GCs incorporated with gelatin and corn starch in different ratios (i.e., 10:0; 9:1; 8:2; 7:3; 0:10). The results showed that as the concentration of starch increased, GCs exhibited higher adhesiveness and stringiness values, which is favorable for GC production. In another study, Pereira et al. [29] evaluated the physical and sensory characteristics of GCs containing 12 and 16% acid-thinned cassava starch compared to control GCs with 8% acid-thinned corn starch. The GCs formulated with 12% cassava starch and 8% corn starch presented similar hardness, chewiness, and sensory attributes, while the incorporation of 16% cassava starch almost doubled textural values and decreased sensory scores. These results showed the possible production of GCs with acid-thinned cassava starch.

Overall, various alternative gelling agents can be successfully utilized in GC production, which promises the production of novel vegan GCs. However, the application of these gelling agents may also have some drawbacks. For instance, some alternative gelling agents require more sugar compared to gelatin to achieve the desired flavor, which may be dangerous for diabetic patients and should be carefully optimized in GC formulation.

2.2. *Alternative Sweeteners*

GCs have been a beloved treat for generations, with traditional recipes using sugar as the primary sweetener. However, with the rise of health concerns related to high sugar consumption, alternative sweeteners (e.g., stevia, erythritol, xylitol, honey, etc.) have become increasingly popular in the food industry. Incorporating alternative sweeteners in the preparation of GCs requires careful consideration of several factors, including the choice of sweetener, recipe modification, and their effects on the texture, flavor, and shelf life of the final product [30].

One of the most popular alternative sweeteners is stevia. It is a plant-based sweetener that is 200 to 400 times sweeter than sucrose but has no calories. When using stevia as a sweetener in GCs, it is important to remember that it has a different flavor profile than sugar and may require additional adjustment of the recipe to achieve the desired flavor. Additionally, because stevia is much sweeter than sugar, less may be required to achieve the same level of sweetness. The primary components responsible for stevia's sweetness are steviol glycosides, particularly rebaudioside A and stevioside [31]. Numerous studies have explored the application of stevia as a sweetener in GCs, demonstrating its feasibility and benefits. For instance, Rivero et al. [32] conducted research to develop stevia-sweetened GCs with an emphasis on the sensory characteristics. The study concluded that the low-calorie GCs formulated with stevia exhibited desirable sensory attributes (e.g., texture, flavor, and sweetness) with high consumer acceptability, indicating the potential of stevia for replacing sugar in GC production. In another study, Aranda-Gonzalez et al. [33] investigated the effects of sugar substitution with 20–100% stevia on the physicochemical and sensory properties of GCs. They found that the GCs with 60% stevia exhibited similar physicochemical properties to those made with 100% sugar. Moreover, the sensory evaluation indicated that stevia-sweetened GCs were well-liked by consumers. In addition, Samakradhamrongthai and Jannu [34] explored the formulation of novel GCs using a combination of stevia and other alternative sweeteners (i.e., xylitol, corn syrup, and erythritol). The study aimed to enhance the sweetness profile while minimizing any potential off-flavors associated with stevia. The results indicated that blending stevia with other sweeteners, such as erythritol or xylitol, improved the taste and overall quality of the gummy candies, offering a promising solution for the GC industry.

Another popular alternative sweetener is erythritol. Erythritol is a sugar alcohol that is derived from corn. It is approximately 70% as sweet as sugar and has a similar texture to sugar. Also, Gan et al. [11] showed that the incorporation of erythritol into GCs can effectively reduce the glycemic index of the final products from 81.9 to 49.9%, indicating its potential to be used in the manufacturing of food supplements for diabetics. Besides, Le et al. [35] evaluated the possible application of erythritol in GC production using three-dimensional (3D) printing technology. The results indicated that when erythritol is regularly added from 0 to 30 g into the printing materials, surface smoothness is dramatically improved. However, the gel structure started to show partial bottom adiposity with the addition of 40 g erythritol, which may be due to the inability of hydrogen bonding between polyols to form a stable gel, thereby weakening gel strength. This research can serve as a basis for the development of 3D printed in the future. Furthermore, it should be noted that erythritol may also have a cooling effect in the mouth when consumed in large quantities, which can affect the overall taste of GCs [36]. Therefore, its proper concentration should be optimized in GC formulation. Xylitol is another commonly used sugar alcohol in GC production. Xylitol has a similar sweetness and texture to sugar and can be used in the same proportion as sugar in most recipes [37]. In this regard, Čižauskaitė et al. [38] conducted a comprehensive analysis of the bioactive compounds in xylitol-sweetened GCs. The study revealed that xylitol not only acted as a sweetener, but also contributed to improve the overall nutritional profile of the GCs. Xylitol supplementation increased the content of polyphenols and other beneficial compounds, potentially offering additional health benefits beyond sweetness. As further research and advancements continue, incorporating xylitol in GC formulations can lead to healthier confectionery options that satisfy consumer cravings while addressing oral health concerns and dietary requirements. However, xylitol can be toxic to pets, so it is important to label the candies properly and keep them out of reach of animals [39].

Honey is another alternative sweetener that can be used in GC recipes. Honey is a natural sweetener with a unique flavor profile that can enhance the sensory attributes of candies. Utilizing honey as an alternative sweetener in GCs presents an opportunity to create confectionery products with natural, nutrient-rich, and potentially beneficial options [40]. However, honey is much sweeter than sugar and can also affect the texture of candies. In this regard, Rivero et al. [41] examined the effect of honey on the glycemic response to GCs. This study compared the glycemic index of honey-sweetened GCs to those sweetened with sucrose. The results indicated that GCs sweetened with honey had a lower glycemic index, which suggests a potentially more favorable effect on blood sugar levels, making them a suitable option for individuals monitoring their glucose levels. Additionally, pure honey is difficult to work with and may require additional adjustments to the recipe to achieve the desired results. For instance, Gunes et al. [5] highlighted that honey's high moisture content could impact the texture and stability of GCs, potentially leading to stickiness or softening. Additional adjustments, such as modifying the gelatin or pectin content, may be necessary to overcome these challenges and achieve the desired GC consistency. However, some confectionery manufacturers have reported that honey may adversely affect the texture and shelf life of GCs and make the final products stickier [42].

Other sweeteners such as isomaltulose, maltitol, and isomaltitol, can also be used in the GC industry. For instance, Gok et al. [43] substituted glucose syrup with maltitol syrup in the formulation of GCs and investigated the effect of different bulking agents (sucrose, mannitol, and soluble wheat fiber) on the physicochemical and sensory attributes of GCs. They found that soluble wheat fiber was the best bulking agent for the GCs with maltitol; while mannitol caused negative changes in the quality of GCs. Besides, Periche et al. [10] demonstrated that the combination of 30% isomaltulose and 70% fructose can be a suitable way to replace traditional sugars in functional GCs with the desired features in terms of optical (e.g., a*, b*, and L* values) and mechanical properties (e.g., springiness, hardness, gumminess, and cohesiveness). Moreover, Jiamjariyatam [44] showed that the toughness and hardness of GCs were significantly decreased by the incorporation of 20, 30, 40, 50, and 100% isomaltulose. Similarly, Ünal and Arslan [45] reported lower hardness, flexibility, and moisture content for isomaltitol-containing GCs than conventional sugar-containing samples. However, the substitution of sugars with isomaltitol negatively affected the sensory acceptability of GCs due to their opaque appearance.

Another approach involved using natural sugar sources like grape, mulberry, and carob molasses to replace sugar syrup and artificial additives in GC production. In this respect, Kurt et al. [46] found that the type of incorporated molasses can significantly affect the texture and appearance of GCs. Grape molasses-sweetened candies had higher thermal stability and lower temperature sensitivity due to their higher total sugar content compared to GCs incorporated with mulberry and carob molasses. On the other hand, sugar can act as a natural preservative, while alternative sweeteners may not have the same effect [36]. Therefore, it may be necessary to add additional preservatives or adjust the packaging of the final products in order to provide sugar-free GCs with a suitable shelf life.

As a result, it is important to adjust the GC recipe according to sweetness and other physicochemical and structural properties of the alternative sweeteners (e.g., melting temperature, particle density, and particle size). The use of alternative sweeteners can also affect the texture and shelf life of GCs. In addition, some alternative sweeteners may have a different mouthfeel or aftertaste than sugar, which could affect the overall flavor of GCs. On the other hand, some studies showed that the consumption of artificial sweeteners (e.g., saccharin, sucralose, aspartame) in food formulations may cause glucose intolerance by inducing compositional and functional changes in the gastrointestinal microbiota [47]. Therefore, it may be necessary to analyze different alternative sweeteners more comprehensively to find one that works best for the GC production.

2.3. Natural Flavors and Colorants

Natural flavors and colorants have become increasingly popular in the food industry as consumers become more conscious of what they are eating. Therefore, incorporating natural ingredients in the preparation of GCs based on novel formulations can enhance the overall flavor and visual appeal of candies [48]. However, careful attention must be paid to the selection and application of natural food flavors and colorants to ensure their compatibility with the conventional recipe and to provide the desired texture and shelf life for GCs.

Fruits can create a wide variety of flavors in GCs, from strawberry and raspberry to orange and lemon. When using fruit flavors, it is important to select natural flavorings derived from the actual fruit rather than artificial or synthetic ones. Whether it is the tangy burst of citrus, the tropical sweetness of mango, or the juicy essence of berries, fruit flavors can add excitement and variety to GC offerings. Generally, consumers appreciate flavors derived from real fruits, and the overall sensory appeal of GCs is enhanced even with the addition of small amounts of fruit flavors. In this regard, Bagautdinov et al. [49] investigated the sensory attributes and consumer preference for GCs flavored with different fruits, including strawberry, orange, and pineapple. The study found that the GCs with fruit flavors were highly favored by consumers, especially strawberry- and orange-containing GCs. Overall, the natural fruit flavors contributed to the overall enjoyment and acceptance of GCs. Moreover, Ali et al. [24] investigated the effects of strawberry as a natural flavoring agent and red beetroot as a natural colorant on the textural and sensory properties of GCs. The study revealed that GCs with natural colors and flavors exhibited variations in texture, such chewiness and firmness. For instance, GC containing a large amount of beetroot was harder than the control sample, which may be related to the increase in fiber and total soluble solids (TSS) of the sample. Also, the incorporation of specific fruits allowed for tailoring the mouthfeel of GCs, enhancing the overall sensory experience for consumers.

Incorporating natural colorants in the preparation of GCs can also enhance their visual appeal. Natural colorants can be derived from various sources, including fruits and vegetables [50]. For example, the extract of the *Clitoria ternatea* L. flower has been used as a natural colorant for the preparation of novel GCs. The study found that the incorporation of 30 g of *C. ternatea* extract improved the color and appearance of GCs, making them more appealing to consumers [51]. In another study, Casas-Forero et al. [52] showed that GCs become darker and more reddish with the addition of freeze-dried blueberry juice due to the presence of anthocyanins. Also, the enriched samples exhibited significantly higher total bioactive compounds (773.3 mg GAE/100 g) and antioxidant activity (4585.4 μM TE/100 g), compared to conventional GCs (233.4 mg GAE/100 g and 782.9 μM TE/100 g, respectively). These results indicate that fruit flavors not only improve the taste, aroma, and appearance of GCs, but also offer potential health benefits through their antioxidant properties. In another study, de Oliveira Nishiyama-Hortense et al. [53] investigated the applicability of grape juice as a natural colorant in GCs and determined their anthocyanin content and sensory properties to evaluate the quality of the GCs. The GCs presented a shiny appearance with a uniform purple color, retention of anthocyanin was 41% and sensory scores were satisfactory.

Nevertheless, many natural colorants, such as betanin, are sensitive to heat and pH, and their color may deteriorate over time. Additionally, their sensitivity to oxidation and low bioaccessibility limits their applications in food products. Encapsulation is a promising approach to overcome the shortcomings of natural colorants in GCs. In this regard, Amjadi et al. [54] loaded betanin in liposomal nanocarriers to improve its stability in GCs. The antioxidant activity and stability of betanin nanoliposomes were considerably higher than those of free betanin in GCs and their sensory parameters did not show significant differences. Consequently, betanin nanoliposomes can be considered a suitable natural colorant for gummy candies.

However, the selection of appropriate natural food flavors and colorants is very critical in order to provide favorable techno-functional properties without negatively affecting the texture of the final products. For instance, the incorporation of some natural ingredients, such as fruit purees, increases the moisture content of the system, which can finally affect the texture of GCs [55]. Therefore, it may be necessary to adjust the recipe to account for these changes by using less liquid or adding additional thickening agents to the formulation of GCs. Another consideration for using natural food flavors and colorants is their potential impact on the shelf life of GCs. Natural ingredients may not have the same preservative properties as artificial ingredients and may reduce the shelf life of GCs [52]. To ensure the best results when incorporating natural food flavors and colorants in the preparation of GCs, it may be necessary to adjust the recipe multiple times to achieve the desired flavor, color, and texture.

In conclusion, incorporating natural food flavors and colorants in the preparation of GCs based on novel formulations can enhance the overall flavor and visual appeal of GCs, as well as their health-promoting properties. However, careful consideration must be given to the selection and application of these natural ingredients to ensure their compatibility with conventional recipes and to maintain the desired texture and shelf life of the GCs.

3. Development of GCs as a Functional Food

Conventional GCs contain high amounts of absorbable carbohydrates and low levels of health-promoting compounds, which have raised many health concerns for consumers. As a result, enrichment with different fibers, proteins, antioxidants, minerals, vitamins, and probiotics is a suitable solution to improve the nutritional value of these products (Figure 2).

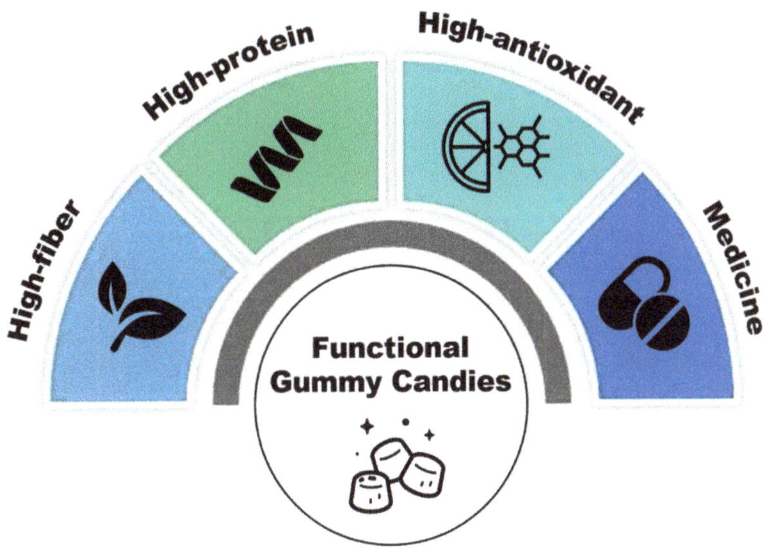

Figure 2. A schematic of GC development as a functional food.

3.1. High-Fiber GCs

Dietary fibers (DFs) include non-digestible forms of carbohydrates, originating from plant-based foods. They can also be categorized into soluble and insoluble fibers according to their solubility in water. The high nutritional value of DF has been well documented over the years and it is widely accepted that DF intake is associated with several health-promoting effects, including protection against cardiovascular and cancer diseases, reduction of blood glucose and cholesterol, regulation of intestinal function, and promotion of gut health [56]. Current DF intake recommendations for adults in most European countries and in the United States are between 30 and 35 g/day for men and between 25 and 32 g/day for women. However, according to the European Nutrition and Health Report, DF intake was below the recommended levels in most European countries [57]. Therefore, the incorporation of DFs into food products is essential to meet such recommendations.

Cappa et al. [58] investigated the effects of adding grape skins with different particle sizes (i.e., small, medium, large) on the physicochemical properties of fruit GCs. The enriched GCs contained more than 60 g fiber per kg product weight, which can be claimed as "high in fiber". Also, fiber-fortified GCs showed better antioxidant and textural features that remained stable within processing. Similarly, Ali et al. [24] showed that the DF content of GCs increased from 0.04 g/100 g to 1.02, 0.92, and 0.90 g/100 g with the addition of 25–75% strawberry and red beetroot. In a further study, Hariadi [59] evaluated the effects of carambola starfruit (*Averrhoa bilimbi*) and papaya (*Carica papaya*) fruits on the physicochemical properties of GCs. The results revealed that carambola starfruit contained a higher amount of DF (2.8 g/100 g of fresh fruit) compared to papaya (1.8 g/100 g of fresh fruit). Therefore, by increasing the proportion of carambola starfruit in the GC formulation from 25 to 75%, the DF content of GCs increased from 0.94 to 0.97 g/100 g. These findings introduce novel fruit-based GCs with favorable fiber content.

Non-starch polysaccharides, such as inulin and konjac glucomannan (KGM), are other important DFs that can be utilized in various food products as non-digestible and health-promoting ingredients. Inulin is composed of fructose units linked by β-(2-1)-D-fructosyl-fructose bonds, improving techno-functional and nutritional properties of foodstuffs [60]. In this regard, Delgado and Bañón [61] investigated the effect of substitution of starch with inulin on the physicochemical, texture, and sensory properties of GCs. Inulin-containing GCs provided a slightly softer, springier, and stickier texture. In addition, the incorporation of inulin can not only improve the technological and sensorial properties of GCs, but also increase the DF of samples with potential prebiotic activity. Similar results were also presented by Cedeño-Pinos et al. [62] with the addition of β-fructan fibers (e.g., chicory inulin and fructooligosaccharides) into the GC formulations as alternative ingredients to starches and sugars. In another study, Fernandes et al. [63] reported that the waist circumference and hunger/appetite of forty-two overweight individuals decreased after the consumption of KGM-enriched GCs for fourteen days. These results indicate that the incorporation of KGM into the GC manufacturing process can significantly prevent the development of cardiovascular diseases in overweight and obese individuals.

3.2. High-Protein GCs

Proteins are important ingredients in the human diet due to their essential role in immune responses, repairing damaged cells, and maintaining muscle mass [64]. However, according to previous studies, more than 10% of people suffer from malnutrition and protein-energy undernutrition (PEU), especially in the developing countries of Asia and Africa. It is estimated that 149.6 million children under the age of five are malnourished, which causes about 50% of child mortality globally [65,66]. Therefore, the development of high-protein food formulations is necessary to increase the protein intake of consumers worldwide.

In this respect, some researchers evaluated the possible application of legume proteins into GCs. Siegwein et al. [67] evaluated the textural, rheological, and sensory properties of starch-based confections enriched with 33, 50, and 66% soy protein isolate (SPI). According to the results, SPI could reduce the hardness, cohesiveness, and gumminess values of GCs by the disruption of starch gel network. In addition, the dynamic oscillatory test showed that with increasing SPI concentration, the viscoelastic parameters (G′ and G″) of the starch network decreased. SPI could also prevent hardening of samples during 20 days of storage. In another study, Bartkiene et al. [68] prepared functional GCs formulated with lacto-fermented Vilciai and Vilniai lupine protein concentrates (LPCs). The *Lactobacillus sakei*-fermented LPCs from Vilniai variety showed the highest protein content (90.11%) and protein digestibility (89.94%), but the lowest trypsin inhibitor activity (19.40%). Furthermore, the incorporation of these proteins into the GC formulation up to 13.0%, not only affects the amino acid profile and genistein content of the samples, but also improves their texture and overall acceptability. In a different study, the possible application of 1, 3, and 5% *Spirulina* biomass as an innovative ingredient in the development of GCs was investigated by Paternina et al. [69]. They reported that the protein and mineral contents of *Spirulina*-enriched GCs increased by 36.7 and 414.3%, respectively, compared to the control sample. In addition, these functional GCs represented relatively high phenolic compounds (0.081 mg GAE/g), antioxidant activity (11.4%), and overall acceptability (80.0%).

Overall, the addition of high-protein ingredients into GC formulations can be a promising way to obtain functional and nutraceutical confectionary products. However, there is still limited information available in this area and it needs to be given much more attention in the future work of food researchers.

3.3. *High-Antioxidant GCs*

Antioxidants are a group of natural or man-made substances that can prevent or slow damage to cells caused by free radicals, which are unstable molecules that the body produces as a reaction to environmental and other pressures. These substances can be found in many foods, including fruits and vegetables, and are also available as dietary supplements, such as vitamins C and E, selenium, and carotenoids. Additionally, it has been observed that numerous polyphenolic compounds present in herbal extracts possess notable antioxidant activities [70–72].

The incorporation of antioxidant substances in GC products has gained a lot of attention in recent years. For instance, Cedeño-Pinos et al. [62] conducted a study to assess the incorporation of rosemary extract into stevia-based GCs, aiming to enhance their antioxidant properties. The researchers found that adding 0.26 g of rosemary extract per kg of GC resulted in a significant increase in polyphenol content from 197 to 411 µg GAE/g and antioxidant capacity from 1.77 to 4.14 µmol Trolox/g. Importantly, this addition did not negatively impact the consumer acceptance of enriched GCs. Similarly, Charoen et al. [73] developed high-antioxidant GCs by utilizing *Psidium guajava* L. leaf extracts obtained at a drying temperature of 50 °C and a pH of 4.0. In another study by Archaina et al. [74], freeze-dried GCs using blackcurrant and yogurt were developed. The results showed that the antioxidant activity of both GC formulations was similar to kiwi and apple fruits (ranging from 4.2 to 4.5 mmol Trolox/kg), demonstrating the possible application of these components in the supplementation of GCs. In a further study, Ali et al. [24] evaluated the antioxidant properties of functional GCs incorporating fresh fruits, such as strawberries and red beetroots. GCs formulated with 75% strawberry and 25% beetroot exhibited the highest levels of bioactive compounds and antioxidant activity (52.55%) with acceptable sensory attributes. In addition, Altamash et al. [75] investigated the possible preparation of GCs using a blend of pineapple and beetroot juice. The GC prepared with a pineapple/beetroot ratio of 90:10 exhibited the best antioxidant and organoleptic qualities, indicating its favorable nutritional and sensory characteristics. These results are in line with the findings of A Abd EL Latif et al. [76], who developed GCs using natural plant extracts from lemons, red roselles, tamarind, peppermint, and dates.

Furthermore, Gramza-Michalowska and Regula [77] utilized different concentrations of tea extracts (*Camelia sinensis*) in GCs as a source of polyphenolic compounds. The researchers observed that the polyphenol content and antioxidant activity of GCs increased with the addition of 1 and 1.5% tea extract, ranging from 245.9 to 1256.5 and 3.2 to 170.1 mg/100g, respectively. In another study, Suman et al. [78] showed a similar trend for GCs enriched with 1–6% ginger powder, 0.02–0.14% ginger oleoresin, and 5–30% ginger juice. In addition, Sarabandi and Mohammadi [79] evaluated the possible use of peppermint extracts in GCs and assessed their physical properties, total phenolic content (TPC), and antioxidant properties. The GC containing 10% peppermint extract exhibited the highest antioxidant capacity in various assays, making it a suitable antioxidant agent in the formulation of functional GCs with desirable texture, antioxidant, and sensory properties. In a further study, Ghendov-Mosanu et al. [80] substituted synthetic dyes with chokeberry extract in GCs. The researchers investigated the impact of different concentrations (i.e., 1.0, 1.5, and 2.0%) of chokeberry extract on the antioxidant and sensory properties of the candies. The results revealed that the extract improved the antioxidant capacity and sensory attributes of GCs at all studied concentrations, especially 1.5%. Therefore, the incorporation of natural antioxidants into the GC production can not only effectively improve the nutraceutical benefits of the final products, but also increase the buying intention of consumers.

3.4. GCs As a Medicine

In ancient times, it was mistakenly believed that the worse a medicine tastes, the better its therapeutic efficacy. Therefore, improving the taste of medicinal products was not considered by manufacturers. Today, the new perception of pharmaceutical technology is that the taste of an oral formulation should be sufficiently pleasant, especially when it is intended for pediatric use [81]. For this reason, several drug delivery systems have been developed in the past few decades. One of the most cost-effective and safe techniques is to incorporate drugs and bioactive compounds into GCs, which are gaining much more attention these days due to their favorable taste and structure, especially for children [82]. However, it is necessary to reformulate GCs according to the organoleptic and physicochemical properties of the active pharmaceutical ingredients, as well as the target consumers. In this regard, Karaiskou et al. [81] evaluated the possible incorporation of metoclopramide hydrochloride into the GCs containing pomegranate juice. According to the results, this technique met the pharmaceutical recovery requirements and can be used for routine consumption. Also, the stability of metoclopramide in the GCs was tested after 6 months of storage at 15 °C with satisfactory results. Such functional GCs can not only be considered for use by children, but also for adults who are facing various diseases such as mental illnesses. In another study, Hosseini et al. [83] successfully designed anti-fever and analgesic GCs based on gelatin and starch with nano-sized acetaminophen. Cumulative drug release of nanoparticles in GCs was approximately 74% after one day. These types of GCs can be the next generation dosage forms of acetaminophen due to their ease of consumption and greater compatibility among children than conventional oral tablets. However, more studies are needed to confirm present findings.

GCs can also be used for delivery of minerals and vitamins. In this respect, Handayani et al. [84] developed chitosan-based GCs loaded with ferrous gluconate (FeG) microparticles in order to increase the daily intake of iron. The authors demonstrated that the incorporation of FeG not only influenced the nutritional properties of GCs, but also affected their structural and textural characteristics during the storage period, in terms of shape, size, hardness and gumminess. In addition, Constantino and Garcia-Rojas [85] evaluated the functional and physicochemical properties of GCs enriched with beta-carotene (β-C) using amaranth carboxymethyl starch (CMS) and lactoferrin (LF). The results indicated that β-C was microencapsulated in CMS/LF complex coacervates up to 98% with favorable thermal and photolytic stabilities. Moreover, the encapsulation of β-C into GCs significantly protected about 66% of the β-C during processing, where temperatures reached over 100 °C. β-C-containing GCs also showed acceptable bioaccessibility (22%) and better textural properties (e.g., lower hardness). Therefore, the fortification of GCs with β-C microcapsules in CMS/LF complex coacervates can be considered as an effective way to overcome vitamin A deficiency worldwide.

Development of GCs with probiotics and prebiotics is another emerging trend in today's confectionery industry. For instance, Lele et al. [86] prepared functional GCs based on agar, gelatin, and apple pomace (as a pectin source) with the incorporation of probiotics (e.g., *Lactobacillus paracasei* LUHS244 and *Lactobacillus plantarum* LUHS135) and prebiotics (e.g., psyllium husk fibers). The antioxidant activities and textural properties of samples were strongly dependent on the strain of LAB and the use of psyllium husk fiber. Also, the enriched-GCs showed excellent antimicrobial properties against a wide range of pathogenic strains including *Escherichia coli, Streptococcus mutans, Streptococcus aureus, Enterococcus faecalis, Salmonella enterica, Proteus mirabilis,* and *Pseudomonas aeruginosa*. These results are in line with the findings of Bartkiene et al. [87], who developed probiotic GCs with the addition of bovine colostrum and various essential oils. The GCs held antimicrobial activities against different pathogenic bacteria such as *Escherichia coli*, *Streptococcus mutans*, *Enterococcus faecalis*, *Salmonella enterica*, *Proteus mirabilis*, *Pseudomonas aeruginosa*, and *Staphylococcus aureus*. They also reported that GCs consisting of 3% fermented bovine colostrum, 0.2% *Thymus vulgaris* L., *Citrus reticulate* L., and *Citrus paradise* L. essential oils represented the desired textural and sensorial properties, revealing the possible production of value-added GCs on an industrial scale. In a further study, Miranda et al. [88] evaluated the functional, physicochemical, and antioxidant properties of novel GCs reformulated with native fruits of the Atlantic Forest (i.e., palm Juçara (*Euterpe edulis* Martius) and passion fruit (*Passiflora edulis*)) and *Bacillus coagulans*. These GCs meet the probiotic survival requirements of laboratory testing recommended by Food and Agriculture Organization (FAO), as populations higher than 6.82 Log CFU/g reached the second viable intestinal phase, indicating the potential of the carrier matrix. These studies demonstrate the potential of probiotic and/or prebiotic GCs in the confectionary market due to their excellent biological and health-promoting benefits.

4. A Summary of Advantages and Disadvantages of Novel Functional GCs

Despite the techno-functional and health benefits of novel functional GCs, the incorporation of unconventional ingredients into formulations may cause some disadvantages, which are summarized in Table 1. However, there is still limited information available in most areas and more attention should be paid in the future.

Table 1. Development of functional GCs with different strategies.

Novel GCs	Functional Ingredients	Advantages	Disadvantages
Gelatin-free	Pectin, guar gum, agar gum, carrageenan, carboxy methyl cellulose (CMC), alginate, corn starch, and cassava starch	□ Suitable for vegan and vegetarian diets □ Can be labeled as halal and kosher-certified products	■ It is more difficult to control environmental factors such as pH compared to gelatin-based GCs ■ Different hydrocolloids, such as pectin and agar gum, show a synergistic effect on the gel network ■ Some alternative gelling agents require more sugar compared to gelatin to achieve the desired flavor
Sugar-free	Stevia, corn syrup, erythritol, xylitol, honey, isomaltulose, maltitol, isomaltitol, and natural sugar sources (e.g., grape, mulberry, and carob molasses)	□ A smaller amount may be needed to achieve the same level of sweetness compared to sugar □ Suitable for anti-diabetic and low-calorie diets	■ Some alternative sweeteners may have a different mouthfeel or aftertaste than sugar ■ Some alternative sweeteners may cause glucose intolerance
Artificial additives-free	- Natural fruit flavors (e.g., strawberry, raspberry, orange, lemon, pineapple, etc.) - Natural fruit colorants (e.g., blueberry, red beetroot, grape, etc.) - Anthocyanin, betanin, etc.	□ Natural additives can enhance the overall flavor and visual appeal of GCs □ Fruit flavors/colorants can also offer potential health benefits through their antioxidant properties	■ Some natural colorants such as betanin are sensitive to heat and pH, and their color may deteriorate over time ■ Natural food flavors and colorants may negatively affect the texture of GCs

Table 1. *Cont.*

Novel GCs	Functional Ingredients	Advantages	Disadvantages
High-fiber	- Fibers derived from fruits (e.g., grape, carambola starfruit, and papaya) - Non-starch polysaccharides (e.g., inulin and konjac glucomannan (KGM))	□ Higher dietary fiber content with potential prebiotic activity □ Better antioxidant and textural features that stay stable within processing	▪ Different dietary fibers may negatively affect the techno-functional properties of GCs
High-protein	Soy protein isolate (SPI), lupine protein concentrate (LPC), and *Spirulina* biomass	□ Better storage stability and overall acceptability □ Improved amino acid profile and genistein content	▪ There is limited information available in this area and it needs more attention in the future
High-antioxidant	Natural plant extracts (e.g., rosemary, *Psidium guajava* L. leaf, tea, peppermint, chokeberry, etc.), fresh fruits (e.g., strawberries and red beetroots), pineapple and beetroot juice, and ginger powder	□ Higher antioxidant activity and polyphenol content □ Natural antioxidants can not only effectively improve the nutraceutical benefits of the final products, but also increase the buying intention of consumers.	▪ There is limited information available in this area and it needs more attention in the future
Medicine	Metoclopramide hydrochloride, acetaminophen, ferrous gluconate (FeG), beta-carotene (β-C) probiotics (e.g., *Lactobacillus paracasei* LUHS244 and *Lactobacillus plantarum* LUHS135), prebiotics (e.g., psyllium husk fibers), and essential oils (e.g., *Thymus vulgaris* L., *Citrus reticulate* L., and *Citrus paradise* L.)	□ Pleasant taste as a drug carrier □ Greater compatibility among children than conventional oral tablets □ Better antimicrobial properties against a wide range of pathogenic strains	▪ There is still limited information available in this area and it needs more attention in the future

5. Conclusions

The popularity of gummy candies (GCs) by children makes it worthwhile to develop more nutritious and health promoting products. Inclusion of healthier alternative ingredients such as natural flavors and colorants, natural and low-calorie sweeteners, plant and algal proteins, and dietary fibers into GCs is a promising approach to improve the low nutritional profile of this confectionary product. Moreover, GCs have potential utility as a delivery system for antioxidant compounds, plant extracts, and some types of medicines. The application of alternative components in GCs may contribute to their texture, color, flavor, sweetness level, and microbiological quality. These parameters should be considered when formulating functional GCs to fulfill consumer demands and expectations. Efforts should be made to develop GCs with superior nutritional and sensory properties.

Funding: This research received no external funding.

Conflicts of Interest: The authors declare no conflicts of interest.

References

1. Khan, R.S.; Grigor, J.; Winger, R.; Win, A. Functional food product development—Opportunities and challenges for food manufacturers. *Trends Food Sci. Technol.* **2013**, *30*, 27–37. [CrossRef]
2. Siro, I.; Kápolna, E.; Kápolna, B.; Lugasi, A. Functional food. Product development, marketing and consumer acceptance—A review. *Appetite* **2008**, *51*, 456–467. [CrossRef]

3. Baker, M.T.; Lu, P.; Parrella, J.A.; Leggette, H.R. Consumer acceptance toward functional foods: A scoping review. *Int. J. Environ. Res. Public Health* **2022**, *19*, 1217. [CrossRef] [PubMed]
4. GrandViewResearch. Confectionery Market Size, Share & Trends Analysis Report by Product (Chocolate, Sugar Confectionery, Cookies, Ice Cream), by Distribution Channel (Offline, Online), by Region, And Segment Forecasts, 2022–2028. 2022. Available online: https://www.grandviewresearch.com/industry-analysis/confectionery-market# (accessed on 16 May 2022).
5. Gunes, R.; Palabiyik, I.; Konar, N.; Toker, O.S. Soft confectionery products: Quality parameters, interactions with processing and ingredients. *Food Chem.* **2022**, *385*, 132735. [CrossRef] [PubMed]
6. GrandViewResearch. Gummy Market Size, Share & Trends Analysis Report by Product (Vitamins, Minerals, Dietary Fibers), by Ingredient (Gelatin, Plant-Based Gelatin Substitutes), by End-Use (Adults, Kids), by Distribution Channel, by Region, And Segment Forecasts, 2023–2030. 2022. Available online: https://www.grandviewresearch.com/industry-analysis/gummy-market-report# (accessed on 17 May 2022).
7. Moghaddas Kia, E.; Ghaderzadeh, S.L.; Langroodi, A.M.; Ghasempour, Z.; Ehsani, A. Red beet extract usage in gelatin/gellan based gummy candy formulation introducing *Salix aegyptiaca* distillate as a flavouring agent. *J. Food Sci. Technol.* **2020**, *57*, 3355–3362. [CrossRef] [PubMed]
8. Delgado, P.; Bañón, S. Determining the minimum drying time of gummy confections based on their mechanical properties. *CyTA J. Food* **2015**, *13*, 329–335. [CrossRef]
9. Tarahi, M.; Fakhr-Davood, M.M.; Ghaedrahmati, S.; Roshanak, S.; Shahidi, F. Physicochemical and Sensory Properties of Vegan Gummy Candies Enriched with High-Fiber Jaban Watermelon Exocarp Powder. *Foods* **2023**, *12*, 1478. [CrossRef]
10. Periche, A.; Heredia, A.; Escriche, I.; Andrés, A.; Castelló, M. Optical, mechanical and sensory properties of based-isomaltulose gummy confections. *Food Biosci.* **2014**, *7*, 37–44. [CrossRef]
11. Gan, D.; Xu, M.; Chen, L.; Cui, S.; Deng, C.; Qiao, Q.; Guan, R.; Zhong, F. Intake of Sugar Substitute Gummy Candies Benefits the Glycemic Response in Healthy Adults: A Prospective Crossover Clinical Trial. *Gels* **2022**, *8*, 642. [CrossRef]
12. Mandura, A.; Šeremet, D.; Ščetar, M.; Cebin, A.V.; Belščak-Cvitanović, A.; Komes, D. Physico-chemical, bioactive, and sensory assessment of white tea-based candies during 4-months storage. *J. Food Process. Preserv.* **2020**, *44*, e14628. [CrossRef]
13. Pegg, A. The application of natural hydrocolloids to foods and beverages. In *Natural Food Additives, Ingredients and Flavourings*; Elsevier: Amsterdam, The Netherlands, 2012; pp. 175–196.
14. DeMars, L.L.; Ziegler, G.R. Texture and structure of gelatin/pectin-based gummy confections. *Food Hydrocoll.* **2001**, *15*, 643–653. [CrossRef]
15. Renaldi, G.; Junsara, K.; Jannu, T.; Sirinupong, N.; Samakradhamrongthai, R.S. Physicochemical, textural, and sensory qualities of pectin/gelatin gummy jelly incorporated with *Garcinia atroviridis* and its consumer acceptability. *Int. J. Gastron. Food Sci.* **2022**, *28*, 100505. [CrossRef]
16. Tahmouzi, S.; Meftahizadeh, H.; Eyshi, S.; Mahmoudzadeh, A.; Alizadeh, B.; Mollakhalili-Meybodi, N.; Hatami, M. Application of guar (*Cyamopsis tetragonoloba* L.) gum in food technologies: A review of properties and mechanisms of action. *Food Sci. Nutr.* **2023**, *11*, 4869–4897. [CrossRef] [PubMed]
17. Zhang, H.; Zhang, F.; Yuan, R. Applications of natural polymer-based hydrogels in the food industry. In *Hydrogels Based on Natural Polymers*; Elsevier: Amsterdam, The Netherlands, 2020; pp. 357–410.
18. Xu, X.; Ye, S.; Zuo, X.; Fang, S. Impact of Guar Gum and Locust Bean Gum Addition on the Pasting, Rheological Properties, and Freeze–Thaw Stability of Rice Starch Gel. *Foods* **2022**, *11*, 2508. [CrossRef] [PubMed]
19. Dinesh Kumar, R.; Sudhakar, V.; Sairagul, G.; Manoj, J.J.B. Studies on the Consistency of Jaggery-Based Hard-Boiled Candy by Incorporating Thickening and Gelling Agents. *Sugar Tech* **2022**, *24*, 1617–1623. [CrossRef]
20. Matos, Â.P.; Novelli, E.; Tribuzi, G. Algae as food and ingredient: From production to consumer acceptance. *Front. Food Sci. Technol.* **2023**, *3*, 1220050. [CrossRef]
21. Ge, H.; Wu, Y.; Woshnak, L.L.; Mitmesser, S.H. Effects of hydrocolloids, acids and nutrients on gelatin network in gummies. *Food Hydrocoll.* **2021**, *113*, 106549. [CrossRef]
22. Song, X.; Chiou, B.S.; Xia, Y.; Chen, M.; Liu, F.; Zhong, F. The improvement of texture properties and storage stability for kappa carrageenan in developing vegan gummy candies. *J. Sci. Food Agric.* **2022**, *102*, 3693–3702. [CrossRef]
23. Hamka, N. Pengaruh Sifat Kimia dan Organoleptik Permen Jelly Buah Naga (*Hylocereus polyrhizus*) dengan Penambahan Karagenan Sebagai Gelling Agent. *Bul. Loupe* **2020**, *16*, 8–13.
24. Ali, M.R.; Mohamed, R.M.; Abedelmaksoud, T.G. Functional strawberry and red beetroot jelly candies rich in fibers and phenolic compounds. *Food Syst.* **2021**, *4*, 82–88. [CrossRef]
25. de Avelar, M.H.M.; Efraim, P. Alginate/pectin cold-set gelation as a potential sustainable method for jelly candy production. *LWT* **2020**, *123*, 109119. [CrossRef]
26. Tarahi, M.; Shahidi, F.; Hedayati, S. A Novel starch from bitter vetch (*Vicia ervilia*) seeds: A comparison of its physicochemical, structural, thermal, rheological and pasting properties with conventional starches. *Int. J. Food Sci. Technol.* **2022**, *57*, 6833–6842. [CrossRef]
27. Tarahi, M.; Shahidi, F.; Hedayati, S. Physicochemical, pasting, and thermal properties of native corn starch–mung bean protein isolate composites. *Gels* **2022**, *8*, 693. [CrossRef] [PubMed]
28. Marfil, P.H.; Anhê, A.C.; Telis, V.R. Texture and microstructure of gelatin/corn starch-based gummy confections. *Food Biophys.* **2012**, *7*, 236–243. [CrossRef]

29. Pereira, D.G.; de Toledo Benassi, M.; Beleia, A.D.P. Gummy candies produced with acid-thinned *Cassava* starch: Physical and sensory evaluation. *J. Food Process. Preserv.* **2022**, *46*, e16661. [CrossRef]
30. Peteliuk, V.; Rybchuk, L.; Bayliak, M.; Storey, K.B.; Lushchak, O. Natural sweetener *Stevia rebaudiana*: Functionalities, health benefits and potential risks. *EXCLI J.* **2021**, *20*, 1412. [PubMed]
31. Li, Z.; An, L.; Zhang, S.; Shi, Z.; Bao, J.; Tuerhong, M.; Abudukeremu, M.; Xu, J.; Guo, Y. Structural elucidation and immunomodulatory evaluation of a polysaccharide from *Stevia rebaudiana* leaves. *Food Chem.* **2021**, *364*, 130310. [CrossRef]
32. Rivero, R.; Archaina, D.; Sosa, N.; Schebor, C. Sensory characterization, acceptance, and stability studies on low calories fruit jelly candies. *J. Food Sci. Technol.* **2023**, *60*, 2204–2212. [CrossRef]
33. Aranda-Gonzalez, I.; Tamayo-Dzul, O.; Barbosa-Martin, E.; Segura-Campos, M.; Moguel-Ordonez, Y.; Betancur-Ancona, D. Development of a gummy candy reduced in calories by sugar substitution with *Stevia rebaudiana* B. *Nutr. Hosp.* **2014**, *31*, 334–340.
34. Samakradhamrongthai, R.S.; Jannu, T. Effect of stevia, xylitol, and corn syrup in the development of velvet tamarind (*Dialium indum* L.) chewy candy. *Food Chem.* **2021**, *352*, 129353. [CrossRef]
35. Le, H.; Wang, X.; Wei, Y.; Zhao, Y.; Zhang, J.; Zhang, L. Making Polyol Gummies by 3D Printing: Effect of Polyols on 3D Printing Characteristics. *Foods* **2022**, *11*, 874. [CrossRef] [PubMed]
36. Wölnerhanssen, B.K.; Meyer-Gerspach, A.C.; Beglinger, C.; Islam, M.S. Metabolic effects of the natural sweeteners xylitol and erythritol: A comprehensive review. *Crit. Rev. Food Sci. Nutr.* **2020**, *60*, 1986–1998. [CrossRef] [PubMed]
37. Gasmi Benahmed, A.; Gasmi, A.; Arshad, M.; Shanaida, M.; Lysiuk, R.; Peana, M.; Pshyk-Titko, I.; Adamiv, S.; Shanaida, Y.; Bjørklund, G. Health benefits of xylitol. *Appl. Microbiol. Biotechnol.* **2020**, *104*, 7225–7237. [CrossRef] [PubMed]
38. Čižauskaitė, U.; Jakubaitytė, G.; Žitkevičius, V.; Kasparavičienė, G. Natural ingredients-based gummy bear composition designed according to texture analysis and sensory evaluation in vivo. *Molecules* **2019**, *24*, 1442. [CrossRef] [PubMed]
39. Ahuja, V.; Macho, M.; Ewe, D.; Singh, M.; Saha, S.; Saurav, K. Biological and pharmacological potential of xylitol: A molecular insight of unique metabolism. *Foods* **2020**, *9*, 1592. [CrossRef] [PubMed]
40. Ranneh, Y.; Akim, A.M.; Hamid, H.A.; Khazaai, H.; Fadel, A.; Zakaria, Z.A.; Albujja, M.; Bakar, M.F.A. Honey and its nutritional and anti-inflammatory value. *BMC Complement. Med. Ther.* **2021**, *21*, 1–17. [CrossRef] [PubMed]
41. Rivero, R.; Archaina, D.; Sosa, N.; Leiva, G.; Coronel, B.B.; Schebor, C. Development of healthy gummy jellies containing honey and propolis. *J. Sci. Food Agric.* **2020**, *100*, 1030–1037. [CrossRef] [PubMed]
42. Iqbal, A.; Kamran, H.; Khalid, S.; Jabeen, S.; Aslam, M. Glycemic response of natural sweeteners like sugarcane juice, honey and jaggery in healthy individuals. *EAS J. Humanit. Cult. Stud.* **2020**, *2*, 279–281. [CrossRef]
43. Gok, S.; Toker, O.S.; Palabiyik, I.; Konar, N. Usage possibility of mannitol and soluble wheat fiber in low calorie gummy candies. *LWT* **2020**, *128*, 109531. [CrossRef]
44. Jiamjariyatam, R. Influence of gelatin and isomaltulose on gummy jelly properties. *Int. Food Res. J.* **2018**, *25*, 776–783.
45. Ünal, M.H.; Arslan, D. Single and combined use of isomalt, polydextrose, and inulin as sugar substitutes in production of pectin jelly. *J. Food Process. Preserv.* **2022**, *46*, e17174. [CrossRef]
46. Kurt, A.; Bursa, K.; Toker, O.S. Gummy candies production with natural sugar source: Effect of molasses types and gelatin ratios. *Food Sci. Technol. Int.* **2022**, *28*, 118–127. [CrossRef] [PubMed]
47. Suez, J.; Korem, T.; Zeevi, D.; Zilberman-Schapira, G.; Thaiss, C.A.; Maza, O.; Israeli, D.; Zmora, N.; Gilad, S.; Weinberger, A. Artificial sweeteners induce glucose intolerance by altering the gut microbiota. *Nature* **2014**, *514*, 181–186. [CrossRef] [PubMed]
48. Otálora, M.C.; de Jesús Barbosa, H.; Perilla, J.E.; Osorio, C.; Nazareno, M.A. Encapsulated betalains (*Opuntia ficus-indica*) as natural colorants. Case study: Gummy candies. *LWT* **2019**, *103*, 222–227. [CrossRef]
49. Bagautdinov, I.; Gusev, A.; Nigmatzyanov, A.; Chernenkov, E.; Kaluzhina, O. Development of Confectionery Products with Functional Properties Using Non-Traditional Plant Raw Materials. *J. Culin. Sci. Technol.* **2023**, 1–20. [CrossRef]
50. Albuquerque, B.R.; Oliveira, M.B.P.; Barros, L.; Ferreira, I.C. Could fruits be a reliable source of food colorants? Pros and cons of these natural additives. *Crit. Rev. Food Sci. Nutr.* **2021**, *61*, 805–835. [CrossRef]
51. Lonez, H.; Banwa, T. Butterfly Pea (*Clitoria ternatea*): A natural colorant for soft candy (Gummy Candy). *Indian. J. Sci. Technol.* **2021**, *14*, 239–244. [CrossRef]
52. Casas-Forero, N.; Trujillo-Mayol, I.; Zúñiga, R.N.; Petzold, G.; Orellana-Palma, P. Effects of cryoconcentrated blueberry juice as functional ingredient for preparation of commercial confectionary hydrogels. *Gels* **2022**, *8*, 217. [CrossRef]
53. de Oliveira Nishiyama-Hortense, Y.P.; de Paula Rossi, M.J.; Shimizu-Marin, V.D.; Janzantti, N.S.; Gómez-Alonso, S.; Da-Silva, R.; Lago-Vanzela, E.S. Jelly candy enriched with BRS Violeta grape juice: Anthocyanin retention and sensory evaluation. *Fut. Foods* **2022**, *6*, 100179. [CrossRef]
54. Amjadi, S.; Ghorbani, M.; Hamishehkar, H.; Roufegarinejad, L. Improvement in the stability of betanin by liposomal nanocarriers: Its application in gummy candy as a food model. *Food Chem.* **2018**, *256*, 156–162. [CrossRef]
55. Ramesh, M.; Muthuraman, A. Flavoring and coloring agents: Health risks and potential problems. In *Natural and Artificial Flavoring Agents and Food Dyes*; Elsevier: Amsterdam, The Netherlands, 2018; pp. 1–28.
56. Barber, T.M.; Kabisch, S.; Pfeiffer, A.F.; Weickert, M.O. The health benefits of dietary fibre. *Nutrients* **2020**, *12*, 3209. [CrossRef] [PubMed]
57. Ktenioudaki, A.; Gallagher, E. Recent advances in the development of high-fibre baked products. *Trends Food Sci. Technol.* **2012**, *28*, 4–14. [CrossRef]

58. Cappa, C.; Lavelli, V.; Mariotti, M. Fruit candies enriched with grape skin powders: Physicochemical properties. *LWT-Food Sci. Technol.* **2015**, *62*, 569–575. [CrossRef]
59. Hariadi, H. The influence of carambola starfruit (*Averrhoa bilimbi*) and Papaya (*Carica papaya*) on the quality of the organoleptic properties, vitamin C content, and fiber at jelly candies. In *IOP Conference Series: Earth and Environmental Science*; IOP Publishing: Bristol, UK, 2020.
60. Mohammadi, F.; Shiri, A.; Tahmouzi, S.; Mollakhalili-Meybodi, N.; Nematollahi, A. Application of inulin in bread: A review of technological properties and factors affecting its stability. *Food Sci. Nutr.* **2023**, *11*, 639–650. [CrossRef] [PubMed]
61. Delgado, P.; Bañón, S. Effects of replacing starch by inulin on the physicochemical, texture and sensory characteristics of gummy jellies. *CyTA J. Food* **2018**, *16*, 1–10. [CrossRef]
62. Cedeño-Pinos, C.; Martínez-Tomé, M.; Murcia, M.A.; Jordán, M.J.; Bañón, S. Assessment of rosemary (*Rosmarinus officinalis* L.) extract as antioxidant in jelly candies made with fructan fibres and stevia. *Antioxidants* **2020**, *9*, 1289. [CrossRef] [PubMed]
63. Fernandes, A.C.S.; Muxfeldt, L.; Motta, N.G.; Skonieski, C.; Fagundes, K.R.; Sandri, G.; de Chaves, D.B.; Suthovski, G.; Gallina, A.L.; Araujo, S.M. Gummies candy enriched with Konjac glucomannan reduces hunger intensity and waist circumference of overweight individuals. *Int. J. Biol. Macromol.* **2023**, *226*, 72–76. [CrossRef]
64. Tarahi, M.; Hedayati, S.; Shahidi, F. Effects of mung bean (*Vigna radiata*) protein isolate on rheological, textural, and structural properties of native corn starch. *Polymers* **2022**, *14*, 3012. [CrossRef]
65. Poniedziałek, B.; Perkowska, K.; Rzymski, P. Food Fortification: What's in It for the Malnourished World? In *Vitamins and Minerals Biofortification of Edible Plants*; John Wiley & Sons Ltd.: Hoboken, NJ, USA, 2020; pp. 27–44.
66. Tarahi, M.; Ahmed, J. Recent advances in legume protein-based colloidal systems. *Legume Sci.* **2023**, e185. [CrossRef]
67. Siegwein, A.M.; Vodovotz, Y.; Fisher, E.L. Concentration of soy protein isolate affects starch-based confections' texture, sensory, and storage properties. *J. Food Sci.* **2011**, *76*, E422–E428. [CrossRef]
68. Bartkiene, E.; Sakiene, V.; Bartkevics, V.; Wiacek, C.; Rusko, J.; Lele, V.; Ruzauskas, M.; Juodeikiene, G.; Klupsaite, D.; Bernatoniene, J. Nutraceuticals in gummy candies form prepared from lacto-fermented lupine protein concentrates, as high-quality protein source, incorporated with *Citrus paradise* L. essential oil and xylitol. *Int. J. Food Sci. Technol.* **2018**, *53*, 2015–2025. [CrossRef]
69. Paternina, L.P.R.; Moraes, L.; Santos, T.D.; de Morais, M.G.; Costa, J.A.V. Spirulina and açai as innovative ingredients in the development of gummy candies. *J. Food Process. Preserv.* **2022**, *46*, e17261. [CrossRef]
70. Carocho, M.; Ferreira, I.C. A review on antioxidants, prooxidants and related controversy: Natural and synthetic compounds, screening and analysis methodologies and future perspectives. *Food Chem. Toxicol.* **2013**, *51*, 15–25. [CrossRef] [PubMed]
71. Rasheed, A.; Azeez, R.F.A. A review on natural antioxidants. In *Traditional and Complementary Medicine*; Intech Open: London, UK, 2019; pp. 1–24.
72. Pokorný, J.; Schmidt, S. Natural antioxidant functionality during food processing. In *Antioxidants in Food: Practical Applications*; Woodhead Publishing: Cambridge, UK, 2001; pp. 331–354.
73. Charoen, R.; Savedboworn, W.; Phuditcharnchnakun, S.; Khuntaweetap, T. Development of antioxidant gummy jelly candy supplemented with *Psidium guajava* leaf extract. *Appl. Sci. Eng. Prog.* **2015**, *8*, 145–151. [CrossRef]
74. Archaina, D.; Sosa, N.; Rivero, R.; Schebor, C. Freeze-dried candies from blackcurrant (*Ribes nigrum* L.) and yoghurt. Physicochemical and sensorial characterization. *LWT* **2019**, *100*, 444–449. [CrossRef]
75. Altamash, A.; Peter, E.S.; Nautiyal, H. Studies on the preparation of gummy candy from a blend of pineapple and beetroot juice. *Pharma Innov. J.* **2022**, *11*, 1708–1713.
76. A Abd EL Latif, M.; A Abd El Aziz, H.; A Kamal El Deen, A. Utilization of some natural plants sources in producing new product (gummy jelly candy). *Int. J. Fam. Stud. Food Sci. Nutr. Health* **2022**, *3*, 40–63. [CrossRef]
77. Gramza-Michalowska, A.; Regula, J. Use of tea extracts (*Camelia sinensis*) in jelly candies as polyphenols sources in human diet. *Asia Pac. J. Clin. Nutr.* **2007**, *16* (Suppl. S1), 43–46.
78. Suman, K.M.; Gupta, A.; Vaidya, D.; Ranjan, K. Standardization of formulation for the preparation of ginger supplemented jelly candies. *Pharma Innov. J.* **2021**, *10*, 608–613.
79. Sarabandi, K.; Mohammadi, A. Stabilization of peppermint polyphenols within crystalline sucrose matrix: Fortification of gummy candy as a food model system. *J. Food Process. Preserv.* **2022**, *46*, e16720. [CrossRef]
80. Ghendov-Mosanu, A.; Cristea, E.; Sturza, R.; Niculaua, M.; Patras, A. Synthetic dye's substitution with chokeberry extract in jelly candies. *J. Food Sci. Technol.* **2020**, *57*, 4383–4394. [CrossRef] [PubMed]
81. Karaiskou, S.G.; Kouskoura, M.G.; Markopoulou, C.K. Modern pediatric formulations of the soft candies in the form of a jelly: Determination of metoclopramide content and dissolution. *Pharm. Dev. Technol.* **2020**, *25*, 20–27. [CrossRef] [PubMed]
82. Crawford, E.B.; Coco, T.; Gaines, L.; Shah, N.; Slattery, A. Pediatric ingestions with gummy formulated medications: A retrospective study. *Clin. Toxicol.* **2021**, *59*, 440–444. [CrossRef] [PubMed]
83. Hosseini, A.; Bagheri, F.; Mohammadi, G.; Azami, M.; Tahvilian, R. Design and preparation of oral jelly candies of acetaminophen and its nanoparticles. *Appl. Nanosci.* **2022**, *12*, 101–107. [CrossRef]
84. Handayani, N.A.; Krisanti, E.; Kartohardjono, S.; Mulia, K. Effect of iron fortification on gummy candies properties: Basic nutrient, microstructure, and texture during the storage period. In *IOP Conference Series: Materials Science and Engineering*; IOP Publishing: Bristol, UK, 2021.
85. Constantino, A.B.T.; Garcia-Rojas, E.E. Microencapsulation of beta-carotene by complex coacervation using amaranth carboxymethyl starch and lactoferrin for application in gummy candies. *Food Hydrocoll.* **2023**, *139*, 108488. [CrossRef]

86. Lele, V.; Ruzauskas, M.; Zavistanaviciute, P.; Laurusiene, R.; Rimene, G.; Kiudulaite, D.; Tomkeviciute, J.; Nemeikstyte, J.; Stankevicius, R.; Bartkiene, E. Development and characterization of the gummy–supplements, enriched with probiotics and prebiotics. *CyTA J. Food* **2018**, *16*, 580–587. [CrossRef]
87. Bartkiene, E.; Ruzauskas, M.; Lele, V.; Zavistanaviciute, P.; Bernatoniene, J.; Jakstas, V.; Ivanauskas, L.; Zadeike, D.; Klupsaite, D.; Viskelis, P. Development of antimicrobial gummy candies with addition of bovine colostrum, essential oils and probiotics. *Int. J. Food Sci. Technol.* **2018**, *53*, 1227–1235. [CrossRef]
88. Miranda, J.S.; BCosta, V.; de Oliveira, I.V.; de Lima, D.C.N.; Martins, E.M.F.; Júnior, B.R.D.C.L.; Benevenuto, W.C.A.D.N.; de Queiroz, I.C.; da Silva, R.R.; Martins, M.L. Probiotic jelly candies enriched with native Atlantic Forest fruits and Bacillus coagulans GBI-30 6086. *LWT* **2020**, *126*, 109275. [CrossRef]

foods

MDPI

Article

Characterization of Băbească Neagră Grape Pomace and Incorporation into Jelly Candy: Evaluation of Phytochemical, Sensory, and Textural Properties

Mariana Spinei [1,2,*] and Mircea Oroian [2]

[1] Integrated Center for Research, Development and Innovation in Advanced Materials, Nanotechnologies, and Distributed Systems for Fabrication and Control (MANSiD), "Ştefan cel Mare" University of Suceava, 13th University Street, 720229 Suceava, Romania

[2] Faculty of Food Engineering, "Ştefan cel Mare" University of Suceava, 13th University Street, 720229 Suceava, Romania; m.oroian@fia.usv.ro

* Correspondence: mariana.spinei@fia.usv.ro

Abstract: The influence of particle size intervals (<125 μm, ≥125–<200 μm, and ≥200–<300 μm) of grape pomace (*Vitis vinifera* var. Băbească Neagră) was analyzed in terms of the proximate composition, functional properties, and physicochemical parameters. The aim was to study the effect of the formulation variables (extract from grape pomace with different particle size intervals and gelatin doses—7, 8.5, and 10 g) on the mechanical properties (hardness, cohesiveness, adhesiveness, springiness, and gumminess), color, and sensorial, microbiological, and phytochemical parameters. The jelly candy formulated with grape pomace extract (<125 μm) and 7 g of gelatin showed the highest total phenolic content (156 mg GAE/g) and antioxidant activity (65.8% inhibition), while grape pomace jellies with a particle size of ≥125–<200 μm and different concentration of gelatin presented the greatest sensory acceptance in terms of sweetness, taste, odor, elasticity, color, and overall acceptability of the resulting jelly. The concluding observation was supported by the microbiological analysis, which also showed that there is no growth in jelly samples except the jelly candies prepared with a ≥200–<300 μm particle size interval of grape pomace extract.

Keywords: grape pomace; fortification; jelly candy; valorization

1. Introduction

The grape has been one of the most planted fruits for a long time, consumed in a variety of ways [1,2], and it is classified as a berry produced by the woody vines of the plant genus *Vitis* [2]. The grapevine, native to Europe and the Mediterranean region, relates to the species *Vitis vinifera*, which consists of approximately 10,000 distinct varieties [3]. Depending on their uses, cultivated grapes can be divided into three different kinds, i.e., wine, table, and dried grapes [2,4]. Table grapes are utilized for the production of jelly, jam, juice, juice concentrate, vinegar, and other products, although grapes can also be dehydrated into raisins [5].

Wine grapes have a significant part in the global economy [6]; in 2022, the world vineyard surface area was estimated to be 7.3 mha [7]. Looking at the European Union (EU) member states, Romania had about 188 kha of vineyard surface in 2022 with 3.9 mhL of the world wine production, excluding musts and juices [7]. The major by-products that are generated after the maceration process are grape pomace, grape seeds, yeast, etc. [1]. Grape pomace or grape marc comprises three different constituents: skins, seeds, and stalks remaining after various stages of wine production, such as crushing, draining, and pressing processes [8]. One of the main destinations of the grape pomace is distillation in order to obtain alcohol [8]. In addition, this by-product is commonly used as fertilizer, bio-fuel, bio-based material, or fodder, but can also be used for the extraction of different compounds, such as phenols, tocopherols, tannins, oil, etc. [1]. Therefore, there is a growing interest in grape pomace as a value-added compound in food products. In recent times, grape pomace has been used as a fortifying agent for a lot of food products, e.g., plant foodstuffs (bread, cookies, extruded cereals, muffins, biscuits, pancakes, noodles, tomato puree, and pasta), meat and fish products (pork sausages, pork burgers, chicken meat, beef frankfurters, minced fish muscles, pork loin marinade, and salmon burgers), and dairy products (cheese, fermented milk, yogurt, and ice cream) [9]. Confectionary products (e.g., nougat, hard and soft jelly candies) are attracting growing interest considering their high organoleptic indicators and low prices in comparison with other types of candy [10]. Jelly candy is consumed in great amounts, especially by children; this consumption can negatively influence health because of coloring agents, contaminants obtained during heat treatment, and artificial flavoring [10].

Romania has a significant range of indigenous varieties (*Vitis vinifera* L.), widely distributed throughout the country, e.g., Fetească Neagră, Băbească Neagră, Busuioacă Neagră, Busuioacă de Bohotin, Fetească Albă, Grasă de Cotnari, Tămâioasă Românească, Galbenă de Odobești, and Riesling de Banat [11]. Băbească Neagră (also Rară Neagră) is a dark-skinned grape variety native to Romania and the Republic of Moldova; the grapes are medium to large, cylindrical–conical with medium-sized and spherical berries which easily accumulate significant amounts of sugar (210–220 g/L) [12]. According to the operational data portal of the Ministry of Agriculture and Rural Development (MARD), Băbească Neagră occupied 2613 ha of the main noble wine grape varieties registered in cultivation in 2020 [13]. Waste minimization, sustainable food production, and valorization of by-products are important to prolong the shelf life of food products, produce new products, improve the nutritional value of products, and make products available out of season [14]. In the scientific literature, there are some studies on the valorization of grape-pomace (Kalecik Karasi cultivar) into soft candies, and their characterization from texture profile analysis, sensory analysis, color, opacity, and rheology testing point of view [15], but the study does not deal with proximate composition, FT-IR analysis, and SEM analysis of the pomace, nor with the phytochemical profile of the finished product. Moreover, the final product is different since soft candies are sweets in which sucrose esters can improve softness and comprise a blend of at least saccharide, vegetable oil, and emulsifier, while jelly candies (hard candies) are a broad category of gelatin-based chewable sweets. To our knowledge, there are no other studies on the valorization of Băbească Neagră grape pomace as a fortifying agent in jelly candies.

Consequently, the aim of this study was to improve knowledge about the fortification process in food products, mainly jellied products, to produce jelly candy, and to evaluate physicochemical (color parameters), phytochemical (total phenolic content and DPPH assay scavenging activity), textural (hardness, cohesiveness, adhesiveness, springiness, and gumminess), sensorial, and microbiological characteristics of the final product.

2. Materials and Methods

2.1. Plant Material and Jelly Candy Preparation

Grape pomace was collected by processing the Băbească Neagră variety (*Vitis vinifera* L.) 2019 harvest. The grape pomace was dried in an air-circulating oven Zhicheng ZRD-A5055 (Zhicheng, Shanghai, China) for 12 h at 50 °C. Then, dried pomace was powdered using a laboratory mill Perten Instruments LM 3310 (PerkinElmer Inc., Waltham, MA, USA) and sieved using a sieve shaker Retsch AS 200 Basic (Retsch GmbH, Haan, Germany) in order to obtain three different particle sizes: <125 μm, ≥125–<200 μm, and ≥200–<300 μm.

The different steps of the by-product-based jelly candy preparation process were:

1. Aqueous extract (*w/v*): 1 g of grape pomace with each particle size was mixed with 30 mL of water; the samples were homogenized with an Ultra-Turrax Homogenizer (Daigger Scientific Inc., Vernon Hills, IL, USA) for 5 min and heated for 1 h at 70 °C in a water bath Precisdig (J.P. Selecta, Barcelona, Spain). At the end of the treatment, the

samples were centrifuged for 10 min at 5000 rpm; the supernatant was filtered and used for jelly candy formulation.

2. Gelatin hydration (*w*/*v*): 7, 8.5, and 10 g of gelatin (to make 100 g of jelly) were hydrated with water for 10 min at 40 °C.
3. Heat treatment and homogenization (*v*/*v*): the final mixture consisted of 12.5 g sweetener (stevia) and 210 mL of grape pomace extract for the products with 7, 8.5, and 10 g of gelatin, respectively (to make 100 g of jelly). These blends were heated for 5 min at 55 °C in a Thermomix device (Vorwerk, Wuppertal, Germany). Finally, the different formulations were established (F1–7, F1–8.5, F1–10, F2–7, F2–8.5, F2–10, F3–7, F3–8.5, and F3–10; F1—grape pomace extract with the particle size of <125 μm; F2—grape pomace extract with the particle size of ≥125–<200 μm; F3—grape pomace extract with the particle size of ≥200–<300 μm; 7—7 g of gelatin; 8.5—8.5 g of gelatin; 10—10 g of gelatin). The molding and maturation stages were performed at 5 °C for 24 h.

The information about the production of grape pomace jelly is presented in Table S1 of the Supplementary Materials.

2.2. Methods

2.2.1. Characterization of Grape Pomace and Grape Pomace Aqueous Extract

Grape pomace of the Băbească Neagră variety (*Vitis vinifera* L.) was analyzed for proximate composition by AOAC methods for moisture content (g/100 g), crude fat (g/100 g), crude protein (g/100 g), ash (g/100 g), total dietary fiber (TDF, g/100 g), soluble dietary fiber (SDF, g/100 g), and insoluble dietary fiber (IDF, g/100 g) [16]. Other carbohydrates were determined according to the difference method by subtracting the other components from 100 (100 − (ash + protein + fat + moisture)).

For a better understanding of jelly candy composition, the phytochemical characteristics (total phenolic content and DPPH assay scavenging activity), moisture content (%), crude protein (g/100 mL), crude fat (g/100 mL), ash (g/100 mL), and total dietary fiber (g/100 mL) of grape pomace aqueous extract were analyzed according to the method described by Spinei and Oroian [17] and the AOAC methods [16]. Other carbohydrates were determined according to the difference method by subtracting the other components from 100 (100 − (ash + protein + fat + moisture)).

2.2.2. Functional Properties of Grape Pomace

Water-holding capacity (WHC, g/g), swelling capacity (SC, mL/g), oil-binding capacity (OBC, g/g), and water solubility index (WSI, %) were determined according to methods previously described by Zhang et al. [18] with slight modifications detailed by Gao et al. [19].

2.2.3. Color of Grape Pomace and Jelly Candy

The color of the grape pomace and jelly candy samples was analyzed in conformity with the method reported by Spinei and Oroian [20], using the granular material attachment CR-A50 (Konica Minolta, Inc., Tokyo, Japan). A glass Petri dish containing grape pomace and jelly candies was placed above the attachment.

2.2.4. Phytochemical Characterization of Grape Pomace Extract

Extract Preparation

The extraction of the phytochemicals from the dried grape pomace was accomplished in accordance with the protocol reported by Spinei and Oroian [17] with slight modifications. Briefly, 1 g of grape pomace of the Băbească Neagră variety with different particle sizes, <125 μm, ≥125–<200 μm, and ≥200–<300 μm, was mixed with 25 mL of ethanol (70%, *w*/*v*), the samples were homogenized using an Ultra-Turrax Homogenizer (Daigger Scientific Inc., Vernon Hills, IL, USA) for 3 min, and heated for 30 min at 70 °C in a water bath Precisdig (J.P. Selecta, Barcelona, Spain). At the end of the treatment, the samples were centrifuged for 10 min at 5000 rpm. The supernatant was filtered and diluted 1:10 with ethanol 70% (*v*/*v*); then, it was described with regard to the total phenolic content (TPC), total flavonoid content (TFC), total monomeric anthocyanin content (TMA content), and antioxidant activity.

Bioactive Compounds and Antioxidant Activity of Grape Pomace Extract

The TPC was determined using the Folin–Ciocalteu method described by Spinei and Oroian [17]. The results were represented as mg of gallic acid equivalents per gram of grape pomace extract. The polyphenols' calibration curve was performed by using gallic acid at concentrations of 1, 2, 3, 4, and 5 mg/L (regression coefficient R^2 = 0.988).

The TMA content was established using the methodology described by Spinei and Oroian [17] and the data were reported as mg cyanidin-3-glucoside per gram of grape pomace extract. The TMA content was calculated using the formula reported by Spinei and Oroian [17].

The TFC was measured utilizing the method reported by Spinei and Oroian [20] and the results were represented as mg quercetin equivalents per gram of grape pomace extract. The flavonoids' calibration curve was performed by using quercetin at concentrations of 1, 2, 3, 4, and 5 mg/L (regression coefficient R^2 = 0.987).

The DPPH assay scavenging activity was estimated using the methodology reported by Spinei and Oroian [17]. The absorbance was determined at 515 nm against a blank. The DPPH was calculated using the formula described by Spinei and Oroian [17].

2.2.5. FT-IR Analysis of Grape Pomace

FT-IR analysis of grape pomace samples was performed by the method used by Spinei and Oroian [20].

2.2.6. SEM Analysis of Grape Pomace

The microstructure of the grape pomace samples was examined by scanning electron microscopy (SEM; SU-70, Hitachi, Tokyo, Japan). Dried grape pomace powder was attached to the sample spot with conductive double-sided adhesive carbon tape and investigated using an accelerating voltage of 30 kV at a magnification of 400×.

2.2.7. Total Phenolic Content and Antioxidant Activity of Jelly Candy

The total phenolic content (TPC) of jelly candies was determined according to Rashad et al. [10] with slight modifications. For extraction, 2 g of jelly candies was homogenized with 20 mL of ethanol (70%, *w*/*v*) under continuous stirring at 400 rpm for 1 h. The extract was filtered in order to obtain a clear solution and TPC was determined using the Folin–Ciocalteu method described by Spinei and Oroian [17].

The DPPH assay scavenging activity of jelly candies was estimated according to Rashad et al. [10] with slight modifications. For extraction, 6 g of jelly candies was homogenized with 75 mL of ethanol (70%, *w*/*v*) under continuous stirring at 400 rpm for 1 h and then filtered. The extract was determined using the method described by Spinei and Oroian [17].

2.2.8. Textural Profile of Jelly Candy

Mechanical properties (hardness (H), cohesiveness (Co), adhesiveness (Ad), springiness (S), and gumminess (G)) were determined with a Perten TVT-6700 device (Perten Instruments, Stockholm, Sweden), by using the method described by Guiné et al. [21] and Figueroa and Genovese [22].

2.2.9. Sensorial Analysis of Jelly Candy

The sensory profile (e.g., color, odor, taste, elasticity, sweetness, appearance, biting, and overall acceptability) of jelly candy samples was determined using the methodology described by Rashad et al. [10]. Sensory characteristics were evaluated by 45 untrained panelists of the Faculty of Food Engineering, Stefan cel Mare University of Suceava (30 females and 15 males, aged 22 to 50 years). A 9-point hedonic scale was used for this aim; the rejection limit is less than 5 [10]. The grape pomace jellies used for the sensory analysis were prepared using silicone molds with 50 teddy bear spaces (size: 19 × 14 cm) and served at room temperature in identical beakers coded with digits.

2.2.10. Microbiological Analysis of Jelly Candy

Grape pomace jellies (10 g of each sample) were homogenized with sodium chloride solution (0.85%, w/v) in order to have a final dilution of 10^{-1} for 5 min in a laboratory blender. Serial decimal dilutions were established according to the method used by Rashad et al. [10]. The total count was determined on nutrient agar at 30 °C after 48 h of incubation, while mold and yeast counts were set out on malt extract agar at 28 °C after 5 days of incubation.

2.3. Statistical Analysis

The results were submitted to analysis of variance (ANOVA) using XLSTAT 2023.2.1413 Statistical Software (Lumivero, Denver, CO, USA). Fisher's least significant difference (LSD) procedure was used at the 95% confidence level. Principal component analysis (PCA) was used in order to correlate color, phytochemical, and textural characteristics (loadings) with different jelly formulations (scores) for easier viewing. The data were arranged in a matrix of 9 lines (jelly candy samples) and 10 columns (color, phytochemical, sensorial, and textural characteristics); the data were centered using variable transformation which is necessary to obtain a more representative variable and normalized by Pareto scaling (the scaling factor used is the square root of the standard deviation). The mean of the samples was used to build the PCA.

3. Results and Discussion

3.1. Influence of the Particle Size on the Proximate Composition of Grape Pomace

The structure of the grape pomace as a by-product depends on the terroir, grape variety, degree of grape ripeness, processing method, and other factors [1,9]. Therefore, all of these factors contribute to the high variability in grape pomace composition and suggest a considerable challenge for the fortification processing of the grape pomace [23]. The results of proximate analysis for grape pomace of Băbească Neagră (BN) variety (*Vitis vinifera* L.) with three different particle sizes, <125 μm, ≥125–<200 μm, and ≥200–<300 μm, are presented in Table 1; thus, the proximate composition was significantly influenced by the particle size of BN grape pomace ($p < 0.0001$).

The grape pomace with the particle size interval of ≥200–<300 μm had the highest content of insoluble dietary fiber (49.4 g/100 g oven-dried pomace powder (ODPP)), with smaller quantities of crude protein (8.4 g/100 g ODPP), crude fat (20.1 g/100 g ODPP), and moisture content (11.5 g/100 g ODPP) in comparison with grape pomace with the granularity of <125 μm (41.4 g/100 g ODPP, 7.7 g/100 g ODPP, 18.3 g/100 g ODPP, and 12 g/100 g ODPP, respectively). The obtained results are consistent with other studies, which related that the predominant constituents of the grape cell wall are lignin, cellulose, hemicellulose, and other polymeric components [1,9]. The relatively high fat content of grape pomace can be ascribed to oil from grape seeds, especially to lipophilic compounds (e.g., unsaturated fatty acids) [1,24]. The total dietary fiber (TDF) value diminished as the particle size consecutively decreased; this may be an effect of the degradation of insoluble dietary fiber (IDF), such as hemicellulose, cellulose, and lignin, in grape pomace into some tiny molecular substances [18]. Moreover, with a decrease in particle size intervals from ≥200–<300 μm to <125 μm, the IDF content decreased from 49.4 g/100 g ODPP to 41.4 g/100 g ODPP, while soluble dietary fiber (SDF) content increased from 6.9 g/100 g ODPP (<125 μm) to 7.8 g/100 g ODPP (≥125–<200 μm) and decreased to 7.6 g/100 g ODPP (≥200–<300 μm); this indicated that smaller particle size led to the redistribution of DF from the insoluble to soluble part [18]. Also, Gao et al. [18] presented the same tendency of a decrease in the SDF content of asparagus pomace with a decrease in the granularity from 109.3 to 3.18 μm. Furthermore, Antonić et al. [9] presented the proximate composition of grape pomace, obtained from studies that comprised an analysis of different grape pomace varieties, in this way: ash of 1.73–9.10 g/100 g, protein about 3.57–14.17 g/100 g, fat around 1.14–13.90 g/100 g, and TDF of 17.28–88.70 g/100 g. On the other hand, Tolve et al. [25] obtained the physicochemical composition of grape pomace (*Vitis vinifera* cv. Corvina) with a particle size of <200 μm as follows: total dietary fiber—52.3 ± 2.1 g/100 g dry matter (DM), crude protein—11.19 ± 0.97 g/100 g DM, and ash—4.17 ± 0.87 g/100 g DM, while Bender et al. [26] reported the following proximate composition of dried grape pomace (*Vitis vinifera* cv. Malbec) with a particle size of <0.59 mm: total dietary fiber—65.56 ± 0.83 g/100 g dry weight (DW), crude protein—14.17 ± 0.08 g/100 g DW, and ash—5.14 ± 0.00 g/100 g DW.

Table 1. Proximate composition, physicochemical, and functional properties of grape pomace.

Parameter	Particle Size (μm)			*p*-Value
	<125	≥125–<200	≥200–<300	
Proximate composition				
Moisture (g/100 g ODPP)	12.0 (0.34) [a]	11.9 (0.26) [b]	11.5 (0.29) [c]	$p < 0.0001$
Crude protein (g/100 g ODPP)	7.7 (0.28) [c]	8.0 (0.33) [b]	8.4 (0.42) [a]	$p < 0.0001$
Crude fat (g/100 g ODPP)	18.3 (0.22) [c]	19.4 (0.25) [b]	20.1 (0.18) [a]	$p < 0.0001$
Ash (g/100 g ODPP)	1.9 (0.14) [a]	1.8 (0.19) [b]	1.6 (0.27) [c]	$p < 0.0001$
Carbohydrates [A] (g/100 g ODPP)	60.0 (0.27) [a]	58.8 (0.21) [b]	58.2 (0.34) [c]	$p < 0.0001$
TDF (g/100 g ODPP)	48.2 (0.36) [c]	55.0 (0.32) [b]	57.0 (0.26) [a]	$p < 0.0001$
IDF (g/100 g ODPP)	41.4 (0.19) [c]	47.1 (0.23) [b]	49.4 (0.28) [a]	$p < 0.0001$
SDF (g/100 g ODPP)	6.9 (0.23) [b]	7.8 (0.15) [a]	7.6 (0.19) [a]	$p < 0.0001$
Functional properties				
WHC (g/g ODPP)	4.0 (0.13) [c]	5.1 (0.17) [b]	7.8 (0.19) [a]	$p < 0.0001$
OBC (g/g ODPP)	2.7 (0.08) [a]	2.1 (0.12) [b]	1.9 (0.07) [c]	$p < 0.0001$
SC (mL/g ODPP)	4.1 (0.28) [c]	6.9 (0.32) [b]	10.1 (0.37) [a]	$p < 0.0001$
WSI (% ODPP)	9.4 (0.11) [a]	8.3 (0.18) [b]	5.2 (0.09) [c]	$p < 0.0001$
Color parameters				
L*	17.6 (0.14) [b]	25.1 (0.11) [a]	29.9 (0.09) [a]	$p > 0.05$
h^*_{ab}	39.9 (0.21) [a]	38.1 (0.28) [b]	38.7 (0.34) [ab]	$p > 0.05$
C^*_{ab}	11.2 (0.17) [b]	14.3 (0.12) [ab]	14.9 (0.07) [a]	$p > 0.05$
Phytochemical characteristics				
TPC (mg GAE/g ODPP)	146 (0.22) [a]	136 (0.32) [b]	135 (0.37) [b]	$p < 0.0001$
TMA content (mg C3G/g ODPP)	11.4 (0.46) [a]	9.2 (0.41) [b]	6.9 (0.38) [c]	$p < 0.0001$
TFC (mg QE/g ODPP)	19.0 (0.13) [b]	21.4 (0.26) [a]	17.5 (0.17) [c]	$p < 0.0001$
DPPH (% inhibition ODPP)	67.1 (0.19) [a]	50.4 (0.54) [b]	48.6 (0.68) [c]	$p < 0.0001$

Mean values (n = 9) and standard deviation are shown in brackets. [a–c]—different letters in the same row indicate significant differences among samples ($p < 0.0001$) in conformity with the Fisher test with $\alpha = 0.05$. [A]—carbohydrate content was determined by the difference method; TDF—total dietary fiber; IDF—insoluble dietary fiber; SDF—soluble dietary fiber; WHC—water-holding capacity; OBC—oil-binding capacity; SC—swelling capacity; WSI—water solubility index; L*—luminosity; h^*_{ab}—hue; C^*_{ab}—chroma; TPC—total phenolic content; TMA—total monomeric anthocyanin content; TFC—total flavonoid content; GAE—gallic acid equivalent; C3G—cyanidin-3-glucoside; QE—quercetin equivalent; ODPP—oven-dried pomace powder.

3.2. Influence of the Particle Size on the Functional Properties of Grape Pomace

Water-holding capacity (WHC) and swelling capacity (SC) are essential hydration characteristics of dietary fibers in food fortification and are outright attributed to the quality and functional features of food products [18]. WHC and SC belong to the insoluble polysaccharides since they can attach water by either surface tension in the pores of the matrix or ionic bonds, hydrophilic interactions, and/or hydrogen bonds [27]. As shown in Table 1, WHC varied between 4.0 (grape pomace with the particle size of <125 μm) and 7.8 g/g ODPP (grape pomace with the particle size of ≥200–<300 μm), while SC had a range from 4.1 (grape pomace with the particle size of <125 μm) to 10.1 mL/g ODPP (grape pomace with the particle size of ≥200–<300 μm); this can be explicated by the fact that the highest particle size interval of BN grape pomace contained a great amount of IDF. These results are in accordance with the data obtained by Zhang et al. [27] and Ma and Mu [28], who found that IDF could bind more water with minimal swelling. Therefore, the WHC obtained for the grape pomace with the particle size of ≥200–<300 μm (7.8 g/g) was higher than that determined for asparagus pomace (4.18–6.12 g/g), black currant pomace (2.78 g/g), cranberry pomace (3.83–3.87 g/g), lingonberry pomace (3.27–3.28 g/g), quince pomace (5.30–7.55 g/g), and sea buckthorn pomace (4.24–4.32 g/g) [18,29,30], but lower than that of grape juice pomace (8.48 g/g), *Dendrobium officinale* pomace (15.30–20.12 g/g), cranberry pomace (10.59 g/g), and deoiled red raspberry pomace (8.97 g/g) [27,31–33]. These data may be a result of the influence of different parameters, e.g., different particle size intervals, preparation procedure, porosity, surface characteristics, and chemical structure of the fiber [33]. Concerning the applications of by-products in the food industry, e.g., dietary fiber with a high value of WHC can be utilized as a functional ingredient in order to modify some characteristics (viscosity and texture) of composed foods, while dietary fiber with low WHC could be utilized as a sugar substitute to produce low-calorie food products (e.g., corn flakes, cookies, extruded snacks, crackers, and jelly) [27].

The water solubility index (WSI) depends on the swelling fluid, swelling power, chemical, and structure composition of the by-product [34]. The influence of particle size on the WSI is presented in Table 1; thus, the WSI was significantly influenced by the particle size of BN grape pomace (*Vitis vinifera* L.) ($p < 0.0001$). The highest value of WSI (9.4% ODPP) was obtained for the particle size interval of <125 μm, while the lowest WSI (5.2% ODPP) for the particle size of ≥200–<300 μm; in this way, the WSI of grape pomace increased significantly as its particle size decreased. Gao et al. [18] reported the same tendency for asparagus pomace samples, in which the WSI varied between 1.32 and 13.25% for 0.12 and 0.005 mm, respectively. This can be explained by the fact that grinding treatment may be capable of destroying the structure of the dietary fiber of grape pomace and releasing the soluble components, resulting in enhanced solubility [18]. Moreover, experimental characteristics, such as stirring, could influence the chemical structure of dietary fiber and the hydration process [18]. Oil-binding capacity (OBC) can be utilized in order to evaluate the functional ingredient's property of preventing fat loss during food processing and decreasing serum cholesterol levels through adsorbing fat in the intestinal lumen [27]. As shown in Table 1, the OBC of the grape pomace varied between 1.9 (grape pomace with the particle size of ≥200–<300 μm) and 2.7 g/g ODPP (grape pomace with the particle size of <125 μm); these data can be related to the presence of lignin (insoluble dietary fiber) which plays a role in oil absorption [18,27]. Certainly, the OBC of the BN grape pomace obtained in this study was relatively higher in comparison with dietary fiber extracted from grape juice pomace (0.65 g/g), asparagus pomace with a particle size of 0.03 mm (2.45 g/g), black currant pomace (1.14 g/g), cranberry pomace (1.57 g/g), lingonberry pomace (1.46 g/g), grape pomace (1.42 g/g), and sea buckthorn pomace (1.09 g/g) [18,26,27,29]. Previous studies reported a higher value of OHC for citrus peel dietary fiber, such as orange dietary fiber—3.62 g/g and grapefruit dietary fiber—8.20 g/g [35]. This resulted most probably as a consequence of fiber composition, surface properties, particle size, hydrophobic nature of particles, and overall charge density [27]. Therefore, the functional properties of grape pomace, e.g., WHC and SC, presented a typical exponential curve with a maximal value for upper-size particles, while OBC showed this curve for lower-size particles.

3.3. *Influence of the Particle Size on the Color of Grape Pomace*

The color of grape pomace, particularly red varieties, is determined by the amount of anthocyanins, localized mostly in the skin [4]. Furthermore, color is affected by different physicochemical parameters, e.g., pH, temperature, particle size of material, and time [4]. The influence of particle size on the color parameters (L^*, C^*_{ab}, and h^*_{ab}) of grape pomace is presented in Table 1; thus, the L*, C^*_{ab}, and h^*_{ab} were not significantly influenced by the granularity of grape pomace ($p > 0.05$). The highest value of L* (29.9) and C^*_{ab} (14.9) was obtained for grape pomace with the granularity of ≥200–<300 μm, while the lowest L* (17.6) and C^*_{ab} (11.2) by <125 μm grape pomace. Therefore, grape pomace samples showed a color ranging from blue-purple to purple in conformity with the CIE chromaticity diagram. This can be explained by the fact that anthocyanins interact with other phenolics to create co-pigments which improve color stability during and after the winemaking process [36]. Moreover, grape pomace is rich in flavonoids, polyphenols, and anthocyanins with a high antioxidant capacity and distinct phenolic compounds [1]. In order to concur with the results presented for color parameters (L*, C^*_{ab}, and h^*_{ab}), images of grape pomace samples with different particle sizes are shown in Figure 1.

Figure 1. Images of Băbească Neagră grape pomace under the influence of different granularities.

3.4. Phytochemical Characterization of Grape Pomace Extract

Vinification by-products consist of a relatively high content of polyphenolic substances, which depends on the grape variety (white or red), processing conditions (e.g., type of maceration), and the part of vegetal material (seeds, skins, etc.) [37,38]. The phytochemical characterization of grape pomace (Băbească Neagră variety) extract was performed by evaluating the following parameters: TPC, TMA content, TFC, and antioxidant activity (Table 1). The highest value of TPC, TMA content, and DPPH was obtained for the granularity of <125 μm (146 mg gallic acid equivalent (GAE)/g ODPP, 11.4 mg cyanidin-3-glucoside (C3G)/g ODPP, and 67.10% ODPP, respectively), while the lowest for the particle size of ≥200–<300 μm (135 mg GAE/g ODPP, 6.9 mg C3G/g ODPP, and 48.6% ODPP, respectively). This can be explained by the fact that particle size intervals significantly influence the extraction of bioactive compounds ($p < 0.0001$). Moreover, the phytochemical characterization (TPC, TMA content, and DPPH) is affected by several factors, such as drying, extraction process, solvent, and extraction temperature [39]. The same results were obtained for Cabernet Franc pomace (153.8 mg GAE/g for TPC and 91.7 mg catechin equivalents (CE)/g for TFC) [40]. On the other hand, the obtained results for grape pomace extract are higher than the TPC reported by Carmona-Jiménez et al. [41] for different grape pomace varieties, such as Tempranillo (49.70 mg GAE/g), Cabernet Sauvignon (44.73 mg GAE/g), Tintilla de Rota (43.85 mg GAE/g), Syrah (44.11 mg GAE/g), and Petit Verdot (46.57 mg GAE/g). Also, Barriga-Sánchez et al. [39] reported a lower TPC (54.36 mg GAE/g) for Black Borgoña (*Vitis labrusca* L.) pomace using ethanol–water as solvent.

3.5. FT-IR Spectra of Grape Pomace

Different organic functional groups, notably N–H, O–H, and C=O, are determined by FT-IR spectroscopy. Consequently, FT-IR spectra were obtained to study the influence of particle size on the chemical structure of Băbească Neagră (BN) grape pomace (*Vitis vinifera* L.). Common absorption bands of phenolic compounds and polysaccharides that are found in the structure of grape pomace are shown in Figure S1 of the Supplementary Materials. The grape pomace samples had stretching bands of O–H and O(6)H···O(3) at 3315–3310 cm^{-1} which were attributed to vibrations of the functional group (–OH) of the phenolic and polysaccharide structures, specifically insoluble dietary fibers, e.g., cellulose and hemicellulose [26]. Furthermore, the C–H stretching vibration peak at 2918 cm^{-1} was assigned to a methylene group of polysaccharides [26]. The band position at 1741 cm^{-1} corresponded to the stretching band of carbonyl group C=O, which indicated the presence of the galloyl group or uronic acid [26,42]. Stretching bands at 1625–1605 cm^{-1} showed the presence of benzene rings in lignin (C–H bonds) [26]. The band positions of 1510 cm^{-1}, 1435 cm^{-1}, and 1377 cm^{-1} corresponded to C=C stretching vibration, antisymmetric, and symmetric in-plane bending of $–CH_3$, respectively [43]. The peaks at 1132 cm^{-1} and 1064 cm^{-1} were related to aromatic C–O–C stretching vibrations (presence of cellulose and hemicellulose) and C–O stretching bands (presence of pyranose ring), respectively [26,44]. Moreover, these band positions are attributed to polysaccharide structures [40]. The absorptive peaks at 905 cm^{-1}, 842 cm^{-1}, and 789 cm^{-1} were ascribed to β-*D*-pyranose, the phenolic compounds, and the aromatic ring breathing manner, respectively [42]. The spectral profile of BN grape pomace with different particle size intervals indicated that the position and number of the characteristic absorption peaks of samples did not change significantly.

3.6. Microstructural Analysis by SEM

SEM is an efficacious technique for analysis of the crystallography, shape, size, composition, and other physicochemical properties of different materials [45]. The morphology of Băbească Neagră (BN) grape pomace was investigated using SEM. As shown in Figure 2, a couple of changes were examined among the grape pomace samples with different particle sizes; the grape pomace with the particle size of <125 μm had a small fibrous, compact, and reticular structure, while other samples presented various rough and irregular surfaces in combination with fibrous structures. According to Pala et al. [46] and Gowman et al. [47], the irregular surface of the samples is attributed to the strong internal bonds in the structure of the lignocellulosic fibers. Moreover, it can be explained by the fact that the grape pomace consisted of seeds, skins, and stems which have different components. The grape pomace skins are flatter, but become irregular after the drying process, while the stems are more fibrous because of the amount of insoluble dietary fibers. The grape pomace seeds are irregular and, like stems, have fibrous and porous surfaces [47]. Therefore, different particle size intervals influenced the surface morphology of grape pomace samples and resulted in different fibrous structures, as presented in the SEM images. The porous surface gives additional space for the formation of hydrogen bonding by water molecule content; this feature could increase the OBC and WHC of the grape pomace samples [48]. The same results were reported by Drevelegka and Goula [49], who established that grape pomace had a solid and relatively organized cell structure. Also, Baldán et al. [48] explained that different particle sizes were related to the hydrophilic components of the grape pomace, e.g., fiber, polysaccharides, protein, and pigments.

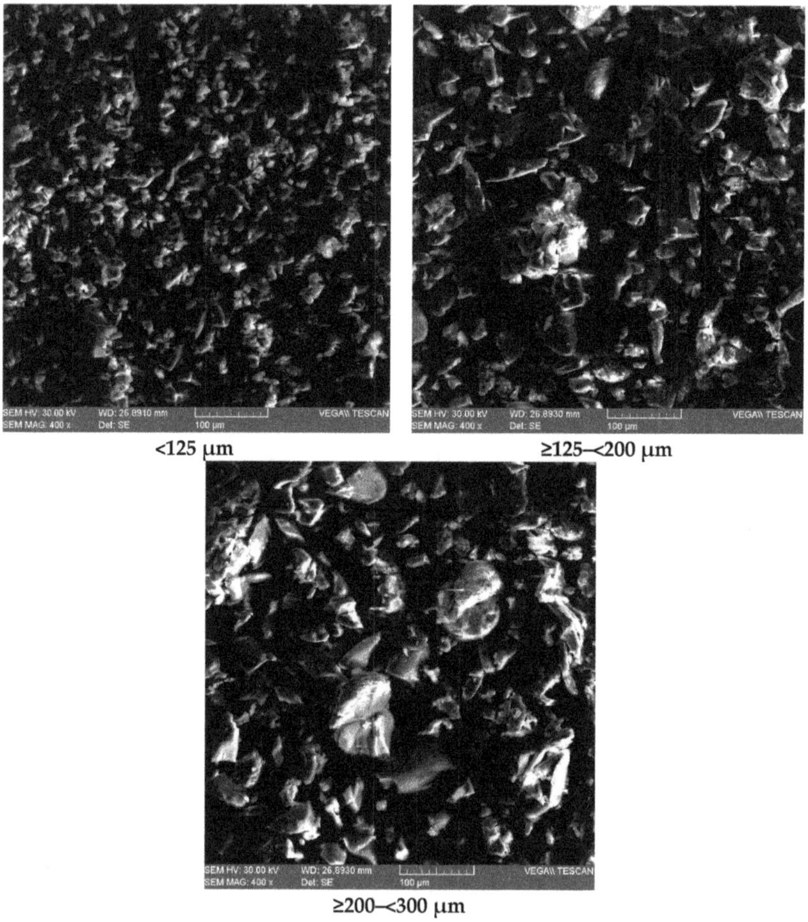

Figure 2. SEM images of Băbească Neagră grape pomace under the influence of different granularities.

3.7. Phytochemical Parameters and Proximate Composition of Grape Pomace Aqueous Extract

The proximate composition and phytochemical parameters (DPPH assay scavenging activity and TPC) of grape pomace aqueous extract are shown in Table 2. The parameters of proximate composition, such as moisture, ash, carbohydrates, and total dietary fiber (TDF), were significantly influenced by the granularity of grape pomace extract ($p < 0.0001$).

Table 2. Proximate composition and phytochemical parameters of grape pomace aqueous extract.

Parameter	Particle Size (μm)			*p*-Value
	<125	≥125–<200	≥200–<300	
Proximate composition				
Moisture (% ODPP)	84.1 (0.13) ^{c}	86.7 (0.16) ^{b}	89.3 (0.14) ^{a}	$p < 0.0001$
Crude protein (g/100 mL ODPP) *	n.d.	n.d.	0.23 (0.01) ^{a}	$p < 0.0001$
Crude fat (g/100 mL ODPP) *	n.d.	n.d.	n.d.	–
Ash (g/100 mL ODPP) *	0.02 (0.02) ^{c}	0.05 (0.02) ^{b}	0.12 (0.01) ^{a}	$p < 0.0001$
Carbohydrates ^{A} (g/100 mL ODPP) *	15.9 (0.06) ^{a}	13.2 (0.08) ^{b}	10.3 (0.11) ^{c}	$p < 0.0001$
TDF (g/100 mL ODPP) *	12.4 (0.11) ^{c}	10.6 (0.12) ^{b}	9.7 (0.16) ^{a}	$p < 0.0001$

Table 2. *Cont.*

Parameter	Particle Size (μm)			*p*-Value
	<125	≥125–<200	≥200–<300	
Phytochemical characteristics				
TPC (mg GAE/mL ODPP)	126 (0.22) [a]	107 (0.32) [c]	109 (0.37) [b]	$p < 0.0001$
DPPH (% inhibition ODPP)	60.8 (0.17) [a]	49.6 (0.25) [b]	42.5 (0.32) [c]	$p < 0.0001$

Mean values (n = 9) and standard deviation are shown in brackets. [a–c]—different letters in the same row indicate significant differences among samples (p < 0.0001) in conformity with the Fisher test with α = 0.05. [A]—carbohydrate content was determined by the difference method; TDF—total dietary fiber; TPC—total phenolic content; GAE—gallic acid equivalent; ODPP—oven-dried pomace powder; *—The data was expressed as g per 100 mL of extract produced for each of the studied fractions from ODPP; n.d.—not detected.

The highest values for moisture content (89.3% ODPP) and ash (0.12 g/100 mL ODPP) were obtained for the grape pomace extract with the granularity of ≥200–<300 μm, while the lowest values were obtained for the grape pomace extract with the granularity of <125 μm (84.1% ODPP and 0.02 g/100 mL ODPP for moisture content and ash, respectively). Crude fat was not found in any sample; this can be explained by the fact that fats did not pass into the water extract and did not have any polar bonds. Moreover, only the grape pomace extract with the granularity of ≥200–<300 μm had crude protein (0.23 g/100 mL ODPP) in its composition. The presence of protein can be attributed to the constituents of grape pomace, especially due to inflorescence architectures and grape seeds which contain an amount of 6.1% and 11% protein, respectively [1]. The highest contents of carbohydrates (15.9 g/100 mL ODPP) and TDF (12.4 g/100 mL ODPP) were obtained for the grape pomace extract with the granularity of <125 μm, while the lowest for the grape pomace extract with the granularity of ≥200–<300 μm (10.3 g/100 mL ODPP and 9.7 g/100 mL ODPP, respectively). Therefore, viable sources of carbohydrates that are presented in grape pomace extract are water-soluble monosaccharides (e.g., fructose and glucose) [1,4].

The grape pomace extracts showed a great quantity of TPC (a range between 107 and 126 mg GAE/mL ODPP) as a result of hydrophilic compounds (e.g., phenolic acids) which are abundant in the solid parts of grape pomace, especially in grape skins and grape seeds [2,4]. The DPPH assay scavenging activity showed a similar tendency to TPC; these results are in accordance with other studies [24,26].

3.8. Physicochemical Parameters of Jelly Candy

The physicochemical parameters, such as color parameters (L^*, C^*_{ab}, and h^*_{ab}), DPPH assay scavenging activity, and TPC were determined for jelly candies (Table 3); thus, the L^*, C^*_{ab}, and h^*_{ab} were significantly influenced by the granularity of grape pomace ($p < 0.0001$).

Table 3. Physicochemical and textural properties of jelly candy.

Parameter	Samples									*p*-Value
	F1–7	F1–8.5	F1–10	F2–7	F2–8.5	F2–10	F3–7	F3–8.5	F3–10	
Color parameters										
L^*	29.9 (0.13) [d]	32.5 (0.17) [c]	39.4 (0.16) [ab]	29.2 (0.14) [d]	33.4 (0.19) [c]	40.2 (0.19) [ab]	33.1 (0.12) [c]	36.2 (0.08) [b]	42.7 (0.11) [a]	$p < 0.0001$
h^*_{ab}	40.7 (0.14) [d]	42.1 (0.17) [cd]	43.6 (0.17) [c]	41.3 (0.13) [cd]	44.7 (0.14) [c]	46.3 (0.15) [bc]	45.6 (0.18) [bc]	47.1 (0.13) [b]	48.9 (0.15) [a]	$p < 0.0001$
C^*_{ab}	15.1 (0.30) [c]	16.2 (0.22) [bc]	16.8 (0.23) [bc]	15.9 (0.17) [c]	16.9 (0.21) [bc]	17.2 (0.16) [b]	16.8 (0.24) [bc]	17.5 (0.29) [b]	18.1 (0.27) [a]	$p < 0.0001$
Phytochemical characteristics										
TPC (mg GAE/g ODPP)	156 (1.42) [a]	148 (1.29) [b]	138 (2.01) [c]	130 (2.13) [d]	129 (1.08) [d]	127 (1.37) [d]	120 (1.23) [e]	118 (0.99) [e]	115 (1.14) [e]	$p < 0.0001$
DPPH (% inhibition ODPP)	65.8 (0.98) [a]	63.3 (1.10) [ab]	58.3 (2.18) [b]	48.1 (1.15) [c]	46.4 (0.57) [c]	46.4 (0.69) [c]	30.1 (1.77) [d]	30.0 (2.13) [d]	29.9 (2.27) [d]	$p < 0.0001$

Table 3. *Cont.*

Parameter	Samples									*p*-Value
	F1–7	F1–8.5	F1–10	F2–7	F2–8.5	F2–10	F3–7	F3–8.5	F3–10	
Textural parameters										
H (N)	6.3 (0.13) ^{c}	11.1 (2.13) ^{b}	16.7 (1.10) ^{a}	6.5 (0.27) ^{c}	10.8 (1.08) ^{b}	17.1 (0.98) ^{a}	6.1 (0.24) ^{c}	10.7 (1.37) ^{b}	16.5 (0.46) ^{a}	$p < 0.0001$
Co (adim.)	0.23 (0.02) ^{a}	0.24 (0.02) ^{a}	0.24 (0.01) ^{a}	0.23 (0.06) ^{a}	0.24 (0.06) ^{a}	0.25 (0.05) ^{a}	0.23 (0.01) ^{a}	0.24 (0.03) ^{a}	0.25 (0.07) ^{a}	$p > 0.05$
Ad (N × s)	−2.9 (0.08) ^{a}	−2.9 (0.10) ^{a}	−3.0 (0.06) ^{a}	−3.0 (0.13) ^{a}	−3.0 (0.08) ^{a}	−3.1 (0.07) ^{a}	−3.2 (0.07) ^{ab}	−3.2 (0.11) ^{ab}	−3.3 (0.06) ^{b}	$p < 0.01$
S (adim.)	0.85 (0.03) ^{b}	0.84 (0.01) ^{b}	0.83 (0.01) ^{b}	0.88 (0.01) ^{a}	0.85 (0.01) ^{b}	0.84 (0.02) ^{b}	0.85 (0.03) ^{b}	0.86 (0.02) ^{b}	0.89 (0.01) ^{a}	$p < 0.01$
G (N)	48.9 (0.12) ^{d}	54.2 (0.18) ^{b}	54.0 (0.17) ^{b}	53.1 (0.15) ^{c}	54.2 (0.12) ^{b}	54.4 (0.16) ^{b}	55.1 (0.18) ^{a}	55.2 (0.12) ^{a}	49.6 (0.11) ^{d}	$p < 0.001$
Microbiological tests										
TP count (CFU/mL)	n.d.	n.d.	n.d.	n.d.	n.d.	n.d.	69.3 (0.14) ^{c}	74.6 (0.17) ^{b}	95.3 (0.23) ^{a}	$p < 0.0001$
MY counts (CFU/mL)	n.d.	n.d.	n.d.	n.d.	n.d.	n.d.	51.6 (0.13) ^{c}	53.3 (0.18) ^{b}	74.1 (0.21) ^{a}	$p < 0.0001$

Mean values ($n = 9$) and standard deviation are shown in brackets. $^{a–e}$—different letters in the same row indicate significant differences among samples ($p < 0.0001$) in conformity with the Fisher test with $\alpha = 0.05$. F1—grape pomace extract with the granularity of <125 μm; F2—grape pomace extract with the granularity of ≥125–<200 μm; F3—grape pomace extract with the granularity of ≥200–<300 μm; 7—7 g of gelatin; 8.5—8.5 g of gelatin; 10—10 g of gelatin; L*—luminosity; h^*_{ab}—hue; C^*_{ab}—chroma; TPC—total phenolic content; GAE—gallic acid equivalent; H—hardness; Co—cohesiveness; Ad—adhesiveness; S—springiness; G—gumminess; TP Count—total plate count; MY Counts—mold and yeast counts; ODPP—oven-dried pomace powder; n.d.—not detected.

The highest values for L* (42.7), C^*_{ab} (18.1), and h^*_{ab} (48.9) were obtained for jelly formulation F3–10 (grape pomace extract with the granularity of ≥200–<300 μm and 10 g of gelatin), while the lowest for F1–7 (grape pomace extract with the granularity of <125 μm and 7 g of gelatin). Thus, jelly samples showed a color ranging from red-purple to purple in conformity with the CIE chromaticity diagram. The color of the jelly candies obtained from the grape pomace extract with the particle size of <125 μm and 7 g of gelatin got darker after the mixing step; this may be explained by the quick dissolution of grape pomace with the diameter of <125 μm, and also, by the presence of polyphenols and other water-soluble compounds (e.g., pigments). The resulting color is also directly correlated to the chemical structures of anthocyanins present in grape pomace extract, the presence of different co-pigmentations, and the chemical reactions implicating these components during the homogenization of the mixture, as well as the concentration of gelling agent (gelatin) and temperature used in that process [50,51]. In order to concur with the results presented for the color parameters (L*, C^*_{ab}, and h^*_{ab}), images of jelly samples are presented in Figure 3.

Figure 3. Jelly samples obtained from grape pomace extract, gelatin, and stevia.

TPC and DDPH assay scavenging activity of jelly candies are greatly dependent on the gelling agent, solvent nature (water and ethanol), extraction method, and formulation procedure [39]. The highest value of TPC was found for jelly samples obtained from the grape pomace extract with the particle size of <125 μm and with the addition of 7 g of gelatin. DPPH values varied from 29.9 to 65.8% for F3–10 and F1–7, respectively. Statistically, DPPH antioxidant activity values showed a similar tendency to TPC. It is well known that polyphenol content is definitely correlated with antioxidant capacity, particularly in the case of fruit, fruit juices, and extracts from different pomaces [50]. Mainly, it can be described that the higher the content of the gelling agent, the lower the functional properties and bioactive compounds (e.g., polyphenols) of jellies [50,52]. On the other hand, the addition of stevia in the jelly formulations increases the antioxidant capacity of the samples because stevia contains higher amounts of antioxidant compounds (hydroxyl radicals, free radicals, and superoxide anion radical scavenging activities) [53]. Similar results for DPPH and TPC were reported by Cano-Lamadrid et al. [50] for jelly candies based on pomegranate juice "Mollar de Elche".

3.9. Textural Parameters of Jelly Candy

Textural analysis was performed using the methodology reported by Garrido et al. [54]. The values obtained for textural parameters, such as hardness, cohesiveness, adhesiveness, springiness, and gumminess are presented in Table 3. Hardness relates to the force necessary to obtain a given deformation [18]. The data obtained for hardness were statistically different among samples ($p < 0.0001$), with the highest value (17.1 N) for F2–10 (grape pomace extract with the particle size of ≥125–<200 μm and 10 g of gelatin) and the lowest (6.1 N) for F3–7 (grape pomace extract with the particle size of ≥200–<300 μm and 7 g of gelatin); these data can be related to the concentration of the gelling agent (gelatin) and the composition of grape pomace extracts with different particle sizes (e.g., protein, carbohydrates, and fiber content). Previous studies reported a higher value of hardness for strawberry and mint jelly gum (26.7–30.3 N), strawberry and anise jelly gum (24.5–26.7 N), raspberry and mint jelly gum (22.8–27.7 N), blueberry and mint (25.1–29.3 N), grape seed soft candies (11.34–47.53 N), and grape skin soft candies (16.79–24.08 N) [15,18]. On the other hand, Garrido et al. [54] obtained a lower value of hardness of apple jelly (0.6–5.2 N). Additionally, the interaction of proteins and carbohydrates which are presented in grape pomace extract has a significant influence on enhancing hardness and other textural parameters, such as springiness and cohesiveness [22].

Cohesiveness refers to the internal bonds of the food that keep the mass cohesive and prevent it from disintegrating [21]. The values of cohesiveness ranged from 0.23 for F2–7 (grape pomace extract with the particle size of ≥125–<200 μm and 7 g of gelatin) to 0.25 for F3–10 (grape pomace extract with the particle size of ≥200–<300 μm and 10 g of gelatin). The fiber content (9.7–12.4 g/100 mL ODPP) of aqueous extract had a positive effect on cohesiveness; fibers cause a significant increase in cohesiveness [22]. Similarly, Akesowan [55] reported a range between 0.202 and 0.322 for the cohesiveness of konjac jelly, while Garrido et al. [54] obtained higher values (0.37–0.53) for apple jelly; this can be explained by the fact that the more gelling agent existed in the gel system, the more cohesive the jelly candy became [55].

Adhesiveness refers to the force needed to eliminate the material that adheres to a surface [21]. The results obtained for adhesiveness were statistically different among samples ($p < 0.01$), with the highest value (-3.3 N × s) for F3–10 (grape pomace extract with the particle size of ≥200–<300 μm and 10 g of gelatin) and the lowest (-3.0 N × s) for F1–7 (grape pomace extract with the particle size of <125 μm and 7 g of gelatin). Several factors, e.g., the amount of sugar naturally present in grape pomace, pectin content, acidity, and concentration of gelling agent may explain the variation in adhesiveness among samples prepared with different particle sizes of grape pomace. On the other hand, Rios de Souza et al. [56] reported higher values ((-12.75)–(-37.50) N × s) for blackberry jellies from different cultivars (Choctaw, Brazos, Cherokee, Comanche, Caingangue, Guarani, and Tupy). In addition, other parameters, such as phytochemical (e.g., TPC) and proximate composition (e.g., moisture content, TDF, and crude protein) of the grape extract can affect the textural characteristics by influencing the yield, cooking time, and water activity of the final jelly candies [56].

Springiness is related to the ability of shape recovery after compression and represents the rate at which a deformed material returns to the initial form after the force is removed [54]. The values of springiness ranged from 0.83 for F1–10 (grape pomace extract with the particle size of <125 μm and 10 g of gelatin) to 0.89 for F3–10 (grape pomace extract with the particle size of ≥200–<300 μm and 10 g of gelatin). Similarly, Rios de Souza et al. [56] reported a range between 0.93 and 0.98 for the springiness of blackberry jellies from different cultivars (Brazos, Caingangue, Cherokee, Choctaw, Comanche, Guarani, and Tupy), while Altınok et al. [15] obtained a lower value (0.484) for grape skin soft candies with a particle size of 100 μm. The results showed that springiness was most reproducible in relation to the other textural responses, e.g., cohesiveness and hardness. Moreover, the highest particle size interval (≥200–<300 μm) influenced the springiness due to the fiber content of grape pomace and gelling agent, which subsequently affects the melting properties [22,55].

Gumminess has been interpreted as the energy involved in disintegrating a semi-solid food product into a state ready for swallowing [50]. The highest value (55.2 N) for F3–8.5 (grape pomace extract with the particle size of ≥200–<300 μm and 8.5 g of gelatin) and the lowest (48.9 N) for F1–7 (grape pomace extract with the particle size of <125 μm and 7 g of gelatin); these data can be related to the gelatin dose, treatment, and the presence of sugars [21]. The increase in the gelatine dose enhanced gumminess, cohesiveness, and hardness; these results are in accordance with the data obtained by Mutlu et al. [57]. Moreover, the results of the texture profile analysis showed that the particle size of plant material (i.e., grape pomace) is a significant factor in the formulation of jelly candy.

3.10. Sensorial Analysis of Jelly Candy

The radar chart of sensory analysis (e.g., color, odor, taste, elasticity, sweetness, appearance, biting, and overall acceptability) of the grape pomace jellies is presented in Figure 4; the quality sensory characteristics of jelly samples were expressed as appealing. Generally, the jelly candies presented good sensory acceptance for all characteristics, with average scores ranging between the hedonic terms "I liked it slightly" and "I liked it very much". The textural parameters (biting and elasticity) were presented as easy to bite and soft by the majority of the panelists. The jelly formulations F3–7 and F3–10 recorded the lowest value of overall acceptability due to the presence of the highest contents of insoluble fiber and fats, which have an unpleasant texture and taste. The highest values for taste, sweetness, odor, and overall acceptability were obtained for jelly formulations F2–7, F2–8.5, and F1–8.5; these results can be attributed to the lower sizes of grape pomace particles which are correlated with a lower content of fats and insoluble fiber. The results showed that grape pomace jellies have a gummy and smooth texture with a gelatinous appearance and sweet taste, which indicates that the formulation and ratios chosen for the gelling agent (gelatin) were adequate and presented a positive influence on the sensory characteristics of the final product. The same tendency of sensorial parameters was reported for jelly candies based on pomegranate juice "Mollar de Elche", strawberry and red beetroot jelly candies, and strawberry jellies from different cultivars (Camino Real, Festival, Camarosa, Oso Grande, Albion, and San Andreas) [10,50,58]. These findings can be highlighted for production by focusing attention on the fact that grape pomace jelly is a product based on stevia (zero-calorie sweetener), does not contain added sugar (only natural sweeteners, i.e., water-soluble monosaccharides present in grape pomace extract), and artificial dyes.

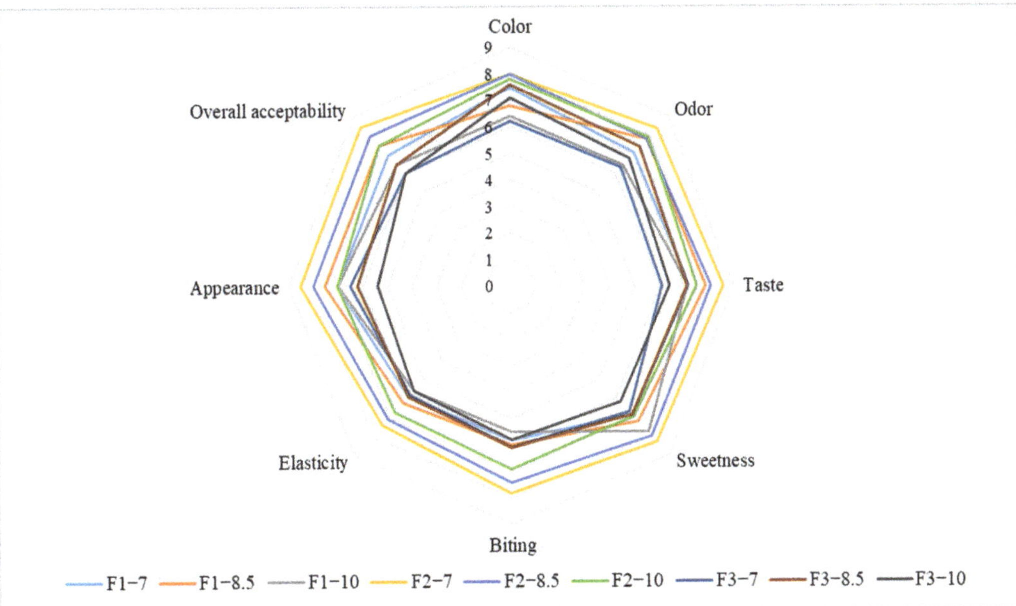

Figure 4. Radar chart of sensory analysis of jelly samples (F1—grape pomace extract with the granularity of <125 μm; F2—grape pomace extract with the granularity of ≥125–<200 μm; F3—grape pomace extract with the granularity of ≥200–<300 μm; 7—7 g of gelatin; 8.5—8.5 g of gelatin; 10—10 g of gelatin).

3.11. Microbiological Analysis

The results of the microbiological analysis (total plate count, mold, and yeast counts) of grape pomace jellies are presented in Table 3. The total plate count (TP count) and mold and yeast counts (MY counts) were significantly influenced by the particle size of BN grape pomace and the ratio of gelatin used for jelly formulation ($p < 0.0001$). The highest values for TP count (95.3 CFU/mL) and MY counts (74.1 CFU/mL) were obtained for jelly formulation F3–10; this can be explained by the fact that F3–10 had the highest contents of fats and proteins. Moreover, gelatin used as the gelling agent may facilitate the formation of mold and yeast due to gelatin consisting of 98–99% protein, which can be a source of microorganisms [59]. According to Thompson [59], jellies have values for water activity ranging between 0.65 and 0.75, which means that jellies are susceptible to spoilage by xerophilic molds and osmophilic yeasts. At the same time, there was no growth in other jelly samples, which may be a result of the high phenolic content of grape pomace. On the other hand, Ismawati et al. [60] reported a higher TP count (2.1×10^2 CFU/mL) and MY counts (8.9×10^1 CFU/mL) for seaweed jelly candy.

3.12. Principal Component Analysis

Principal component analysis (PCA) of the experimental findings was utilized to decompose the dataset into scores (jelly candies) and loadings (color parameters, phytochemical characteristics, and sensorial and textural parameters). As presented in Figure 5, the principal components described the better part of variability, with a total variance contribution of 67.13% (F1—1: 46.12%, F2—2: 21.01%). The biplot presents that the most of obtained characteristics demonstrated a significant influence on the distinctions among jelly samples. The jelly samples F2–7, F2–8.5, and F2–10 were correlated with sensorial parameters (e.g., sweetness, taste, odor, elasticity, color, and overall acceptability), gumminess, and h^*_{ab}, while formulations F3–8.5 and F3–10 were associated with hardness, cohesiveness, L*, and C^*_{ab}. The arrangement of the F1–7 and F2–8.5 samples more to the right was related to their high values of adhesiveness, and phytochemical characteristics (TPC and DPPH), while F3–7 and F1–10 jelly candies were not correlated with any of the studied parameters.

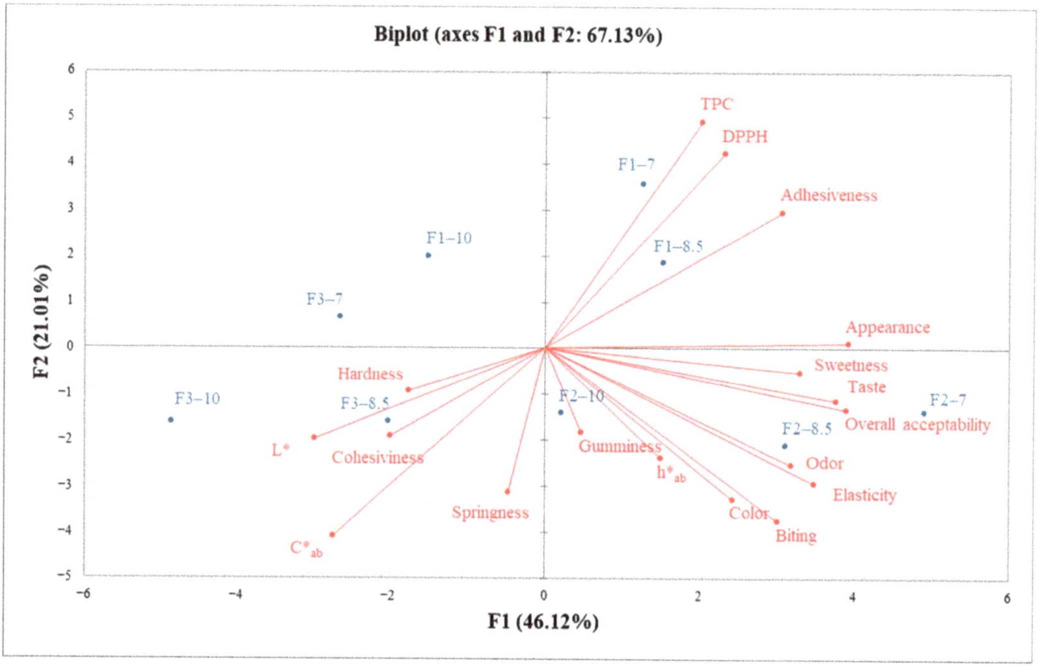

Figure 5. PCA biplot showing the correlation between scores (jelly samples, •) and loadings (color parameters, phytochemical characteristics, sensorial, and textural parameters, •); F1—grape pomace extract with the granularity of <125 µm; F2—grape pomace extract with the granularity of ≥125–<200 µm; F3—grape pomace extract with the granularity of ≥200–<300 µm; 7—7 g of gelatin; 8.5—8.5 g of gelatin; 10—10 g of gelatin; L*—luminosity; h^*_{ab}—hue; C^*_{ab}—chroma; TPC—total phenolic content.

4. Conclusions

In recent times, grape pomace has been used as a fortifying agent for a lot of food products (plant foodstuffs, dairy products, meat, and fish products). Concerning the valorization of by-products, it is very important to improve the nutritional value of products and obtain new products. Therefore, Băbească Neagră grape pomace was used as a fortifying agent in jelly candy. For a better understanding of how to fortify grape pomace, the proximate composition, functional properties, and physicochemical characteristics were analyzed. The obtained results were in agreement with FT-IR and SEM analysis of grape pomace. Furthermore, the grape pomace extract was analyzed in terms of phytochemical parameters and proximate composition; the results contributed to a more complete explanation of the jelly formulation and characterization of the final product. So, by-product-based jelly candies were formulated and had the following composition: grape pomace extract with different particle sizes (<125 µm, ≥125–<200 µm, and ≥200–<300 µm), gelatin (7, 8.5, and 10 g), and stevia. The obtained results showed that grape pomace jellies with the particle size of ≥125–<200 µm and different concentrations of gelatin had the greatest acceptance in terms of sensorial properties (e.g., sweetness, taste, odor, elasticity, color, and overall acceptability), while the highest value of phytochemical characteristics (156 mg GAE/g for total phenolic content and 65.8% inhibition of antioxidant activity) was acquired for jelly candy formulated with grape pomace extract (<125 μm) and 7 g of gelatin. The application of grape pomace extract as a natural ingredient in the formulation of jelly candy resulted in an improvement with a great concentration of polyphenols, along with uniform purple color, pleasant taste, and a high acceptance in terms of textural parameters (i.e., hardness, cohesiveness, adhesiveness, springiness, and gumminess). Thereby, all the data indicate the potential of Băbească Neagră grape pomace for use as a fortifying agent in jelly candies.

Supplementary Materials: The following supporting information can be downloaded at: https://www.mdpi.com/article/10.3390/foods13010098/s1, Table S1. The list and amount of ingredients used for jelly formulation; Figure S1. FT-IR spectra of Băbească Neagră grape pomace under the influence of different granularity.

Author Contributions: Conceptualization, M.S. and M.O.; methodology, M.S.; software, M.S.; validation, M.S. and M.O.; formal analysis, M.S. and M.O.; investigation, M.S.; resources, M.S. and M.O.; data curation, M.O.; writing—original draft preparation, M.S.; writing—review and editing, M.S. and M.O.; visualization, M.O.; supervision, M.O. All authors have read and agreed to the published version of the manuscript.

Funding: Mariana Spinei was funded by the Romanian National Council for Higher Education Funding, CNFIS, project number CNFIS-FDI-2023-F-0579. Mircea Oroian was funded by the Ministry of Research, Innovation, and Digitalization within Program 1—Development of National Research and Development System, Subprogram 1.2—Institutional Performance—RDI Excellence Funding projects, under contract no. 10PFE/2021.

Institutional Review Board Statement: The study was conducted in accordance with the Declaration of Helsinki, and approved by the Ethics Committee of Stefan cel Mare University of Suceava (protocol code PO-04, 6 December 2016).

Informed Consent Statement: Informed consent was obtained from all subjects involved in the study.

Data Availability Statement: Data is contained within the article or supplementary material.

Acknowledgments: Mariana Spinei was supported by the Romanian National Council for Higher Education Funding, CNFIS, project number CNFIS-FDI-2023-F-0579. Mircea Oroian was supported by the Ministry of Research, Innovation, and Digitalization within Program 1—Development of National Research and Development System, Subprogram 1.2—Institutional Perfor-mance—RDI Excellence Funding projects, under contract no. 10PFE/2021.

Conflicts of Interest: The authors declare no conflicts of interest.

References

1. Spinei, M.; Oroian, M. The Potential of Grape Pomace Varieties as a Dietary Source of Pectic Substances. *Foods* **2021**, *10*, 867. [CrossRef] [PubMed]
2. Kokkinomagoulos, E.; Kandylis, P. Grape Pomace, an Undervalued By-Product: Industrial Reutilization within a Circular Economy Vision. *Rev. Environ. Sci. Bio/Technol.* **2023**, *22*, 739–773. [CrossRef]
3. Markus, K. *The Science of Grapevines*; Elsevier: Amsterdam, The Netherlands, 2020; ISBN 9780128163658.
4. Castellanos-Gallo, L.; Ballinas-Casarrubias, L.; Espinoza-Hicks, J.C.; Hernández-Ochoa, L.R.; Muñoz-Castellanos, L.N.; Zermeño-Ortega, M.R.; Borrego-Loya, A.; Salas, E. Grape Pomace Valorization by Extraction of Phenolic Polymeric Pigments: A Review. *Processes* **2022**, *10*, 469. [CrossRef]
5. Pisciotta, A.; Barone, E.; Di Lorenzo, R. Table-Grape Cultivation in Soil-Less Systems: A Review. *Horticulturae* **2022**, *8*, 553. [CrossRef]
6. Alston, J.M.; Sambucci, O. Grapes in the World Economy. In *The Grape Genome*; Springer: Cham, Switzerland, 2019; pp. 1–24.
7. OIV. *State of the World Vine and Wine Sector in 2022*; OIV: Paris, France, 2022.
8. OIV. *Managing By-Products of Vitivinicultural Origin*; OIV: Paris, France, 2018.
9. Antonić, B.; Jančíková, S.; Dordević, D.; Tremlová, B. Grape Pomace Valorization: A Systematic Review and Meta-Analysis. *Foods* **2020**, *9*, 1627. [CrossRef] [PubMed]
10. Rashad, M.; Mahgoub, R.; Abedelmaksoud, T.G. Functional Strawberry and Red Beetroot Jelly Candies Rich in Fibers and Phenolic Compounds. *Food Syst.* **2021**, *4*, 82–88.

11. Nicolescu, C.M.; Bumbac, M.; Radulescu, C.; Buruleanu, L.C.; Olteanu, R.L.; Gorghiu, L.M.; Teodorescu, G.; Holban, C.G. Romanian Organic and Conventional Red Grapes Vineyards as Potential Sources of High Value-Added Products, in a Circular Economy Approach. In *Grapes and Wine*; IntechOpen: London, UK, 2022; ISBN 978-1-83969-642-8.
12. Indreaş, A.; Vişan, L. *Principalele Soiuri de Struguri de Vin Cultivate În România*; Ceres: Bucharest, Romania, 2001.
13. MARD Operational Data Portal of Ministry of Agriculture and Rural Development, Romania. Available online: https://www.madr.ro/horticultura/viticultura-vinificatie.html (accessed on 20 October 2023).
14. Capanoglu, E.; Tomás-Barberán, F.A. Introduction to Novel Approaches in the Valorization of Agricultural Wastes and Their Applications. *J. Agric. Food Chem.* **2022**, *70*, 6785–6786. [CrossRef]
15. Altınok, E.; Palabiyik, I.; Gunes, R.; Toker, O.S.; Konar, N.; Kurultay, S. Valorisation of Grape By-Products as a Bulking Agent in Soft Candies: Effect of Particle Size. *LWT* **2020**, *118*, 108776. [CrossRef]
16. AOAC. *Official Methods of Analysis of the AOAC International*; AOAC: Gaithersburg, MD, USA, 2006.
17. Spinei, M.; Oroian, M. The Influence of Osmotic Treatment Assisted by Ultrasound on the Physico-Chemical Characteristics of Blueberries (*Vaccinium Myrtillus* L.). *Ultrasonics* **2021**, *110*, 106298. [CrossRef]
18. Zhang, Z.; Song, H.; Peng, Z.; Luo, Q.; Ming, J.; Zhao, G. Characterization of Stipe and Cap Powders of Mushroom (*Lentinus edodes*) Prepared by Different Grinding Methods. *J. Food Eng.* **2012**, *109*, 406–413. [CrossRef]
19. Gao, W.; Chen, F.; Zhang, L.; Meng, Q. Effects of Superfine Grinding on Asparagus Pomace. Part I: Changes on Physicochemical and Functional Properties. *J. Food Sci.* **2020**, *85*, 1827–1833. [CrossRef] [PubMed]
20. Spinei, M.; Oroian, M. The Influence of Extraction Conditions on the Yield and Physico-Chemical Parameters of Pectin from Grape Pomace. *Polymers* **2022**, *14*, 1378. [CrossRef] [PubMed]
21. Guiné, R.P.F.; Correia, P.M.R.; Reis, C.; Florença, S.G. Evaluation of Texture in Jelly Gums Incorporating Berries and Aromatic Plants. *Open Agric.* **2020**, *5*, 450–461. [CrossRef]
22. Figueroa, L.E.; Genovese, D.B. Fruit Jellies Enriched with Dietary Fibre: Development and Characterization of a Novel Functional Food Product. *LWT* **2019**, *111*, 423–428. [CrossRef]
23. Milincic, D.; Kostic, A.; Spirovic-Trifunovic, B.; Tesic, Z.; Tosti, T.; Dramicanin, A.; Barac, M.; Pesic, M. Grape Seed Flour of Different Grape Pomaces: Fatty Acid Profile, Soluble Sugar Profile and Nutritional Value. *J. Serb. Chem. Soc.* **2020**, *85*, 305–319. [CrossRef]
24. Konuskan, D.B.; Kamiloglu, O.; Demirkeser, O. Fatty Acid Composition, Total Phenolic Content and Antioxidamt Activity of Grape Seed Oils Obtained by Cold-Pressed and Solvent Extraction. *Indian J. Pharm. Educ. Res.* **2019**, *53*, 144–150. [CrossRef]
25. Tolve, R.; Simonato, B.; Rainero, G.; Bianchi, F.; Rizzi, C.; Cervini, M.; Giuberti, G. Wheat Bread Fortification by Grape Pomace Powder: Nutritional, Technological, Antioxidant, and Sensory Properties. *Foods* **2021**, *10*, 75. [CrossRef]
26. Bender, A.B.B.; Speroni, C.S.; Moro, K.I.B.; Morisso, F.D.P.; dos Santos, D.R.; da Silva, L.P.; Penna, N.G. Effects of Micronization on Dietary Fiber Composition, Physicochemical Properties, Phenolic Compounds, and Antioxidant Capacity of Grape Pomace and Its Dietary Fiber Concentrate. *LWT* **2020**, *117*, 108652. [CrossRef]
27. Zhang, L.; Zhu, M.; Shi, T.; Guo, C.; Huang, Y.; Chen, Y.; Xie, M. Recovery of Dietary Fiber and Polyphenol from Grape Juice Pomace and Evaluation of Their Functional Properties and Polyphenol Compositions. *Food Funct.* **2017**, *8*, 341–351. [CrossRef]
28. Ma, M.; Mu, T. Effects of Extraction Methods and Particle Size Distribution on the Structural, Physicochemical, and Functional Properties of Dietary Fiber from Deoiled Cumin. *Food Chem.* **2016**, *194*, 237–246. [CrossRef]
29. Jurevičiūtė, I.; Keršienė, M.; Bašinskienė, L.; Leskauskaitė, D.; Jasutienė, I. Characterization of Berry Pomace Powders as Dietary Fiber-Rich Food Ingredients with Functional Properties. *Foods* **2022**, *11*, 716. [CrossRef] [PubMed]
30. Dimitriu, L.; Preda, D.; Constantinescu-Aruxandei, D.; Oancea, F. Quince Pomace: A Source of Fiber Products and Polyphenols. In Proceedings of the NeXT-Chem 2023. *Chem. Proc.* **2023**, *13*, 6.
31. Liu, W.; Jing, H.; Ma, C.; Liu, C.; Lv, W.; Wang, H. Microstructure, Physicochemical and Functional Properties of *Dendrobium officinale* Pomace and Its Total Dietary Fiber. *J. Food Meas. Charact.* **2023**, *17*, 3927–3937. [CrossRef]
32. Jagelaviciute, J.; Basinskiene, L.; Cizeikiene, D.; Syrpas, M. Technological Properties and Composition of Enzymatically Modified Cranberry Pomace. *Foods* **2022**, *11*, 2321. [CrossRef] [PubMed]
33. Li, M.; Liu, Y.; Yang, G.; Sun, L.; Song, X.; Chen, Q.; Bao, Y.; Luo, T.; Wang, J. Microstructure, Physicochemical Properties, and Adsorption Capacity of Deoiled Red Raspberry Pomace and Its Total Dietary Fiber. *LWT* **2022**, *153*, 112478. [CrossRef]
34. Oladiran, D.A.; Emmambux, N.M. Nutritional and Functional Properties of Extruded Cassava-soy Composite with Grape Pomace. *Starch—Stärke* **2018**, *70*, 1700298. [CrossRef]
35. Wang, L.; Xu, H.; Yuan, F.; Pan, Q.; Fan, R.; Gao, Y. Physicochemical Characterization of Five Types of Citrus Dietary Fibers. *Biocatal. Agric. Biotechnol.* **2015**, *4*, 250–258. [CrossRef]
36. Wojdyło, A.; Samoticha, J.; Chmielewska, J. Effect of Different Pre-Treatment Maceration Techniques on the Content of Phenolic Compounds and Color of Dornfelder Wines Elaborated in Cold Climate. *Food Chem.* **2021**, *339*, 127888. [CrossRef]
37. Makris, D.P.; Kefalas, P. Characterization of Polyphenolicphytochemicals in Red Grape Pomace. *Int. J. Waste Resour.* **2013**, *3*, 126. [CrossRef]
38. Dhua, S.; Kumar, K.; Sharanagat, V.S.; Nema, P.K. Bioactive Compounds and Its Optimization from Food Waste: Review on Novel Extraction Techniques. *Nutr. Food Sci.* **2022**, *52*, 1270–1288. [CrossRef]

39. Barriga-Sánchez, M.; Hiparraguirre, H.C.; Rosales-Hartshorn, M. Chemical Composition and Mineral Content of Black Borgoña (*Vitis Labrusca* L.) Grapes, Pomace and Seeds, and Effects of Conventional and Non-Conventional Extraction Methods on Their Antioxidant Properties. *Food Sci. Technol.* **2022**, *42*, e120021. [CrossRef]
40. Xu, Y.; Burton, S.; Kim, C.; Sismour, E. Phenolic Compounds, Antioxidant, and Antibacterial Properties of Pomace Extracts from Four Virginia-grown Grape Varieties. *Food Sci. Nutr.* **2016**, *4*, 125–133. [CrossRef] [PubMed]
41. Carmona-Jiménez, Y.; García-Moreno, M.V.; García-Barroso, C. Effect of Drying on the Phenolic Content and Antioxidant Activity of Red Grape Pomace. *Plant Foods Hum. Nutr.* **2018**, *73*, 74–81. [CrossRef] [PubMed]
42. Zhao, X.; Zhu, H.; Zhang, G.; Tang, W. Effect of Superfine Grinding on the Physicochemical Properties and Antioxidant Activity of Red Grape Pomace Powders. *Powder Technol.* **2015**, *286*, 838–844. [CrossRef]
43. Fu, C.; Yang, X.; Lai, S.; Liu, C.; Huang, S.; Yang, H. Structure, Antioxidant and α-Amylase Inhibitory Activities of Longan Pericarp Proanthocyanidins. *J. Funct. Foods* **2015**, *14*, 23–32. [CrossRef]
44. Lucarini, M.; Durazzo, A.; Kiefer, J.; Santini, A.; Lombardi-Boccia, G.; Souto, E.; Romani, A.; Lampe, A.; Ferrari Nicoli, S.; Gabrielli, P.; et al. Grape Seeds: Chromatographic Profile of Fatty Acids and Phenolic Compounds and Qualitative Analysis by FTIR-ATR Spectroscopy. *Foods* **2019**, *9*, 10. [CrossRef] [PubMed]
45. Mohammed, A.; Avin, A. Scanning Electron Microscopy (SEM): A Review. In Proceedings of the 2018 International Conference on Hydraulics and Pneumatics—HERVEX, Băile Govora, Romania, 7–9 November 2018; pp. 77–85.
46. Pala, M.; Kantarli, I.C.; Buyukisik, H.B.; Yanik, J. Hydrothermal Carbonization and Torrefaction of Grape Pomace: A Comparative Evaluation. *Bioresour. Technol.* **2014**, *161*, 255–262. [CrossRef]
47. Gowman, A.C.; Picard, M.C.; Rodriguez-Uribe, A.; Misra, M.; Khalil, H.; Thimmanagari, M.; Mohanty, A.K. Physicochemical Analysis of Apple and Grape Pomaces. *BioResources* **2019**, *14*, 3210–3230. [CrossRef]
48. Baldán, Y.; Riveros, M.; Fabani, M.P.; Rodriguez, R. Grape Pomace Powder Valorization: A Novel Ingredient to Improve the Nutritional Quality of Gluten-Free Muffins. *Biomass Convers. Biorefinery* **2023**, *13*, 9997–10009. [CrossRef]
49. Drevelegka, I.; Goula, A.M. Recovery of Grape Pomace Phenolic Compounds through Optimized Extraction and Adsorption Processes. *Chem. Eng. Process.—Process Intensif.* **2020**, *149*, 107845. [CrossRef]
50. Cano-Lamadrid, M.; Calín-Sánchez, Á.; Clemente-Villalba, J.; Hernández, F.; Carbonell-Barrachina, Á.A.; Sendra, E.; Wojdyło, A. Quality Parameters and Consumer Acceptance of Jelly Candies Based on Pomegranate Juice "Mollar de Elche". *Foods* **2020**, *9*, 516. [CrossRef] [PubMed]
51. De Nishiyama-Hortense, Y.P.O.; Rossi, M.J.d.P.; Shimizu-Marin, V.D.; Janzantti, N.S.; Gómez-Alonso, S.; Da-Silva, R.; Lago-Vanzela, E.S. Jelly Candy Enriched with BRS Violeta Grape Juice: Anthocyanin Retention and Sensory Evaluation. *Future Foods* **2022**, *6*, 100179. [CrossRef]
52. Hani, N.M.; Romli, S.R.; Ahmad, M. Influences of Red Pitaya Fruit Puree and Gelling Agents on the Physico-mechanical Properties and Quality Changes of Gummy Confections. *Int. J. Food Sci. Technol.* **2015**, *50*, 331–339. [CrossRef]
53. Shahu, R.; Jobby, R.; Patil, S.; Bhori, M.; Tungare, K.; Jha, P. Phytochemical Content and Antioxidant Activity of Different Varieties of *Stevia rebaudiana*. *Hortic. Environ. Biotechnol.* **2022**, *63*, 935–948. [CrossRef]
54. Garrido, J.I.; Lozano, J.E.; Genovese, D.B. Effect of Formulation Variables on Rheology, Texture, Colour, and Acceptability of Apple Jelly: Modelling and Optimization. *LWT—Food Sci. Technol.* **2015**, *62*, 325–332. [CrossRef]
55. Akesowan, A. Optimization of Textural Properties of Konjac Gels Formed with κ-Carrageenan or Xanthan and Xylitol as Ingredients in Jelly Drink Processing. *J. Food Process. Preserv.* **2015**, *39*, 1735–1743. [CrossRef]
56. Rios de Souza, V.; Aparecida Pimenta Pereira, P.; Carla Marques Pinheiro, A.; Carlos de Oliveira Lima, L.; Pio, R.; Queiroz, F. Analysis of the Subtropical Blackberry Cultivar Potential in Jelly Processing. *J. Food Sci.* **2014**, *79*, S1776–S1781. [CrossRef]
57. Mutlu, C.; Tontul, S.A.; Erbaş, M. Production of a Minimally Processed Jelly Candy for Children Using Honey Instead of Sugar. *LWT* **2018**, *93*, 499–505. [CrossRef]
58. Curi, P.N.; de Sousa Tavares, B.; de Almeida, A.B.; Pio, R.; Peche, P.M.; de Souza, V.R. Influence of Subtropical Region Strawberry Cultivars on Jelly Characteristics. *J. Food Sci.* **2016**, *81*, S1515–S1520. [CrossRef]
59. Thompson, S. Microbiological Spoilage of High-Sugar Products. In *Compendium of the Microbiological Spoilage of Foods and Beverages*; Sperber, W.H., Doyle, M.P., Eds.; Springer: New York, NY, USA, 2009; pp. 301–324.
60. Ismawati, R.; Romadhoni, I.F.; Nurul, Q.T.; Ratna, S.T. Storability of Seaweed Jelly Candy Based on Chemical, Physical and Microbiology Characteristics. In Proceedings of the IOP Conference Series: Earth and Environmental Science; IOP: Bristol, UK, 2019; Volume 347, p. 012021.

foods

MDPI

Article

The Impact of Oil Type on the Performance of β-Amyrin-Based Oleogels: Formation, Physicochemical Properties, and Potential Correlation Analysis

Shuxian Su [1], Si Qin [1], Huiping Xia [1], Peiwang Li [2], Haiyan Li [1], Chenjia Li [1], Shiyin Guo [1,3,*] and Chaoxi Zeng [1,*]

[1] Department of Food Science and Technology, College of Food Science and Technology, Hunan Agricultural University, No. 1 Nongda Road, Furong District, Changsha 410128, China; sushuxian1118@stu.hunau.edu.cn (S.S.); qinsiman@hunau.edu.cn (S.Q.); hpxia@hunau.edu.cn (H.X.); 13578374900@stu.hunau.edu.cn (H.L.); chenjiali@stu.hunau.edu.cn (C.L.)

[2] State Key Laboratory of Utilization of Woody Oil Resource, Hunan Academy of Forestry, 658 Shaoshan South Road, Tianxin District, Changsha 410128, China; 17184655@163.com

[3] College of Life Sciences and Health, Hunan University of Science and Technology, No. 1 Taoyuan Road, Xiangtan 411201, China

* Correspondence: gsy@hunau.edu.cn (S.G.); chaoxizeng@hunau.edu.cn (C.Z.); Tel.: +86-15074909106 (S.G.); +86-18670047526(C.Z.)

Abstract: Pentacyclic triterpenes show potential as oleogelators, but their combination with various vegetable oils has limited research. This study selected linseed, rapeseed, sunflower, coconut, and palm oils to combine with the triterpenoid compound β-amyrin for the preparation of oleogels. The stability, crystal network structure, and other properties of each oleogel were evaluated. The correlation between different oil types and the properties of corresponding oleogels was explored. The results showed that β-amyrin formed stable oleogels with five vegetable oils under suitable temperature conditions, wherein especially the LO-based oleogel not only exhibited higher oil-binding capacity and hardness, but also demonstrated excellent stability at the microscopic level and notable rheological properties. Further analysis revealed a close correlation between the physicochemical properties of the oleogels and lipid characteristics, indicating that oleogels prepared from long-chain highly unsaturated fatty acids exhibit high stability. The above results indicate that β-amyrin can be a novel candidate oleogelator and that the oil type can modify the properties of β-amyrin-based oleogels. This study provides the latest reference for the application of pentacyclic triterpenoids in food.

Keywords: β-amyrin; oleogel; fatty acid properties; stability of the oleogels; correlational analysis

1. Introduction

Foods made of oils and fats provide an important energy source for the human body. A network of fat crystals composed of trans fats and saturated fats produced by hydrogenation, transesterification, or blending comprises a large market share of margarine, chocolate, and other foods rich in solid fats [1,2]. Due to excessive intake of saturated fatty acids and trans fatty acids, cardiovascular disease is induced by increased levels of low-density lipoprotein (LDL), and the risk of type II diabetes mellitus is increased by inducing insulin resistance [3–5]. To be more in line with the contemporary concept of healthy eating and to promote the progress of the energy-saving food market, new technologies or methods should be used to prepare products that are comparable to traditional solid fats.

Oleogels preparation, a technique that imparts solid characteristics to liquid vegetable oils, has emerged as a novel and efficient approach to prepare solid fats. It can enhance not only the product's qualities such as shelf life, but also improve the quality of food nutrition and nutritional value [4,6,7]. The oleogelator used to make oleogels can be classified into two categories: small molecule structuring agents and large molecule structuring agents, depending on the relative molecular mass of the oleogelator. The first category comprises natural waxes, glycerol esters, fatty acids, fatty alcohols, lecithin, mono- and diglycerides of glycerol, and similar substances [1,8,9]. The second category includes polymers and proteins and their derivative structural agents such as proteins, ethyl cellulose, hydroxypropyl methyl cellulose, methyl cellulose, chitosan, chitin, and so on [10–12]. Presently, numerous researchers have investigated diverse applications of oleogels in food production, such as for alternatives to margarine [13], as well as the creation of functional chocolates [14], ice creams [15], cakes [16], and more. Nevertheless, the existing research on oleogels falls short of fully addressing the present requirements for practical use, mostly due to the limited variety of oleogelators that are currently available. Consequently, the current focus in oleogel research is the active investigation of novel oleogelators.

Prior research has indicated that pentacyclic triterpenoids, specifically ursolic acid and oleanolic acid, possess the capability to induce gelation in organic solvents, resulting in the formation of organogels [17,18]. To the best of our knowledge, pentacyclic triterpenoids have not been used to prepare food-grade oleogels. In contrast, β-amyrin, as a class of pentacyclic triterpenoids, is an effective functional ingredient in many food products and traditional Chinese herbal medicines, as well as a characteristic triterpenoid in camellia oil [19]. β-amyrin has anticancer, anti-inflammatory, anti-tumor, anti-diabetic, anti-dyslipidemia, and antidepressant effects [20,21]. It has a wide range of potential uses for healthy food and drug development. Therefore, in this manuscript, we chose to use β-amyrin as an oleogelator for the preparation of oleogels from different vegetable oils to investigate its effect on the properties of the formed oleogels.

The formation of an oleogel is a three-dimensional network structure formed by the interaction of oil with liquid oil during the cooling process, such as hydrogen bonding, van der Waals forces, and dipole–dipole forces. The characteristics of the network structure formed are related to the chain length and unsaturation of the liquid oil [1,22]. Therefore, a large number of studies have examined the effect of different oils on the properties of oleogels, and the results indicate that different oil types are one of the main factors affecting the physicochemical properties of oleogels [23–25]. Higher unsaturation increases the crystallinity and network stability of oil-based oleogels, according to Han et al. [1]. Pang et al. found that the degree of unsaturation of the oil phase of the prepared oleogel was proportional to the hardness of the oleogel [13]. Liu et al. discovered that variability in oil polarity due to oil types also affects oleogel performance [26]. Valoppi et al. and Chai et al. pointed out that different types of oils may also affect the formation of the network and thus the final structure of the oleogel due to differences in viscosity [27]. Many studies have analyzed oleogels made from different cooking oils and examined how oil type affects them. However, the effects of different oil types on the properties and mechanisms of the oleogel formed based on triterpenoids such as β-amyrin are still unknown.

This study aimed to investigate the effect of different oil types on the oil properties of β-amyrin gels at different preparation temperatures and to reveal the correlation relationship between different types of oils and the physical properties of the oleogel. In this study, we first examined the possibility of preparing oleogels from β-amyrin and then investigated the probable relationships and correlations that exist between oil types and the physicochemical structure of the oleogels as well as the potential molecular roles during the gelling process of the organogels. This work helps to provide insight into the preparation and properties of pentacyclic triterpenoid oleogels.

2. Materials and Methods

2.1. Materials

Linseed oil, rapeseed oil, coconut oil, and sunflower oil were purchased from the local market, palm oil was purchased from Yihai Kerry Group, and β-amyrin (purity > 98%) was purchased from Shanxi Jinkangtai Biotechnology Co., Ltd. (Xianyang, China), which can be used directly without purification. All other chemical reagents were analytically pure.

Fatty acid composition was analyzed using gas chromatography. Gas chromatography analysis conditions: using Agilent 7890A gas chromatograph; Agilent J&WDB-FATWAX ultra-high inert column; detector FID; inlet temperature 270 °C, detector temperature 280 °C. Chromatographic column heating program: initial temperature 100 °C, keep 13 min, 10 °C/min to 180 °C, keep 6 min, then 1 °C/min to 200 °C, keep 20 min, then 4 °C/min to 230 °C, keep 10.5 min, until the analysis is complete. The carrier gas is pure nitrogen at a flow rate of 30 cm/s, constant flow mode, hydrogen flow rate of 40 mL/min; air flow rate: 400 mL/min, the tail gas is nitrogen at a flow rate of 25 mL/min, the injection volume is 1 μL, and the sample is injected by shunt mode, the shunt ratio is 100:1. The qualitative analysis of fatty acids in the five samples was conducted by comparing the retention times of fatty acids in the standard reference solution. The quantitative analysis of fatty acid content was performed using the area normalization method in the Agilent Chemstation software, which provided the relative percentage of each fatty acid. The analysis was performed in triplicate for each sample.

2.2. Oleogel Preparation

Eight grams of linseed oil, rapeseed oil, palm oil, coconut oil, and sunflower oil were weighed and added with 4% β-amyrin, respectively. The mixture was heated and stirred for 30 min at certain temperatures (60 °C, 80 °C, 100 °C, 120 °C, or 140 °C) with a speed of 500 rpm. All samples were stored at room temperature (25 ± 2 °C) for 24 h before detection. The gelability of the samples was determined by the inverted tube method. Oleogel samples with β-amyrin concentrations of 1–5% (*w*/*w*) were also prepared to determine the critical gel concentrations for different oil phases.

2.3. Oil Binding Capacity

The Oil-Binding Capacity (OBC) of the oleogel was analyzed according to the method described by Da et al. [28]. The oleogel samples prepared under various temperature conditions were collected in 1.5 mL centrifuge tubes. Following high-speed centrifugation at 13,000 rpm for 10 min, the tubes were inverted onto filter paper and allowed to stand for 30 min to ensure complete drainage of any remaining liquid oil. The OBC was calculated using the provided formula 1, and the experiment was repeated three times with the average of the results taken. Additionally, the oleogel samples were kept at room temperature for 0, 14, and 28 days to compare the differences in OBC among the different samples.

$$\mathrm{OBC} = \frac{c-a}{b-a} \times 100\% \tag{1}$$

where c denotes the weight of the oleogel and tube after centrifugation (g), b denotes the weight of the oleogel and tube before centrifugation (g), and a denotes the weight of the empty tube (g).

2.4. Texture Analysis

A texture analyzer (TA. XT Plus, Surrey, UK) test was performed to determine the textural properties of the oleogels [29]. The samples prepared at 80 °C, 100 °C, and 120 °C were left at 25 °C for 24 h before measurement. The texture analyzer was set with the parameters of a test speed of 1 mm/s and a compression distance of 10.00 mm. Each sample was tested three times, and the average value was taken. The oleogel samples were then kept at 25 °C for 0, 14, and 28 days, followed by hardness measurements under the same conditions.

2.5. Polarized Light Microscopy

The microstructure of the oleogel was observed using a polarized microscope (Olympus, BX53M, Tokyo, Japan). To observe the crystalline morphology of the oleogel at different heating temperatures, the samples were photographed at 25 °C using a 100× magnification (10× eyepiece and 10× objective) under the polarizing microscope. An appropriate amount of the oleogel was taken and placed on a clean slide. The slide was then covered to ensure even distribution of the oleogel.

ImageJ software (NIH, Bethesda, MD, USA) with the FracLac plugin was used to calculate fractal dimension (Db), average length (L), and lacunarity (A) in two-dimensional space for quantitative analysis. The box-counting method was employed to calculate these parameters. The formula for the box-counting method is as follows:

$$\mathrm{Db} = -\frac{\mathrm{logN}\varepsilon}{\mathrm{log}\varepsilon} \quad (2)$$

2.6. X-ray Diffraction Analysis

The crystalline shape of the oleogel samples was analyzed by the X-diffraction technique (XRD-6000 type, Shimadzu, Kyoto, Japan) by referring to the method of Blake et al. [30]. The instrument test conditions were set as follows: Cu target as the radiation source (wavelength λ = 1.5401 Å), operating voltage 40 kV, operating current 40 mA, scanning step size 0.02°, scanning rate 2°/min, emitting and antireflecting slit 1.0 mm, receiving slit 0.1 mm, test temperature 25 °C, and scanning range of 5.0–55.0° at 2θ angle.

2.7. Differential Scanning Calorimetry

Samples were prepared for differential scanning calorimetry (DSC-822, Mettler Toledo, Zurich, Switzerland) measurements by placing 5 mg of oleogel in an aluminum pan. Each sample was kept at 120 °C for 5 min until completely melted to eliminate crystal memory. The temperature was reduced to −65 °C of cooling rates of 5 °C/min and 10 °C/min. After holding for 5 min, the temperature was raised to 30 °C at the same heating rate. Melting and crystallization curves of the samples under different oil treatments were recorded. The onset dissolution temperature (Ton-m), the melting peak temperature (Tm), the onset crystallization temperature (Ton-c), and the crystallization peak temperature (Tc) of the samples were measured utilizing the DSC curves provided by the TA Instruments Universal Analysis Software. All measurements were repeated three times.

2.8. The Crystallization Kinetics of Oleogel

The heat flow rate obtained during the crystallization process during DSC measurements was integrated against time using Origin software. The Avrami equation was used to study the crystallization kinetics of the oleogel under isothermal conditions, and the relative crystallinity X(t) at time t was obtained, which is of the following form:

$$\mathrm{X(t)} = 1 - \exp(-\mathrm{kt}^{\mathrm{n}}) \quad (3)$$

where X(t) is the crystallinity at time t; t is the crystallization time; k is the complex crystallization rate constant; and n is the Avrami index.

Equation (1) was obtained by deforming the sample according to its relative crystallinity X(t):

$$\ln[\ln(1 - \mathrm{Xt})] = \ln\mathrm{k} + \mathrm{n}\ln\mathrm{k} \quad (4)$$

2.9. Rheological Measurement

Analysis of the rheological behavior of the oleogels was performed using a rotational rheometer (Kinexus Pro+ Rheometer, Malvern, London, UK). Strain scanning experiments were carried out on 40.0 nm diameter flat plates with a plate spacing of 1 mm, a constant frequency of 1 Hz, and a strain range of 0.01–10% at room temperature to determine the linear viscoelastic region. Shear viscosity experiments were carried out at shear rates between 0.1–100 s^{-1}. Frequency scanning experiments were carried out at frequencies between 0.01 and 10 Hz with a fixed strain and the rheological behavior of the oleogel was analyzed using the power law equation.

$$G' = a\,(\omega)^b \quad (5)$$

where G′ is the storage modulus (Pa), a (Pa·b) is a parameter used to describe the rheological behavior, ω is the frequency (rad/s), and b is a dimensionless flow behavior index.

2.10. Statistical Analysis

Significant differences between means were determined using Duncan's analysis of variance (ANOVA), and multiple comparisons between treatments were analyzed using ANOVA with significant differences at the 5% level ($p < 0.05$). Each experiment was repeated thrice, and results were expressed as mean ± standard deviation. Statistical analysis was performed using Origin 2021.

3. Results and Discussion

3.1. Concentration Phase Diagrams

This study investigated how different types of oils influenced the oil phase transition content in β-amyrin oleogel at various preparation temperatures. The results showed that β-amyrin was successfully used to prepare oleogels at high and appropriate temperatures. Additionally, various oils exhibited similar oleogel formation trends at different temperatures (Figure 1a). When the preparation temperature is below 80 °C, the samples did not exhibit any thickening or gelation, and the formation of a gel-to-oil phase was not feasible. This lack of oleogel formation can be attributed to the fact that β-amyrin is unable to establish a stable three-dimensional network structure at lower temperatures, thereby struggling to impede the flow of liquid oil [31,32]. As the preparation temperature rises, the intermolecular forces lead to the aggregation of β-amyrin particles, forming a three-dimensional network structure. This network structure hinders the flow of lipophilic liquid, resulting in the gelation of the entire system [33]. However, when the temperature reaches 140 °C, the increase in temperature causes the unsaturated fatty acids in the vegetable oil to be oxidized and degraded [34].

The minimum concentration required to form an oleogel reflects the gelling ability of β-amyrin in different types of oils. The temperature during the preparation process affects the critical concentration required to form an oleogel. At a preparation temperature of 80 °C (Figure 1b), the critical concentration of β-amyrin for LO-based oleogel is 3%, while for SO and RO, it is 3.2% and 3.5%, respectively. The highest critical concentration is observed for CO at 3.7%. With an increase in the preparation temperature, there is an overall decreasing trend in the critical concentration of gel-forming oil. Consequently, within a specific temperature range, higher temperatures necessitate lower critical concentrations for the formation of an oleogel, implying a reduced quantity of gelating agent necessary to achieve gelation.

The type of oil significantly influences the critical concentration required for the formation of oleogels. At a preparation temperature of 100 °C, compared to other oil types, the critical concentration for oleogels formation in RO and CO are reduced to 3.0% and 3.4%, respectively. LO has the lowest β-amyrin concentration, with the essential concentration for a stable oleogel being 2.4%. From 100 °C to 120 °C, the viscous sol-gel state distribution range rose substantially. Other than for LO, which showed no change, the other four oils showed a slightly decreasing trend in critical concentration, but LO still had the lowest gelling concentration and PO and CO had greater critical concentrations. This result is consistent with the findings of Han et al. [1] and may be attributed to the higher content of linoleic acid in LO, which has the highest degree of unsaturation. This high unsaturation level is favorable for β-amyrin to undergo one-dimensional growth, helix formation, and twisting, resulting in a more stable and robust three-dimensional network structure [33,35–37]. On the other hand, CO and PO are rich in short-chain saturated fatty acids, which overall decrease the ability to bind oils and lower the critical concentration for gel formation. Therefore, the gel network structure formed by the self-assembly of oils rich in long-chain unsaturated fatty acids and β-amyrin in a binary system exhibits enhanced stability at the appropriate preparation temperature.

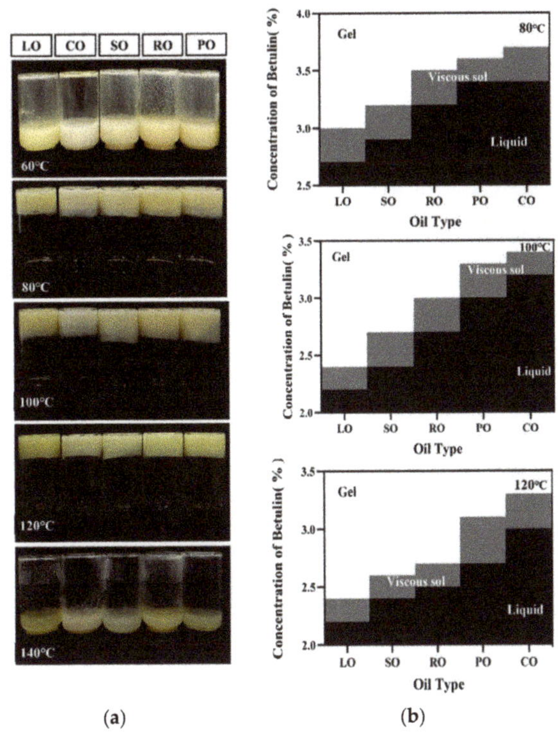

Figure 1. Gel property analysis of β-amyrin-based oleogel. (**a**) The appearance of 4% β-amyrin-based oleogels at different preparation temperatures; (**b**) shows the qualitative phase diagram of different concentrations of β-amyrin-based oleogel under different oil conditions: liquid (no gelation occurred), viscous sol (viscous solution, with fluidity), and gel (i.e., free-standing gel).

3.2. Oil Binding Capacity and Hardness Analysis

Hardness and Oil Binding Capacity (OBC) are used to evaluate the ability of oleogels to achieve structural properties similar to solid fat using a minimal amount of oleogelators, without significantly impacting the sensory characteristics of the product [38]. As temperature rises, oleogel OBC increases significantly (Figure 2a–c). The network structure of an oleogel may catch more liquid oil as the temperature rises, increasing OBC. Figure 2d–f indicates a significant increase in hardness as the preparation temperature rises. Particularly, at a preparation temperature of 80 °C, the CO-based oleogel, which is rich in saturated fatty acids, exhibits a low hardness value of 4.35 g. However, at a preparation temperature of 120 °C, the hardness value of the CO-based oleogel increases to 17.44 g. In addition, according to Figure S1, it can be concluded that the correlation coefficient between temperature and hardness values is greater than 0.98, showing a linear positive correlation. The reason for this could be that as the temperature increases, during the same cooling process, non-covalent bonds spontaneously assemble into aggregates, increasing the entanglement of the three-dimensional network formed by β-amyrin. Therefore, the flow of liquid oil is restricted, leading to the appearance of solid-like properties.

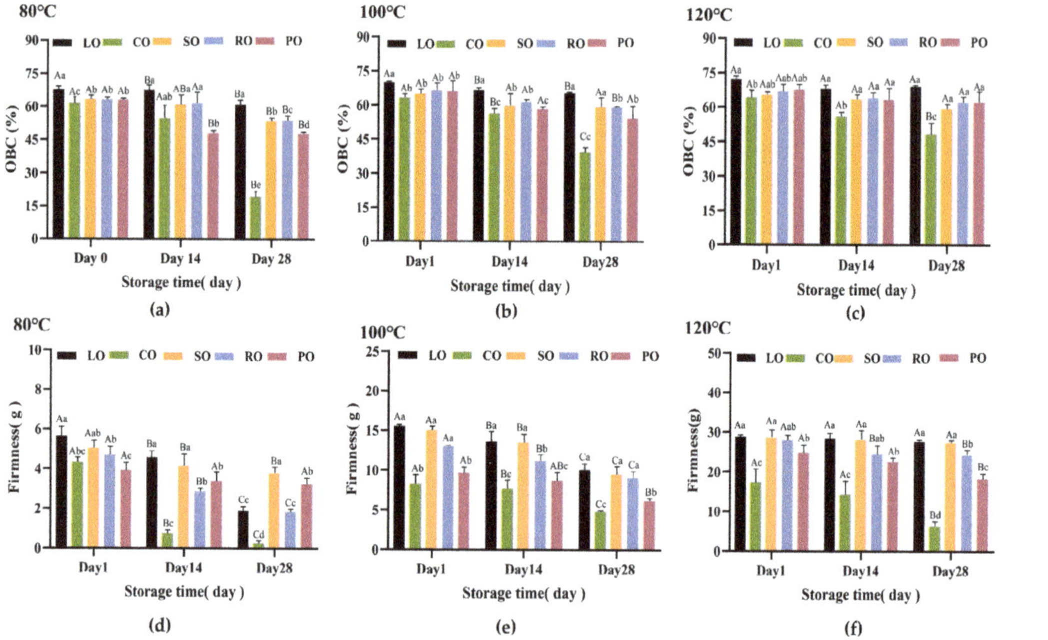

Figure 2. Physical stability of oleogels. (**a–c**) The changes in oil binding capacity values of oleogels prepared from 4% β-amyrin with LO, SO, RO, PO, and CO at temperatures of 80 °C, 100 °C, and 120 °C for 0, 14, and 28 days. (**d–f**) Hardness values of oleogels prepared from 4% β-amyrin with LO, SO, RO, PO, and CO at 80–120 °C temperatures on days 1, 14, and 28. Different capital letters indicate significant differences between the same samples at different stor-age times ($p < 0.05$). Different lowercase letters indicate significant differences between samples of different oil types for the same storage time ($p < 0.05$).

The hardness and OBC of oleogels are both significantly influenced by the unsaturation and the fatty acid chain length of the oils [32]. At a preparation temperature of 120 °C, the hardness values of oleogels prepared with LO, SO, and CO are 28.79 g, 28.73 g, and 17.44 g, respectively. The corresponding OBC values are 72.08%, 65.47%, and 59.59%, respectively. According to Table S1, the oleogel formed by C18:3-dominated LO is more tightly structured compared to oleogels prepared from other oil types and thus has significantly higher hardness [39]. This can be attributed to the unique conformation of the C18:3, which enables it to be more tightly encapsulated within the β-amyrin network. Additionally, the presence of polyunsaturated fatty acids promotes chain–chain interactions between the β-amyrin molecules, enhancing the structural and physical stability of the oleogel [4].

On the other hand, the oil type can also affect the physical properties of the oleogel through polarity differences. According to Table S2, it can be concluded that the total polar component (TPC) content is higher in LO at 16% [23], while it is only 4% in CO [40]. That finding provides evidence of a positive correlation between TPC and the degree of oil unsaturation. It may be because oils enriched with higher levels of polar compounds increase the thermally induced oxidation process, promoting the formation of hydrogen bonds between the solvent and the oleogelator. This contributes to the structural integrity of the gel matrix, thereby improving the stability of the textural morphology of the oleogel.

Oil types can affect the storage stability of oleogels. As the storage time increases, the OBC and hardness exhibit a decline at 80 °C and 100 °C. Notably, the OBC of the CO-based oleogel decreases significantly, potentially due to its higher content of saturated fatty acids [41]. At a preparation temperature of 120 °C, the OBC and hardness of the

LO-based oleogel remain relatively stable as the storage time increases. This finding aligns with the reported literature, which suggests a positive correlation between the degree of unsaturation in oils and the stability of oleogel hardness [33,42]. In conclusion, the OBC and hardness tended to be consistent over a certain storage time, and the fatty crystal network structure of the oleogel prepared from the long-chain oil type with high unsaturation had the best mechanical strength and better storage stability.

3.3. Polymorphism of Different Oleogels

The microscopic images of the oleogel were obtained using polarized light microscopy in Figure 3, aiming to explore the impact of different oil types on the microstructure of the β-amyrin oleogel's crystal network. As shown in the image, oleogel crystal shape and size differ at 80 °C. The CO-based oleogel has irregular long needle-like fibrous aggregated crystals, while the PO-based oleogel has bigger spherical crystals. The RO-based oleogel has a chrysanthemum shape, and the SO and LO-based oleogels have fine needle-like aggregated structures like those described by Frolova et al. [33]. Interestingly, at a preparation temperature of 100 °C, there is a significant alteration in the crystal morphology of the PO-based oleogel, with a predominant presence of uneven fibrous aggregates. As the preparation temperature is raised to 120 °C, all of the oleogels, except for the CO-based oleogel, exhibit similar features in terms of crystal morphology and arrangement. They undergo a transition from relatively dispersed crystals to a dense and uniformly stable crystal network, with consistent spacing between the crystals. In conclusion, as the preparation temperature increases, the crystal types of different oleogels transform into needle-like and fine fibrous structures, with the crystals gradually distributed more uniformly.

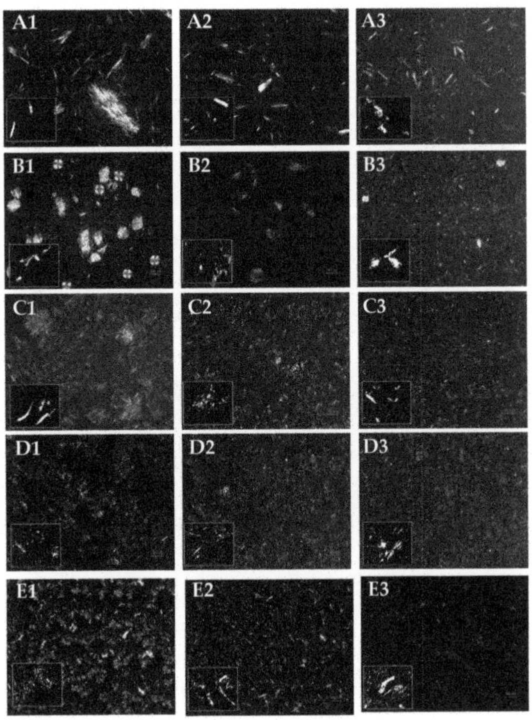

Figure 3. Microphotographs of oleogels prepared at different temperature conditions (80 °C, 100 °C, 120 °C). (**A1–A3**) CO-based oleogel, (**B1–B3**) PO-based oleogel, (**C1–C3**) RO-based oleogel, (**D1–D3**) SO-based oleogel, (**E1–E3**) LO-based oleogel. Scale bar: 100 μm.

Table 1 analyzes the fractal dimension (Db), average length (L), and lacunarity (A) of microscopic structures in different oleogels, investigating how oil types influence the uniformity of crystal network distribution in the oleogels. Smaller L-values mean better OBC of oleogel internal crystals and more oil-binding surface area. Larger Db values allow for quick nucleation and production of more crystals inside of the oleogel, resulting in smaller, denser, uniformly distributed crystal clusters that confine more liquid oil to a denser, more uniform crystal network. When the Db value is smaller and the A value is larger, crystals tend to aggregate, resulting in wider crystal spacing. This hinders the connectivity of the crystal network and results in weak interparticle interactions between β-amyrin particles [27,43]. Over time, this may decrease the hardness of the oleogel, decrease the OBC, and lead to phase separation of the oleogel [44,45].

Table 1. Average length fractal dimension (Db), average length (L), and lacunarity (A) of microstructures in oleogels prepared with different types of oils. Different lowercase letters within the same column indicate significant differences between samples of different oil types at the same preparation temperature ($p < 0.05$). Different uppercase letters within the same column indicate significant differences between samples of the same oil type at different temperatures ($p < 0.05$).

Types	Temperature (°C)	Db	L (μm)	A
LO	80	1.27 ± 0.03 Ca	35.08 ± 1.15 Ad	0.57 ± 0.00 Ab
	100	1.38 ± 0.05 Ba	30.96 ± 1.70 Ad	0.25 ± 0.02 Bd
	120	1.46 ± 0.02 Aa	21.09 ± 6.56 Bc	0.08 ± 0.00 Cd
CO	80	0.82 ± 0.03 Cc	61.10 ± 2.82 Aa	0.86 ± 0.02 Aa
	100	0.97 ± 0.04 Bd	52.10 ± 6.52 Aa	0.85 ± 0.01 Aa
	120	1.07 ± 0.02 Ad	50.54 ± 6.14 Aa	0.78 ± 0.03 Ba
SO	80	1.02 ± 0.01 Ca	42.54 ± 1.21 Ac	0.59 ± 0.04 Ab
	100	1.27 ± 0.05 Ab	35.05 ± 0.72 Bcd	0.36 ± 0.02 Bcd
	120	1.23 ± 0.12 Bbc	31.23 ± 1.28 Bbc	0.13 ± 0.04 Cc
RO	80	1.07 ± 0.04 Bb	45.59 ± 6.96 Abc	0.84 ± 0.02 Aa
	100	1.11 ± 0.02 Ac	41.08 ± 1.55 ABbc	0.45 ± 0.02 Bc
	120	1.29 ± 0.06 Ab	30.35 ± 5.83 Bbc	0.29 ± 0.03 Cbc
PO	80	0.85 ± 0.04 Bc	50.74 ± 2.67 Ab	0.86 ± 0.02 Aa
	100	1.06 ± 0.07 Acd	46.89 ± 2.34 ABab	0.68 ± 0.08 Bb
	120	1.14 ± 0.23 Ac	36.43 ± 8.57 Bb	0.28 ± 0.02 Cb

Based on the results in Table 1, the internal arrangement of crystals in the oleogels is temperature dependent. As the preparation temperature increases, the Db value increases gradually, while the A and L values decrease. This can be attributed to variations in the van der Waals forces between oils and β-amyrin molecules, resulting in differences in the size and structure of the β-amyrin crystals [1]. When comparing oleogels formed by different oil types at the same preparation temperature, it was observed that the LO-based oleogel had the smallest A and L and the largest Db value. This can be explained by the fact that, compared to oleogels prepared with other oils, the LO-based oleogel contains a higher amount of polyunsaturated fatty acids. This leads to smaller and more uniformly distributed crystal clusters in the oleogel, making it easier to achieve a stable gel effect. This further supports the reason for the changes in storage stability results.

Comparing the XRD spectra of β-amyrin (Figure S2), different oils, and β-amyrin gels based on various oil types (Figure S3), it is evident that the long-spacing peaks of β-amyrin diminish or vanish when it is dissolved in different edible oils. This can be attributed to the rearrangement and recrystallization of β-amyrin crystals as they aggregate in the presence of oils. Comparing the oleogel results of different oil types, it was observed that the amorphous peak areas were similar among the five oleogels in the wide-angle range ($2\theta > 15°$). However, the d-spacing of the oleogels varied, with the LO-based oleogel having the shortest d-spacing and the CO-based oleogel having the highest d-spacing. The observed phenomenon may be attributed to the distorted spatial structure of unsaturated fatty acid molecules in long chains, which affects the electron absorption capacity of the carbonyl group. As a result, a higher number of methyl groups exhibit asymmetric stretching and vibration, leading to the LO-based oleogel having stronger van der Waals forces and forming a stable oleogel [46]. Crystal morphology analysis revealed that different oleogels displayed a prominent diffraction peak at 2θ = 19.93 (corresponding to a d-spacing of 4.42 Å–4.6 Å). This suggests that the triglycerides in the oleogels primarily adopt a β-form monoclinic subcell structure [29]. It is likely that the interaction between the triglyceride molecules and dissolved β-amyrin in the oil induces a modification of the crystal morphology [47]. By comparing the relative crystallinity (RC) of the oleogels at different preparation temperatures (Table S2), it is evident that, except for the CO-based oleogel, which shows insignificant changes in RC, the relative crystallinity of the other oleogels tends to increase. This observation suggests that the difference in crystallinity is primarily due to the degree of exposure of alkyl groups in the liquid oil, which may be related to the spatial structure of the molecules formed through crystal cross-linking and entanglement during the gelation process [41]. Additionally, at a preparation temperature of 120 °C, the relative crystallinity of 4 wt% CO (70.87%) and PO (77.85%) is lower compared to 4 wt% LO (86.91%) and 4 wt% RO (79.21%). This can be attributed to the higher degree of unsaturation in the liquid oils, resulting in a higher degree of crystallinity, which is in line with faster crystal growth rates in crystallization kinetics.

In summary, the various characteristic indices reflected by XRD patterns are influenced by different oil types and preparation temperatures in oleogels. At a certain preparation temperature, the higher the degree of unsaturation of long-chain fatty acids in different vegetable oils, the stronger the effect of the bent spatial arrangement, resulting in greater mechanical strength of the formed oleogels.

3.4. Thermal Behaviour of Oleogels

DSC was used to investigate the thermal behavior of oleogels prepared from different oil types at various cooling rates. The results from Figure 4 and Table S3 indicate that all five oleogels exhibited broad melting and crystallization characteristics. When comparing the Ton values under different crystallization rate conditions, it was observed that both the melting and crystallization peaks became narrower with increasing temperature rates. Additionally, the Tm values under melting conditions shifted towards higher temperatures as the temperature rate increased. This shift is likely due to the rapid melting of crystals at higher temperature rates, consistent with the findings of Han et al. [1]. Moreover, oleogels derived from long-chain unsaturated fatty acids have a broad range of dissolution temperatures. This is attributed to the gradual and continuous transition from a gel to a sol state, rather than an abrupt transformation.

On the other hand, from the crystallization graph, it can be observed that long-chain unsaturated fatty acids exhibit a wide range of crystallization peaks with relatively low peak heights. This is likely due to the β-amyrin molecules behaving as uniformly dispersed amorphous entities within the oil. Interestingly, by comparing the ΔHm (enthalpy of melting) and ΔHc (enthalpy of crystallization) values of oleogels from different oil types, it can be inferred that the thermal stability of oleogels based on LO is poor. This could be attributed to the potential interactions between minor components and unsaturated fatty acids present in the oil formulation, which might affect the melting and crystallization processes.

The crystallization process can be further described using the Avrami model. By fitting the crystallization data to the Avrami model, the values of the Avrami exponent (n) and apparent rate constant (k) were determined, along with the correlation coefficient (R). Based on Table 2, the applied model could adequately describe the experimental data ($R^2 > 0.85$), and both the n and k values were influenced by the cooling rate and the type of oil used. The n reflects a parameter related to the time dependence of nucleation and the dimensionality of crystal growth. The n for the oleogels ranged from 2 to 4, indicating a transition from two-dimensional plate-like growth to instantaneous nucleation and spherical crystal growth for the majority of the crystals. Comparing different oleogels, the LO-based oleogel shows the highest n value, indicating an increasing trend with longer chain length. This suggests a transition from spherical to rod-like crystal growth in this oleogel. Additionally, the variation in the value of k indicates changes in the crystal growth rate and nucleation process within the oleogel, resulting in changes in crystal size and type [24,29]. As the cooling rate increases, the k also increases. Specifically, for the LO-based oleogel, k increases from 0.03 to 0.07. This implies that, in a single-component system, samples crystallize faster at higher supercooling temperatures. These results indicate that a higher content of unsaturated fatty acids in the oil phase promotes the formation of more and finer crystals. However, the rate of oleogel formation is not directly correlated with the strength of oleogel formation.

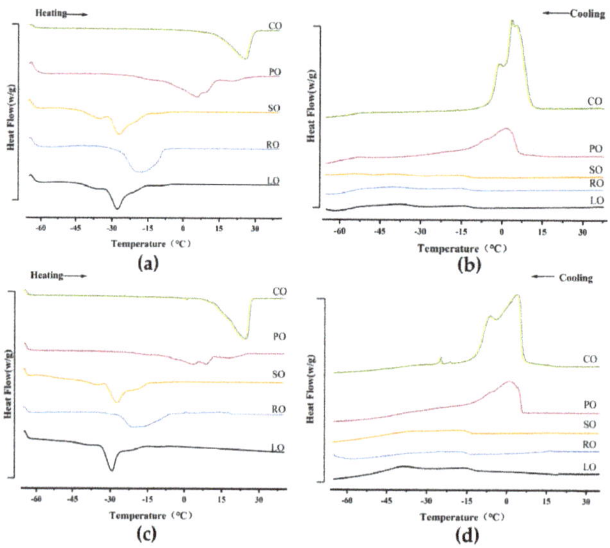

Figure 4. Melting and crystallization curves of different oleogels at 120 °C preparation temperature. (**a**,**b**) Melting and crystallization curves of LO—based oleogels, CO—based oleogels, SO—based oleogels, RO—based oleogels, and PO—based oleogels at 10 °C/min. (**c**,**d**) Melting and crystallization curves of five different oleogels at 5 °C/min.

Table 2. Modified Avrami model parameters of oleogels formed by different oil types and β-amyrin as a function of their crystallization cooling rates.

	5 °C min^{-1}			10 °C min^{-1}		
Type	n	K (min^{-1})	R^2	n	K (min^{-1})	R^2
LO	2.96 ± 0.13	0.03 ± 0.02	0.95 ± 0.02	3.03 ± 0.47	0.07 ± 0.03	0.98 ± 0.01
CO	2.75 ± 0.19	ND	0.85 ± 0.02	2.80 ± 0.02	0.02 ± 0.01	0.87 ± 0.00
SO	2.59 ± 0.18	0.01 ± 0.00	0.96 ± 0.01	2.66 ± 0.32	0.03 ± 0.02	0.98 ± 0.01
RO	2.79 ± 0.07	0.02 ± 0.00	0.98 ± 0.02	2.97 ± 0.30	0.02 ± 0.01	0.97 ± 0.02
PO	2.75 ± 0.12	ND	0.97 ± 0.01	2.31 ± 0.07	0.01 ± 0.01	0.90 ± 0.11

3.5. The Rheological Properties of Oloegels

Rheological analysis was conducted on the oleogels to ascertain their relevant properties. As depicted in Figure 5a, within the LVR region, the storage modulus (G′) consistently exceeds the loss modulus (G″), demonstrating that the oleogel prepared with β-amyrin is predominantly a "solid"-dominant sample. Comparing the LO-based oleogel with the CO-based oleogel, the LO-based oleogel exhibits a wider linear viscoelastic region. Additionally, the LO-based oleogel has a higher G′ value, indicating that it possesses greater strain stability. This suggests that the gel network formed by intermolecular interactions in the LO-based oleogel is stronger, leading to the extension of the linear viscoelastic region. The point at which the transition occurs from the elastic deformation region to the plastic deformation region, as the strain value increases, is known as the linear yield stress ($\gamma 0$). For the LO-based oleogel, $\gamma 0$ is 0.04, while for the CO-based oleogel, it is 0.02 (Table S3). This indicates that the LO-based oleogel is less susceptible to changes in stress values, suggesting a stronger gel strength than other oleogels. As the stress values exceed the LVR range, both the average G' and G'' values decrease, indicating a gradual breakdown of the gel's network structure. Once the stress value surpasses the critical stress value (λc), where $G' = G''$, all samples exhibit elastic softening and viscosity thinning, resulting in a plastic flow. This is consistent with the results of Figure S4. The λc values for the LO-based and CO-based oleogels are 19.95 and 1.26, respectively (Table S3). This suggests that oleogels formed by oils rich in saturated fatty acids are more prone to fracture under lower strain conditions [48]. Interestingly, for the CO-based oleogel, the G' value suddenly decreases under high stress. This suggests that the oleogel exhibits excellent spreadability, making it an ideal choice for novel edible coatings [49]. In addition, comparing the stress–strain curves of oleogels prepared from different types of oils reveals that the strain value increases with an increase in the degree of unsaturation.

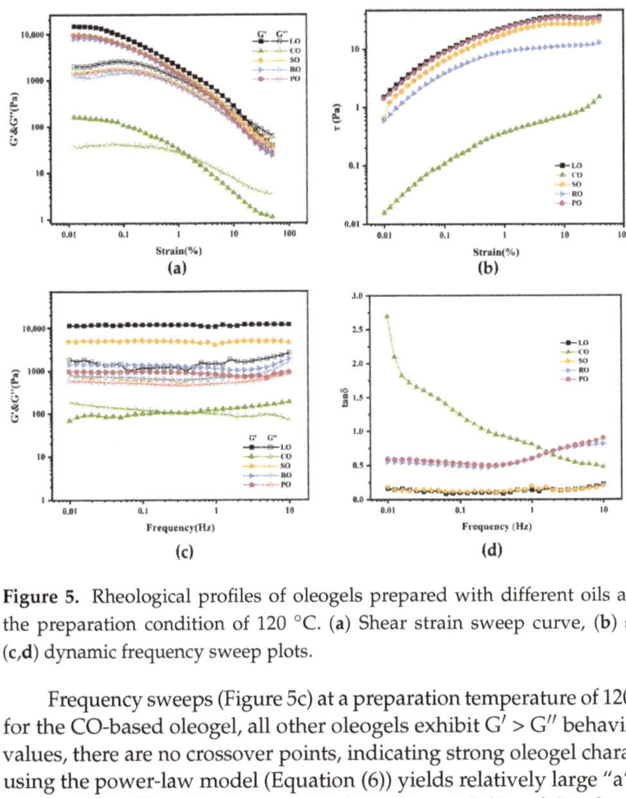

Figure 5. Rheological profiles of oleogels prepared with different oils and 4% β-amyrin under the preparation condition of 120 °C. (**a**) Shear strain sweep curve, (**b**) stress-strain curve, and (**c**,**d**) dynamic frequency sweep plots.

Frequency sweeps (Figure 5c) at a preparation temperature of 120 °C reveal that, except for the CO-based oleogel, all other oleogels exhibit $G' > G''$ behavior. At high-frequency values, there are no crossover points, indicating strong oleogel characteristics. Calculation using the power-law model (Equation (6)) yields relatively large "a" values and small "b" values (Table S4), reflecting good deformation stability of the oleogels. The G' values of the oleogels remain unaffected by frequency. However, the CO-based oleogel is the only one showing frequency dependence and a high "b" value, indicating the weakest oleogel structure. Moreover, computing the loss tangent ($\tan\delta = G''/G'$) can provide additional evidence for the overall viscoelastic behavior of the oleogels (Figure 5d). At a frequency of 0.1 Hz, the $\tan\delta$ values for the LO-based and CO-based oleogels are 0.099 and 1.23, respectively. This indicates that a higher degree of unsaturation and a greater content of polyunsaturated fatty acids (PUFAs) can enhance the strength of the crystal network within the oleogel [50]. These findings further support the idea that the internal network of the oleogels consists of non-covalent interactions, forming a physically crosslinked gel network. Additionally, the CO-based oleogel exhibits the lowest G′ value, which can be attributed to the formation of a rod-like crystal structure through the combination of saturated fatty acids and β-amyrin. This observation aligns with the micrographs of the oleogel and the relative crystallinity values obtained from XRD analysis (Table S2).

3.6. Correlation Analysis of Oleogels

Through correlation analysis, a deeper understanding of the influence of different oil types on the characteristics of oleogels can be gained. From the perspective of fatty acids, C18:3 shows a positive correlation with OBC (R = 0.97), RC (R = 0.83), and G′ (R = 0.92). Additionally, the increase in hardness, OBC, RC, and G′ value is positively correlated with the degree of lipid unsaturation (Figure 6). On the other hand, C12:0 exhibits a negative correlation with ΔHm (R = −0.98) and hardness (R = 0.71). This result indicates that oleogels prepared from longer-chain and highly unsaturated oils exhibit higher strength, whereas oleogels prepared from oils with shorter chains and higher saturation levels have lower strength. The potential reason is that an LO-based oleogel rich in long-chain and unsaturated fatty acids can facilitate interactions between β-amyrin particles [45]. Simultaneously, the result implies that oleogels with higher rheological properties exhibit strong gel strength, characterized by high stability and a solid texture [51]. On the other hand, there is a strong negative correlation between L and RC (R = −0.94), OBC (R = −0.81), and hardness (R = −0.89). Conversely, L shows a positive correlation with A (R = 0.94). These findings suggest that crystal size plays a role in influencing the ability of β-amyrin to bind liquid oil, thereby impacting the texture of the oleogel. An increase in the void fraction is associated with a larger crystal size in the oleogel. The negative correlation between crystal size and gel strength in oleogels has been substantiated in various research studies [33].

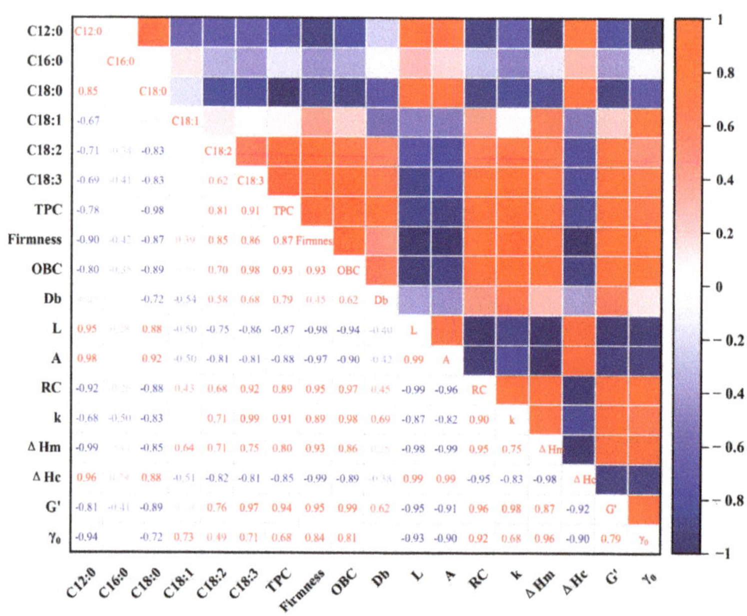

Figure 6. Correlation plots of key parameters (blue indicates negative correlation, red indicates positive correlation).

4. Conclusions

This study represents the first utilization of β-amyrin as a gelator to construct plant oil-based oleogels, demonstrating the feasibility of using β-amyrin alone as a gelator. Additionally, we investigated the physicochemical properties of β-amyrin-based oleogels prepared at different temperatures and with different oil types, aiming to evaluate the influence of oil type and preparation temperature on gel characteristics. The results revealed that β-amyrin has the capability to impart specific structure to oils and form oleogels, with a critical concentration as low as 2–4%. The gel properties were found to be influenced by the preparation temperature as well as the chain length and degree of unsaturation of the lipids. Especially at higher suitable preparation temperatures, lipids with longer-chain unsaturated fatty acids can form stable oleogels when combined with β-amyrin. Through polarized light microscopy, storage stability, physical properties, and rheological results, it is evident that LO-based oleogels exhibit small and dense β-amyrin crystals, resulting in the highest gel strength and stronger storage stability. In contrast, in CO-based oleogels, crystal aggregation leads to weaker gel strength and poorer stability. Correlation analysis indicates a positive correlation between C18:3 and hardness, OBC, and G', while C12:0 shows a negative correlation with G'. These findings hold promising practical applications as they allow for the design of stable organic oleogels with varying mechanical properties by modifying the preparation conditions and lipid types. This provides the latest insights for the utilization of pentacyclic triterpenes as bioactive compounds in food applications.

Supplementary Materials: The following supporting information can be downloaded at: https://www.mdpi.com/article/10.3390/foods13060876/s1, Figure S1: Relationship between hardness and preparation temperature, Figure S2: XRD pattern of β-amyrin, Figure S3: XRD patterns of oleogels. (A) XRD patterns of different edible oils, (B–D) prepa-ration temperatures of 80 °C, 100 °C and 120 °C respectively, Figure S4: Viscosity profiles of β-amyrin-based oleogels prepared with different oil types; Table S1: FO, CO, SO, RO, PO (g/100 g) Major fatty acid composition and total polar fraction (TPC%), Table S2: Relative crystallinity (%) of oleogels at different preparation temperatures, Table S3: Thermal behavioral characteristics of oleogels prepared with different oil types with 4% β-amyrin, Table S4: Physical properties of the oleogels. Including linear ultimate stress ($\gamma 0$), supercritical strain values (λc, stress values for $G' = G''$), and ($G' = a\ \omega b$) under the power-law exponent where a(Pa·b) and b are dimensionless flow behavior exponent.

Author Contributions: Conceptualization, C.Z. and S.S.; methodology and investigation, S.S., C.L. and H.L.; software, P.L.; validation, S.Q.; C.Z. and S.G. writing—original draft preparation and writing—review and editing, S.S. and H.X.; funding acquisition, S.Q. All authors have read and agreed to the published version of the manuscript.

Funding: This study was funded by the Key Research and Development Project of Hunan Province (2022NK2033) and the Natural Science Foundation of Hunan Province (grant number 2021JJ30321).

Informed Consent Statement: Not applicable.

Data Availability Statement: The original contributions presented in the study are included in the article/Supplementary Material, further inquiries can be directed to the corresponding authors.

Acknowledgments: We would like to thank the lab team teachers for their guidance and help, as well as all the students who participated in the experiments.

Conflicts of Interest: The authors declare no conflicts of interest.

References

1. Han, W.; Chai, X.; Liu, Y.; Xu, Y.; Tan, C.-P. Crystal network structure and stability of beeswax-based oleogels with different polyunsaturated fatty acid oils. *Food Chem.* **2022**, *381*, 131745. [CrossRef]
2. Li, L.; Liu, G.; Bogojevic, O.; Pedersen, J.N.; Guo, Z. Edible oleogels as solid fat alternatives: Composition and oleogelation mechanism implications. *Compr. Rev. Food Sci. Food Saf.* **2022**, *21*, 2077–2104. [CrossRef]
3. McClements, D.J. Future foods: A manifesto for research priorities in structural design of foods. *Food Funct.* **2020**, *11*, 1933–1945. [CrossRef] [PubMed]

4. Millao, S.; Iturra, N.; Contardo, I.; Morales, E.; Quilaqueo, M.; Rubilar, M. Structuring of oils with high PUFA content: Evaluation of the formulation conditions on the oxidative stability and structural properties of ethylcellulose oleogels. *Food Chem.* **2023**, *405*, 134772. [CrossRef]
5. Zhu, Y.; Bo, Y.; Liu, Y. Dietary total fat, fatty acids intake, and risk of cardiovascular disease: A dose-response meta-analysis of cohort studies. *Lipids Health Dis.* **2019**, *18*, 91. [CrossRef] [PubMed]
6. Chen, K.; Zhang, H. Fabrication of oleogels via a facile method by oil absorption in the aerogel templates of protein–polysaccharide conjugates. *ACS Appl. Mater. Interfaces* **2020**, *12*, 7795–7804. [CrossRef] [PubMed]
7. Pan, J.; Tang, L.; Dong, Q.; Li, Y.; Zhang, H. Effect of oleogelation on physical properties and oxidative stability of camellia oil-based oleogels and oleogel emulsions. *Food Res. Int.* **2021**, *140*, 110057. [CrossRef]
8. Gao, Y.; Wu, S. Development and evaluation of a novel oleogel system based on starch–water–wax–oil. *Food Funct.* **2020**, *11*, 7727–7735. [CrossRef] [PubMed]
9. Palla, C.A.; Dominguez, M.; Carrín, M.E. An overview of structure engineering to tailor the functionality of monoglyceride oleogels. *Compr. Rev. Food Sci. Food Saf.* **2022**, *21*, 2587–2614. [CrossRef]
10. Gravelle, A.J.; Davidovich-Pinhas, M.; Zetzl, A.K.; Barbut, S.; Marangoni, A.G. Influence of solvent quality on the mechanical strength of ethylcellulose oleogels. *Carbohydr. Polym.* **2016**, *135*, 169–179. [CrossRef]
11. Oh, I.; Lee, J.; Lee, H.G.; Lee, S. Feasibility of hydroxypropyl methylcellulose oleogel as an animal fat replacer for meat patties. *Food Res. Int.* **2019**, *122*, 566–572. [CrossRef] [PubMed]
12. Rogers, M.A.; Strober, T.; Bot, A.; Toro-Vazquez, J.F.; Stortz, T.; Marangoni, A.G. Edible oleogels in molecular gastronomy. *Int. J. Gastron. Food Sci.* **2014**, *2*, 22–31. [CrossRef]
13. Pang, M.; Wang, X.; Cao, L.; Shi, Z.; Lei, Z.; Jiang, S. Structure and thermal properties of β-sitosterol-beeswax-sunflower oleogels. *Int. J. Food Sci. Technol.* **2020**, *55*, 1900–1908. [CrossRef]
14. Sun, P.; Xia, B.; Ni, Z.-J.; Wang, Y.; Elam, E.; Thakur, K.; Ma, Y.-L.; Wei, Z.-J. Characterization of functional chocolate formulated using oleogels derived from β-sitosterol with γ-oryzanol/lecithin/stearic acid. *Food Chem.* **2021**, *360*, 130017. [CrossRef]
15. Silva-Avellaneda, E.; Bauer-Estrada, K.; Prieto-Correa, R.; Quintanilla-Carvajal, M. The effect of composition, microfluidization and process parameters on formation of oleogels for ice cream applications. *Sci. Rep.* **2021**, *11*, 7161. [CrossRef] [PubMed]
16. Tang, Y.R.; Ghosh, S. Canola protein thermal denaturation improved emulsion-templated oleogelation and its cake-baking application. *RSC Adv.* **2021**, *11*, 25141–25157. [CrossRef]
17. Yu, X.; Chen, H.; Shi, X.; Albouy, P.-A.; Guo, J.; Hu, J.; Li, M.-H. Liquid crystal gelators with photo-responsive and AIE properties. *Mater. Chem. Front.* **2018**, *2*, 2245–2253. [CrossRef]
18. Liu, Y.; Xia, H.; Guo, S.; Lu, X.; Zeng, C. Development and characterization of a novel naturally occurring pentacyclic triterpene self-stabilized pickering emulsion. *Colloids Surf. A Physicochem. Eng. Asp.* **2022**, *634*, 127908. [CrossRef]
19. Li, Z.; Liu, A.; Du, Q.; Zhu, W.; Liu, H.; Naeem, A.; Guan, Y.; Chen, L.; Ming, L. Bioactive substances and therapeutic potential of camellia oil: An overview. *Food Biosci.* **2022**, *49*, 101855. [CrossRef]
20. Amiri, S.; Dastghaib, S.; Ahmadi, M.; Mehrbod, P.; Khadem, F.; Behrouj, H.; Aghanoori, M.-R.; Machaj, F.; Ghamsari, M.; Rosik, J. Betulin and its derivatives as novel compounds with different pharmacological effects. *Biotechnol. Adv.* **2020**, *38*, 107409. [CrossRef]
21. Bildziukevich, U.; Rárová, L.; Šaman, D.; Wimmer, Z. Picolyl amides of betulinic acid as antitumor agents causing tumor cell apoptosis. *Eur. J. Med. Chem.* **2018**, *145*, 41–50. [CrossRef]
22. Shi, Z.; Cao, L.; Kang, S.; Jiang, S.; Pang, M. Influence of wax type on characteristics of oleogels from camellia oil and medium chain triglycerides. *Int. J. Food Sci. Technol.* **2021**, *57*, 2003–2014. [CrossRef]
23. Bascuas, S.; Hernando, I.; Moraga, G.; Quiles, A. Structure and stability of edible oleogels prepared with different unsaturated oils and hydrocolloids. *Int. J. Food Sci. Technol.* **2020**, *55*, 1458–1467. [CrossRef]
24. de Vries, A.; Gomez, Y.L.; van der Linden, E.; Scholten, E. The effect of oil type on network formation by protein aggregates into oleogels. *RSC Adv.* **2017**, *7*, 11803–11812. [CrossRef]
25. Doan, C.D.; Tavernier, I.; Okuro, P.K.; Dewettinck, K. Internal and external factors affecting the crystallization, gelation and applicability of wax-based oleogels in food industry. *Innov. Food Sci. Emerg. Technol.* **2018**, *45*, 42–52. [CrossRef]
26. Liu, N.; Lu, Y.; Zhang, Y.; Gao, Y.; Mao, L. Surfactant addition to modify the structures of ethylcellulose oleogels for higher solubility and stability of curcumin. *Int. J. Biol. Macromol.* **2020**, *165*, 2286–2294. [CrossRef] [PubMed]
27. Chai, X.; Meng, Z.; Liu, Y. Comparation of micro-viscosity of liquid oil in different colloidal fat crystal networks using molecular rotors. *Food Chem.* **2020**, *317*, 126382. [CrossRef] [PubMed]
28. Davidovich-Pinhas, M.; Barbut, S.; Marangoni, A. Development, characterization, and utilization of food-grade polymer oleogels. *Annu. Rev. Food Sci. Technol.* **2016**, *7*, 65–91. [CrossRef]
29. Lu, Y.C.; Li, J.L.; Ding, J.; Nie, X.H.; Yu, N.X.; Meng, X.H. Comparison of diosgenin-vegetable oils oleogels with various unsaturated fatty acids: Physicochemical properties, in-vitro digestion, and potential mechanism. *Food Chem.* **2023**, *413*, 135663. [CrossRef]
30. Blake, A.I.; Co, E.D.; Marangoni, A.G. Structure and physical properties of plant wax crystal networks and their relationship to oil binding capacity. *J. Am. Oil Chem. Soc.* **2014**, *91*, 885–903. [CrossRef]
31. Okuro, P.K.; Tavernier, I.; Bin Sintang, M.D.; Skirtach, A.G.; Vicente, A.A.; Dewettinck, K.; Cunha, R.L. Synergistic interactions between lecithin and fruit wax in oleogel formation. *Food Funct.* **2018**, *9*, 1755–1767. [CrossRef]

32. Wang, L.; Wen, Y.; Su, C.; Gao, Y.; Li, Q.; Du, S.; Yu, X. Effect of water content on the physical properties and structure of walnut oleogels. *RSC Adv.* **2022**, *12*, 8987–8995. [CrossRef] [PubMed]
33. Frolova, Y.; Sarkisyan, V.; Sobolev, R.; Makarenko, M.; Semin, M.; Kochetkova, A. The influence of edible oils' composition on the properties of beeswax-based oleogels. *Gels* **2022**, *8*, 48. [CrossRef] [PubMed]
34. Li, P.; Wu, G.; Yang, D.; Zhang, H.; Qi, X.; Jin, Q.; Wang, X. Effect of multistage process on the quality, water and oil distribution and microstructure of French fries. *Food Res. Int.* **2020**, *137*, 109229. [CrossRef]
35. Borriello, A.; Antonella Miele, N.; Masi, P.; Aiello, A.; Cavella, S. Effect of fatty acid composition of vegetable oils on crystallization and gelation kinetics of oleogels based on natural wax. *Food Chem.* **2022**, *375*, 131805. [CrossRef] [PubMed]
36. Franco, D.; Martins, A.J.; López-Pedrouso, M.; Purriños, L.; Cerqueira, M.A.; Vicente, A.A.; Pastrana, L.M.; Zapata, C.; Lorenzo, J.M. Strategy towards replacing pork backfat with a linseed oleogel in frankfurter sausages and its evaluation on physicochemical, nutritional, and sensory characteristics. *Foods* **2019**, *8*, 366. [CrossRef]
37. Gilbert, E.P. Building blocks of β-sitosterol-γ-oryzanol gels revealed by small-angle neutron scattering and real space modelling. *Food Funct.* **2022**, *13*, 7123–7131. [CrossRef] [PubMed]
38. Flöter, E.; Wettlaufer, T.; Conty, V.; Scharfe, M. Oleogels—Their applicability and methods of characterization. *Molecules* **2021**, *26*, 1673. [CrossRef]
39. Laredo, T.; Barbut, S.; Marangoni, A.G. Molecular interactions of polymer oleogelation. *Soft Matter* **2011**, *7*, 2734–2743. [CrossRef]
40. Li, X.; Li, J.; Wang, Y.; Cao, P.; Liu, Y. Effects of frying oils' fatty acids profile on the formation of polar lipids components and their retention in French fries over deep-frying process. *Food Chem.* **2017**, *237*, 98–105. [CrossRef]
41. Fayaz, G.; Goli, S.A.H.; Kadivar, M. A novel propolis wax-based organogel: Effect of oil type on its formation, crystal structure and thermal properties. *J. Am. Oil Chem. Soc.* **2017**, *94*, 47–55. [CrossRef]
42. Scharfe, M.; Flöter, E. Oleogelation: From Scientific Feasibility to Applicability in Food Products. *Eur. J. Lipid Sci. Technol.* **2020**, *122*, 24. [CrossRef]
43. Cerqueira, M.A.; Fasolin, L.H.; Picone, C.S.; Pastrana, L.M.; Cunha, R.L.; Vicente, A.A. Structural and mechanical properties of organogels: Role of oil and gelator molecular structure. *Food Res. Int.* **2017**, *96*, 161–170. [CrossRef]
44. Palla, C.; de Vicente, J.; Carrin, M.E.; Ruiz, M.J.G. Effects of cooling temperature profiles on the monoglycerides oleogel properties: A rheo-microscopy study. *Food Res. Int.* **2019**, *125*, 108613. [CrossRef]
45. Perta-Crisan, S.; Ursachi, C.S.; Chereji, B.D.; Tolan, I.; Munteanu, F.D. Food-Grade Oleogels: Trends in Analysis, Characterization, and Applicability. *Gels* **2023**, *9*, 386. [CrossRef]
46. Wang, Y.; Chi, Y.; Zhang, W.; Yang, Q.; Yang, S.; Su, C.; Lin, Z.; Gu, J.; Hu, C. Structural Diversity of Diosgenin Hydrates: Effect of Initial Concentration, Water Volume Fraction, and Solvent on Crystallization. *Cryst. Growth Des.* **2016**, *16*, 1492–1501. [CrossRef]
47. Guo, Y.; Cai, Z.; Xie, Y.; Ma, A.; Zhang, H.; Rao, P.; Wang, Q. Synthesis, physicochemical properties, and health aspects of structured lipids: A review. *Compr. Rev. Food Sci. Food Saf.* **2020**, *19*, 759–800. [CrossRef] [PubMed]
48. Mohanan, A.; Nickerson, M.T.; Ghosh, S. The effect of addition of high-melting monoacylglycerol and candelilla wax on pea and faba bean protein foam-templated oleogelation. *J. Am. Oil Chem. Soc.* **2020**, *97*, 1319–1333. [CrossRef]
49. Joyner, H.S. *Rheology of Semisolid Foods*; Springer: Berlin/Heidelberg, Germany, 2019.
50. Holey, S.A.; Sekhar, K.P.; Mishra, S.S.; Kanjilal, S.; Nayak, R.R. Effect of oil unsaturation and wax composition on stability, properties and food applicability of oleogels. *J. Am. Oil Chem. Soc.* **2021**, *98*, 1189–1203. [CrossRef]
51. Pehlivanoğlu, H.; Demirci, M.; Toker, O.S.; Konar, N.; Karasu, S.; Sagdic, O. Oleogels, a promising structured oil for decreasing saturated fatty acid concentrations: Production and food-based applications. *Crit. Rev. Food Sci. Nutr.* **2018**, *58*, 1330–1341. [CrossRef] [PubMed]

foods

MDPI

Article

Ready-to-Use Supplementary-Food Biscuit Production with Low-Cost Ingredients for Malnourished Children in Sub-Saharan Africa

Davide Reggi [1,*], Gaia Alessio [1], Andrè Ndereyimana [1], Andrea Minuti [1], Giorgia Spigno [2] and Giuseppe Bertoni [1]

[1] Department of Animal, Nutrition and Food Science (DiANA), Università Cattolica del Sacro Cuore, Via Emilia Parmense 84, 29122 Piacenza, Italy; gaia_95pv@hotmail.it (G.A.); ndereya79@gmail.com (A.N.); andrea.minuti@unicatt.it (A.M.); giuseppe.bertoni@unicatt.it (G.B.)

[2] Department for Sustainable Food Process (DiSTAS), Università Cattolica del Sacro Cuore, Via Emilia Parmense 84, 29122 Piacenza, Italy; giorgia.spigno@unicatt.it

* Correspondence: davide.reggi@unicatt.it

Abstract: In Africa, the number of children under 5 years old who suffer from stunting and wasting are, respectively, 61.4 and 12.1 million, and to manage situations like these, emergency food products like RUTF and RUSF (ready-to-use therapeutic/supplementary food) are very useful. The aim of this study was to develop an RUSF biscuit using the low-cost food resources usually present in Sub-Saharan Africa (Burundi and the DRCongo in our case study); we conducted chemical characterization, nutritional evaluation, and a stability trial simulating the usual storage conditions in a rural context to demonstrate that RUSF can be functional also using low-cost ingredients and a simple method of production. The obtained recipes showed good potential in supplying protein integration—17.81% (BUR) and 16.77% (CON) (% as food) were the protein contents—and the protein digestibility values were very high (BUR: 91.72%; CON: 92.01%). Moreover, 30% of the daily requirement was achieved with less than 50 g of both recipes in all the considered ages. Finally, a good shelf-life was demonstrated during the 35-day testing period at 30 °C, considering moisture, texture, and lipid oxidation evolution. Recipes like these, with appropriate changes, could be very useful in all contexts where child malnutrition is a serious problem.

Keywords: child malnutrition; RUSF; protein biscuit; food security; low-income countries; low-cost integration

1. Introduction

Malnutrition is one of the biggest problems in low-income countries. A total of 45% of the 5.9 million deaths of children under 5 years worldwide can be linked to undernutrition, which is often connected in turn to other causes like infections and parasitosis [1]. The risk of death for a malnourished child is 11 times higher than for a well-nourished child [2]. It is known that malnutrition compromises not only physical development, but also mental/brain development, which is fundamental for the integral development of any human being [3]. For this reason, Black et al. [4] consider the problem of child malnutrition in developing countries not just related to mortality, but particularly to the negative influence on their development, especially regarding the brain; therefore, malnutrition in the first few years of life affects cognitive development, as well as motor and social–emotional development [5]. Other effects of undernutrition, in the longer term, include mental illness, hypertension, diabetes, and impaired working capacity, with consequences for individual health and the standard of living of the affected individuals throughout their life [6,7].

Malnutrition is often linked to household food insecurity (HFI) [8], which is defined as the lack of physical, social, and economic access to sufficient, safe, and nutritious food to meet the needs and food preferences for an active and healthy life [9].

The problem of HFI, despite some improvements, continues to be very worrying, overall, in Africa. In fact, the FAO declared that three-quarters of the population in Africa do not have access to a healthy diet in which appropriate quantities of all needed nutrients are supplied [10]. Moreover, their diets are characterized by monotony and a lack of food rich in nutrients, which are particularly critical for children [11]. In fact, UNICEF et al. [12] indicated that 6.7% (45.4 million) of children under five years old suffer from wasting, and 22% (149.6 million) of them suffer from stunting globally.

Stunting is a result of malnutrition indicated by a low height-for-age (HAZ) level; wasting, on the other hand, is indicated by a low weight-for-height (WHZ) level; and, lastly, severe acute malnutrition (SAM) is a type of malnutrition defined by the WHO and UNICEF [13] as causing at least one of the following: a low WHZ, a mid-upper arm circumference of < 115 mm, or clinical signs of bilateral oedema.

On the African continent, the number of children under five years old that suffer from wasting and stunting are, respectively, 12.1 million and 64.1 million, with 3.1 million that suffer from SAM. Considering Burundi, the percentages for stunting and wasting in the same category of children are 57.6% and 4.8%; in the Democratic Republic of Congo (DRCongo), stunting affects 40.8% of children, with a wasting percentage of 6.4%. These numbers are particularly worrying if the critical levels (very high) for children in a population are considered: 30% for stunting (40% for alarming) and 15% for wasting (medium: 5–10%, low: 2.5–5%) [12].

To manage situations like those already described, it is appositely thought that high-energy and high-nutrient foods can be a good help. Emergency food products, to be useful, must be ready without preparation requirements when consumed, have a long shelf-life, be easy to distribute everywhere, and, obviously, have high energy and nutrient contents [14]; they must also cover the nutritional needs of different ages, supplying energy content as protein (10–15%), as fat (35–45%), and as carbohydrates (40–50%) [15]. For these reasons, depending on the different situations, ready-to-use nutritional food (RUF) can be divided into ready-to-use therapeutic food (RUTF) and ready-to-use supplementary food (RUSF): the first supplies all of the energy and nutrients necessary for malnourished people's growth to rapidly catch up; the second supplies only part of a proper diet [16,17]. The classic example of RUTF is Plumpy'Nut, a paste composed of peanut flour, sugar, plant fat, and milk powder and enriched with minerals and vitamins [18]. RUSF products are very diverse, and often they are also made with some local food sources, particularly those rich in protein and fat with relatively lower costs [19]. For example, a bread enriched with different concentrations of oyster mushroom powder (*Pleurotus plumonarius*), a rich source of high-quality proteins and essential amino acids, was developed in Nigeria by Okafor et al. [20]. Agrahar et al. [21] evaluated the enrichment with chickpea flour of the recipe for laddoo, a traditional lemon-sized ball-shaped sweet composed of roasted cereal or legume flour, fat, and sugar, which is very well accepted by people in India and the Middle East. A very common shape to present a therapeutic or supplementary food in is that of a biscuit, since this food shape is easy to be appreciated and consumed by children [22]. High-energy biscuits (HEBs) are very well documented as supplements (because they can easily have protein-rich ingredients and vitamin–mineral premixes added in the formulation) for emergency feeding programs in situations of nutritional deficiencies or a lack of access to basic facilities [23]. The WHO [24] developed the BP-100 biscuit, a milk-based RUTF used by humanitarian agencies and governmental and non-governmental organizations to treat SAM in any cultural setting. Varghese and Srivastav [25] developed a low-cost, high-energy, and nutritious cookie (RUSF) prepared with soy protein isolate, moringa powder, roasted jackfruit seed powder, and a vitamin–mineral premix with the aim of satisfying 30% of the Recommended Dietary Allowance (RDA) requirement for undernourished adolescents in India. A comparison with Plumpy'Nut and two cookies prepared with peanuts, pistachios, maize, and moringa was carried out in Benin for protein and mineral content by Kouton et al. [26]. A study conducted by Fetriyuna et al. [27] evaluated the introduction of peanut into some RUSF biscuits as a protein and lipid source, in conjunction with mung bean, taro flour, and banana nagka.

The objective of this study was to develop an RUSF, using the relatively low-cost food resources that are readily available in Sub-Saharan Africa, with particular focus on Burundi and the Democratic Republic of Congo. In this context, Project C3S (Production of Appropriate Food: Sufficient, Safe and Sustainable), promoted by Università Cattolica del Sacro Cuore (Piacenza), has established pilot centers with the aim of fostering agricultural, technological, and nutritional progress in rural communities of low-income countries.

The initial recipes were developed based on a previous study [28], where three prototypes, with the same objective as this paper, were prepared and characterized. The protein content (12–13%) was deemed insufficiently high, and thus, in the new recipe of the current study, only whole chicken eggs and peanut flour were retained as essential ingredients. The incorporation of raw soy flour and Niebè flour (Vigna unguiculata) was driven by the necessity of enhancing the protein content, while the substitution of sunflower seed oil with palm oil in both recipes and the elimination of wheat flour in the DRCongo recipe were implemented in order to reduce costs. The newly developed biscuits (one for Burundi, and one for the DRCongo) were characterized for their chemical and nutritional composition and tested for stability (with reference to moisture and peroxide content, and texture), simulating typical storage conditions in the target rural context. The new RUSFs are designed to have an impact on children's nutrition (1–10 years), taking into account their consumption possibilities and the local diets, as well as to be low-cost. The concept has significant potential, contingent upon the implementation of appropriate alterations to the ingredients, in contexts characterized by poverty and rurality where a low-cost RUSF is a necessity for children's nutritional support, provided that the aforementioned trials in the countries cited above are successful.

2. Materials and Methods

2.1. Experimental Plan

The development of the RUSF biscuits included the following steps.

- Ingredient selection and preparation.
- Recipe development, with "La Nuova Dieta Ragionata—2003 (3.6.0)" software.
- Biscuit preparation.
- Chemical characterization and nutritional evaluation of two final prototypes.
- Sample production of two final prototypes for shelf-life trial.
- Monitoring time evolution for moisture, peroxide content, and texture.
- Evaluation of economic aspects.

2.2. Ingredient Selection and Preparation

All the ingredients were sourced from the ethnic market of Piacenza (Italy), with particular attention paid to their availability in Burundi and the DRCongo, particularly in the areas where Project C3S has active projects. Additionally, their costs and traditional varieties and uses were considered. This information was obtained from numerous local informal dietary questionnaires (which do not require ethical approval) within the project [29] (Table 1). Soy flour and Niebè flour (Vigna unguiculata) were prepared by milling the raw grains in "Model 4—Laboratory Mill—Thomas—Wiley" to a particle size of 2 mm.

Table 1. Brand of selected ingredients and their prices (USD/kg) in Burundi and DRCongo (April 2021).

Product	Brand (In Piacenza Market)	Price Burundi (USD/kg)	Price DRCongo (USD/kg)
Whole chicken egg	Terra (Lidl)	2.80	2.93
Raw soy seeds	Calleris	0.73	0.66
Raw Niebè (*Vigna unguiculata*)	Calleris	0.67	0.43
Peanut flour	Kava Foods	1.90	0.58
Rice flour	Adea	0.84	1.08
Wheat flour (type 1)	Cerealpuglia	0.73	2.00
Brown sugar	Tate & Lyle Demerara	0.88	1.40
Palm oil	Blue Bay	0.91	1.00
Baking powder	Paneangeli	1.48	2.11

2.3. Recipe Development and Preparation

Using "La Nuova Dieta Ragionata—2003", a professional software for dieticians, an evaluation was conducted in accordance with Mourey [30] for data pertaining to the nutritional requirements of children. The software database was utilized to assess the chemical characteristics of ingredients, except for Niebè chemical characteristics, which were assessed through laboratory analysis. This evaluation led to the conclusion that a recipe with a protein content of at least 16% and a fat content of the same percentage should be included in the composition of the food. In order to adhere to the aforementioned requirements while maintaining a low price, certain ingredients were substituted, added, or omitted in comparison to the Reggi [28] recipe. Sorghum flour, sweet potato flour, egg yolk, and wheat flour (the latter only present in the DRCongo recipe) were removed. Sunflower seed oil was substituted with palm oil, and raw soy flour and raw Niebè flour were incorporated. In accordance with the aforementioned requirements and utilizing the software, a series of preliminary recipes were devised and subjected to evaluation in order to ascertain the optimal combination of taste and texture. The preparation of prototypes involved the formation of dough through the manual kneading of the dry and wet ingredients for a period of 5 min. Subsequently, the biscuits were shaped (diameter: 3 cm, height: 0.5 cm) and baked in a convection oven (180 °C for 11 min), resulting in biscuits with an approximate weight of 7.1 g each. According to the results of all evaluations, two final biscuits were obtained, one for Burundi and one for the DRCongo (Table 2).

Table 2. Composition of biscuits for Burundi and DRCongo (% of wet matter).

Ingredients	BUR (Burundi)	CON (DRCongo)
Whole chicken egg	20	22
Raw soy flour	13	10
Niebè flour	10	15
Peanut flour	16	20
Rice flour	10	10
Wheat flour (type 1)	8	--
Brown sugar	15	15
Palm oil	6	6
Baking powder	2	2

2.4. Chemical and Nutritional Characterization

The proximate composition analyses of the two final prototypes were conducted using the following methods: AOAC Official Method 942.05 for ashes, Method ISO 712 for dry matter, AOAC Official Method 920.29 for ether extract (fat), AOAC Official Method 996.11 (using dosage kit K-TSTA, Megazyme International, Bray, Ireland) for starch, AOAC Official Method 2009.01 for fiber, AOAC Official Method 990.03 for proteins, and the method described by Directive 71/250/EEC, 15/6/1971, for reducing sugars [31–33]. For the nutritional characterization, the following methods were employed, respectively, for protein digestibility and urease activity: AOAC Official Method 971.09 and Directive 71/250/EEC, 15/6/1971 [31,33]. Once the quantity of biscuits required to fulfill 30% of the protein requirement for children aged between 1 and 10 years had been estimated, the software "La Nuova Dieta Ragionata—2003" was employed, utilizing the requirements suggested by Mourey [30], for a theoretical evaluation of the biscuits' potential to meet the vitamin A, Fe, and Zn requirements.

2.5. Shelf-Life Trial

In order to assess the quality stability of the two biscuits under standard climate conditions, a room-temperature storage test was conducted in Burundi and the DRC using a simple plastic bag packaging. For both recipes, 6 samples of biscuits packed in plastic (PE) lab bags closed with rubber bands were stored in a thermostat at 30 °C in darkness and analyzed at time 0 and after 7, 14, 21, 28, and 35 days. At each time, the biscuits were analyzed for their shelf-life in terms of moisture and peroxide contents, as well as their texture. The moisture content was analyzed using the ISO 712 method [32], following milling with a kitchen mixer. The peroxide content of the fat extracted was evaluated using the method described by Calligaris et al. [34], in accordance with the procedure described in Annex III to Regulation 2568/91/EEC [35]. The peroxide contents are reported as meq O_2/kg oil. The texture of the biscuits was evaluated according to the method described by Kaur et al. [36], utilizing the Texture Analyzer TVT 6700 with the following settings: height of sample—10 mm, starting distance between sensor and sample—5 mm, starting speed—1 mm/s, test speed—3 mm/s, starting force—50 g, data rate—333 pps. Two parameters were evaluated for texture: hardness (g strength) and fracturability (mm).

2.6. Data Analysis

The data for chemical and nutritional parameters (average of triplicate analyses) were compared using the Student test, while the data for stability were compared by ANOVA, with the influence of time and recipe factors also analyzed. Tukey test was employed for post hoc analysis to differentiate the averages. All analyses were conducted using IBM SPSS Statistics (Version 25, SPSS Inc., Chicago, IL, USA). GraphPad Prism 8.0.1 was used for the shelf-life graphs. The results are presented as mean values ± sd, $p < 0.05$.

3. Results

3.1. Chemical and Nutritional Characterization

The results presented in Table 3 indicate no significant differences between the two recipes with regard to all the parameters. Moreover, it can be observed that a relatively high content of protein (17.81 and 16.77) and fat (19.69 and 20.06) was achieved. The urease activity was found to be very low for both biscuit samples.

Table 3. Proximate composition (% as food), protein digestibility (%), and urease (mg N/g min at 30 °C) of the experimental biscuits. Values reported as mean ± sd.

Nutrients	BUR	CON
Dry matter	94.35 ± 0.05	94.17 ± 0.02
Ash	3.06 ± 0.002	3.05 ± 0.004
Protein	17.81 ± 0.03	16.77 ± 0.12
Fat	19.69 ± 0.04	20.06 ± 0.02
Starch	26.37 ± 0.06	22.16 ± 0.01
Fiber	6.63 ± 0.04	6.95 ± 0.02
Soluble fiber	1.56 ± 0.05	1.11 ± 0.02
Insoluble fiber	5.07 ± 0.01	5.84 ± 0.04
Redox sugars	0.19 ± 0.02	0.34 ± 0.03
Protein digestibility	91.72 ± 0.01	92.01 ± 0.03
Urease	0.028 ± 0.003	0.020 ± 0.001

The weight of a single biscuit is 7.1 ± 0.1 g, and the mass of biscuits needed to cover 30% of daily protein is shown in Table 4.

Table 4. Biscuit weight needed to cover 30% of the daily protein requirement for children 1 to 10 years old (as defined by Mourey [30]).

Children's Age	Protein Requirement (g/d)	BUR (g/d)	BUR (Number of Biscuits)	CON (g/d)	CON (Number of Biscuits)
1–2 years	4.3	24.14	3.44	25.64	3.61
2–3 years	4.58	25.72	3.62	27.31	3.85
3–4 years	5.39	30.26	4.31	32.14	4.53
4–5 years	5.24	29.42	4.14	31.25	4.4
5–6 years	6.27	35.20	4.96	37.39	5.27
6–7 years	6.21	34.87	4.91	37.03	5.22
7–10 years	8.18	45.93	6.47	48.78	6.87

Table 5 shows the calculated fulfilment of some micronutrients, according to the daily intake of biscuits.

Table 5. Calculated fulfilment of micronutrients (vitamin A, Zn, Fe) by the biscuit amounts shown in Table 4 (requirements by Mourey [30]; micronutrients supplied by "La Nuova Dieta Ragionata—2003").

Children's Age	Vit. A Needs (μg/Day)	Vit. A Covered (%)		Zn Needs (mg/Day)	Zn Covered (%)		Fe Needs (mg/Day)	Fe Covered (%)	
		BUR	CON		BUR	CON		BUR	CON
1–3 years	400	6.19	7.19	4	16	17.75	3.9	24.61	27.95
4–6 years	400	7.67	8.94	6	13.16	14.66	4.2	28.33	32.14
7–10 years	500	8.51	9.9	7	15.57	17.43	5.9	27.96	31.69

3.2. Economic Considerations

In terms of economic considerations, it is essential to recognize that the analysis presented here has solely focused on the quantity and the retail price/kg of individual ingredients utilized in two recipes, without an evaluation of all the other production costs (machineries, energy, staff, and packaging) and the possible weight loss or processing waste. Subsequently, the prices of the ingredients were collected during the study (April 2021), and the production costs associated with the ingredients purchased were calculated and compared for the two biscuits in both countries. In Burundi, the cost of ingredients for the recipes was found to be 1.471 USD/kg and 1.386 USD/kg, and in the DRCongo 1.312 USD/kg and 1.388 USD/kg, respectively. However, the cost of the two biscuits would be slightly higher in Burundi, with a difference between the two recipes. In Burundi, BUR would be the cheapest (−5.78%), while in the DRCongo it would be the most expensive (+5.48%).

3.3. Shelf-Life Trials

With regard to moisture content (Figure 1), the two recipes demonstrated no significant differences over the entire duration of the study, with the same storage times. There were no relevant differences within the same recipe in terms of evolution over time. The sole exception was observed at 21 days for both samples. However, this discrepancy is likely due to a packaging issue with the PE bags used for the 21-day samples.

It can be reasonably assumed that the closure problem also affects the peroxide content (Figure 2), which shows a faster oxidation for BUR compared with CON. In fact, the BUR samples exhibited an increase between 14 and 21 days, while the CON samples demonstrated a similar increase but between 28 and 35 days. Nevertheless, the peroxide values for BUR remained consistently low, suggesting that this trend could be considered almost constant with an oxidation level slightly lower than that of CON.

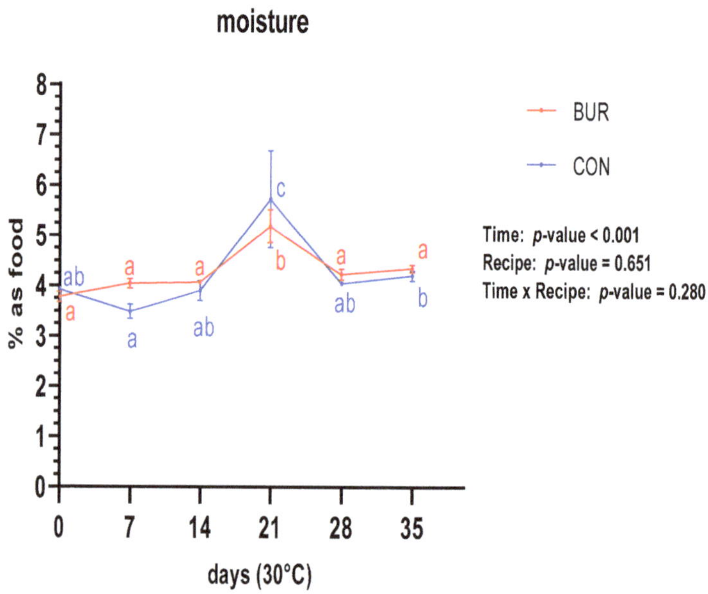

Figure 1. Behavior of moisture content for both recipes during shelf-life trial. a, ab, c—significant differences in the same recipe are indicated with these different letters.

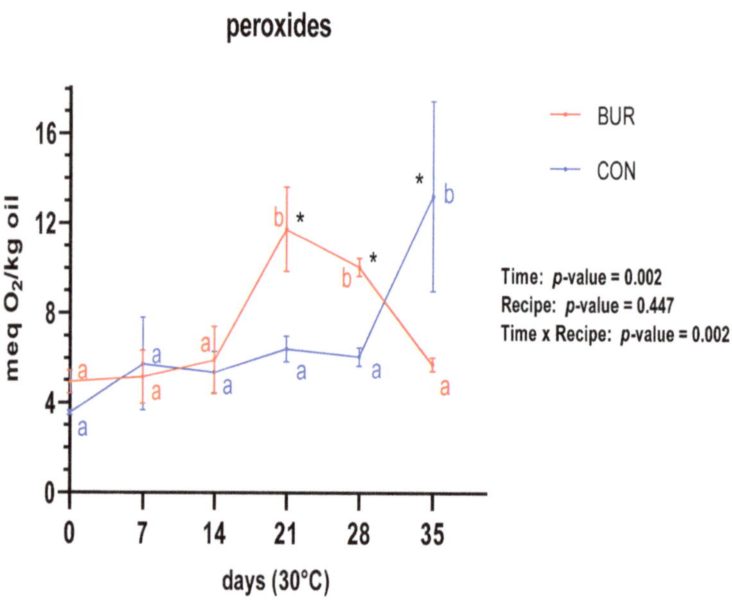

Figure 2. Behavior of peroxide content for both recipes during shelf-life trial. *, a, b—significant differences in the same recipe are indicated with different letters, and significant differences between recipes are indicated with *.

The texture analysis yielded two distinct parameters: fracturability (Figure 3) and hardness (Figure 4). The fracturability did not significantly vary between the two recipes throughout the entirety of the trial period, aligning with the observed trend in moisture content. In contrast, a significant difference was observed between the two biscuits, with a lower fracturability (higher value in mm run by the sensor before breaking) for BUR than for CON. Concordant trends were observed for the hardness, with constant values over time and higher values for BUR.

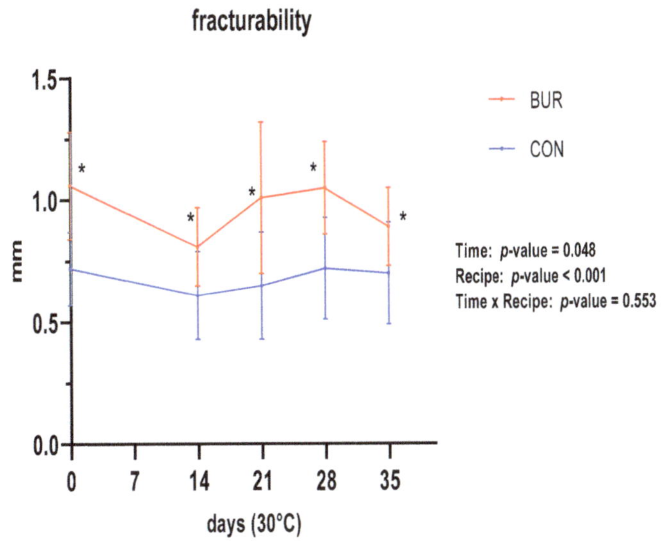

Figure 3. Trends of fracturability for both recipes during shelf-life trial. Significant differences between recipes are indicated with *.

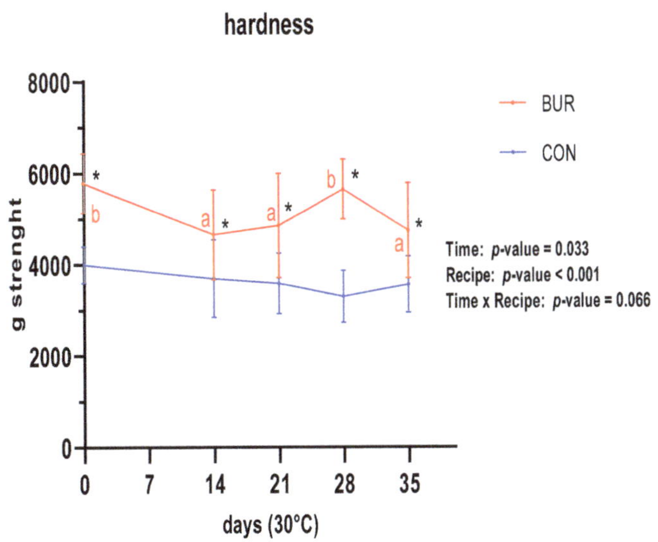

Figure 4. Trends of hardness for both recipes during shelf-life trial. *, a, b—significant differences in the same recipe are indicated with different letters, and significant differences between recipes are indicated with *.

4. Discussion

The age group most affected by malnutrition, defined as a lack of proper foods (and thus nutrients), is 3–5 years old. During this period, children require more specific dietary components than at other ages, while often, in the Sub-Saharan Africa context, the weaning transition due to slow digestive tract adaptation and the lack of available baby foods can result in inadequate nutrition. Such deficiencies in childhood can have adverse consequences, including an increased incidence of infections; developmental impairment, particularly in the brain and cognitive functions; a heightened risk of poor academic and future occupational performance; and an elevated prevalence of chronic diseases [37]. In an attempt to address the issue of malnutrition, many different therapeutic foods have been developed. One such example is Plumpy'Nut, which is currently being used for child nutrition care in a significant number of developing countries [18]. However, this type of product has certain limitations. Its availability and subsequent distribution are irregular, which hampers a good nutritional recovery process. Furthermore, its characteristics do not align with the typical eating habits of the population, increasing the risk of rejection [38]. One potential solution could be the use of RUSF and RUTF prepared with local available ingredients [26]. This, in fact, can stimulate local economies and increase sustainability, which also has a more beneficial social impact [39], like the possibility of household production and selling preventive food in local markets [40].

In light of the aforementioned considerations, the selected ingredients for the two biscuit recipes are the best compromise between cost effectiveness and nutritional value. The objective is to produce a biscuit that is both rich in nutrients and affordable, with the aim of providing 30% of the daily protein requirement for children. The cost of RUSF and RUTF represents a significant challenge for their availability in low-income countries [41]. For these reasons, milk and other dairy products, and their protein supply, were replaced with the use of egg, soy flour, Niebè flour, and peanut flour. Milk powder represents the most expensive ingredient used in BP-100 [25], a conclusion that was also reached by Sandige et al. [42] and Dibari et al. [43].

Our economic evaluation demonstrates that two distinct recipes are necessary to maintain the lowest possible cost in Burundi and the DRCongo. For instance, the incorporation of wheat flour into the DRCongo recipe would have increased the cost by 5.48%.

Our nutritional evaluation of the biscuits has confirmed the possibility of producing an RUSF suitable to meet 30% of the daily protein requirement with a daily biscuit portion of a maximum of 50 g. This quantity is low enough to be easily consumed and at an affordable cost, allowing the biscuits to be used as a convenient snack between normal meals consumed throughout the day [16,17].

With regard to macronutrients, the BUR and CON biscuits exhibited protein contents of 17.81% and 16.77%, respectively, which were both higher than those of Plumpy'Nut (16.35% [26]) and BP-100 (12.3% [25]), two of the most prevalent and utilized RUFs in low-income countries. This is also true of the biscuits produced by Varghese and Srivastav [25] (13.46%) and Kouton et al. [26] (16.18%; 16.12%). Our driving criterion was the use of low-cost ingredients to enhance the protein content, while also considering the amino acid composition of soy and egg, which is regarded as highly beneficial. The Biological Value (BV) and Protein Digestibility-Corrected Amino Acid Score (PDCAAS) of these two products are, respectively, 100 and 1.00 for egg, and 74 and 1.00 for soy [44].

In terms of fat content, the values for BUR and CON (19.69%; 20.06%) were lower than those observed in BP-100 (24.8%), cookies (25.97%) developed by Varghese and Srivastav [25], and T1 and T2 (respectively, 37.35% and 43.25%) developed by Kouton et al. [26]. In any case, the fat content of our two prototypes is clearly above the 12% of fat defined as standard for high-energy biscuits by USAID [45]. The high lipid values in the recipes are intended to satisfy dietary needs, particularly those pertaining to polyunsaturated fatty acid, given the paucity of such nutrients in routine diets. The relatively low values observed in BUR and CON can be justified by their supportive and non-therapeutic nature.

With regard to the fiber content, the two recipes are distinguished by a remarkably high level of fiber (BUR: 6.63%; CON: 6.95%), which is above the standard maximum defined by USAID [45] for high-energy biscuits (2.3%). This outcome can be attributed to the incorporation of pulse ingredients in the recipes.

With regard to the ash content (BUR: 3.06%; CON: 3.05%), both biscuits exhibited a slightly lower level than the minimum (3.5%) recommended by USAID [45], but this discrepancy can be attributed to the absence of additional minerals not readily available in local markets. The ash content of the two recipes is also lower in comparison with the three biscuits based on mushroom powder (3.51%; 3.89%; 4.23%) of Cornelia and Chandra [14].

The analysis of urease activity revealed values of, respectively, 0.028 (BUR) and 0.02 (CON) mg N/g min at 30 °C, which are ten times lower than the limit defined as 0.23 mg N/g min 30 °C [46]. This evidence demonstrates the capacity of biscuit baking to denature antinutritional factors, such as trypsin inhibitors [47], which are often present in legumes.

The protein digestibility of the biscuits was found to be very high (BUR: 91.72%; CON: 92.01%), which is a significant advantage in terms of suitability for feeding malnourished children.

With regard to the estimated micronutrient content, the two prototypes were not comparable with the similar products presented by Varghese and Srivastav [25] or Fetriyuna et al. [27], who used vitamin and mineral premix in the formulation. It is evident that the two biscuits are unable to fulfil the daily requirements of vitamin A, Zn, and iron for the different age groups. In fact, they can only cover 6–10% of the vitamin A daily need, 13–18% of the Zn daily need, and 25–32% of the Fe daily need. A comparison of the two recipes revealed that CON is the richest one in terms of micronutrients.

With regard to the stability trial, the moisture content of both prototypes remained almost constant, without significant increase, throughout the 35-day period at 30 °C. There were also no significant differences between the two recipes. A comparison with Romani et al. [48] reveals a comparable trend in the temporal evolution of moisture content at similar temperatures, although the data indicate a higher level (3.9–4.1%) in BUR and CON. The humidity in the prototypes also remained below the maximum levels defined by USAID [45] for high-energy biscuits (4.5%), as well as the levels observed in six prototypes created by Beshir et al. [49] with lupin flour (5.32%; 5.50%; 6.38%; 5.75%; 5.44%; 5.37%).

With regard to the peroxide content, the closure problem influenced the trends for the two prototypes, which demonstrated a discernible difference in the rate of lipidic oxidation. This was evidenced by the fact that BUR reached a level higher than 10 Meq O_2/kg oil at a faster rate than CON, which is considered the starting point for an increase in consumer rejection of similar baking products by Calligaris et al. [34,50]. The latter study [50] defines the rejection area for 70% of the consumers as being between 13.7 and 19.7 Meq O_2/kg oil, which are levels that were never achieved by the two biscuits during the course of the study. In addition, a comparison with Romani et al. [48] indicates that the maximum peroxide levels in BUR and CON are lower than the maximum levels achieved in similar products tested by these authors (14 Meq O_2/kg oil).

The final parameter analyzed to assess the product shelf-life was the texture. Two parameters were identified that permit a more detailed explanation of the texture profile: hardness, expressed in g strength, and fracturability, expressed in mm walked by the sensor from minimum to maximum strength (hence, a higher mm indicates a lower fracturability). During the aforementioned period, neither of these two texture parameters exhibited any significant change in either biscuit. However, BUR was found to be harder and, correspondingly, less fracturable than CON. This difference, which does not affect the perception of palatability, can be attributed to the inclusion, in the first prototype, of wheat flour, which represents the sole significant difference between the two formulations. The formation of the gluten network can significantly contribute to the hardness of the biscuit. Similar results were obtained by Lages Rodrigues et al. [51], who conducted a similar texture trial.

5. Conclusions

The two biscuit recipes developed for a low-cost ready-to-use supplementary food (RUSF) in the areas followed by Project C3S of Università Cattolica del Sacro Cuore (Piacenza) demonstrate promising results in terms of protein and lipidic integration for the prevention and care of malnutrition in children. The shelf-life test demonstrated that biscuits packaged in a simple PE plastic bag would remain stable for up to 35 days at 30 °C, a temperature compatible with the environmental conditions of Sub-Saharan Africa. The biscuits can be produced with an accessible recipe and with locally available, affordable ingredients, allowing for straightforward distribution, consumption, and further household production in the event that the context does not permit the establishment of industrial facilities.

The subsequent stage of research may entail a feeding trial, with the objective of evaluating the impact of these products on malnourished children.

Author Contributions: Conceptualization, A.N.; methodology, D.R. and G.A.; formal analysis, D.R. and G.A.; investigation, D.R. and G.A.; resources, G.S. and G.B.; writing—original draft preparation, D.R.; writing—review and editing, A.N., G.S. and G.B.; visualization, D.R. and A.M.; supervision, G.B.; project administration, G.B.; funding acquisition, G.S. All authors have read and agreed to the published version of the manuscript.

Funding: This work was supported by the Doctoral School on the Agro-Food System (Agrisystem), Project C3S (Production of Appropriate Food: Sufficient, Safe and Sustainable) funded by Fondazione Invernizzi, and 5 × 1000 of the Università Cattolica del Sacro Cuore (Italy).

Institutional Review Board Statement: Ethical review and approval were waived for this study due to informality of dietary questionnaires used to understand the most used food resources in local contexts.

Informed Consent Statement: Informed consent was obtained from all subjects involved in the study.

Data Availability Statement: The original contributions presented in the study are included in the article, further inquiries can be directed to the corresponding author.

Acknowledgments: Riccardo Dioli and Samantha Sigolo—lab technicians in Department of Animal, Nutrition and Food Science (DiANA), Università Cattolica del Sacro Cuore, Via Emilia Parmense 84, 29122 Piacenza, Italy.

Conflicts of Interest: The authors declare no conflicts of interest. The funders had no role in the design of the study; in the collection, analyses, or interpretation of data; in the writing of the manuscript; or in the decision to publish the results.

References

1. WHO. *Children: Reducing Mortality*; WHO: Geneva, Switzerland, 2016; p. 178.
2. Mc Donald, C.M.; Olofin, I.; Flaxman, S.; Fawzi, W.W.; Spiegelman, D.; Caufield, L.E.; Black, R.E.; Ezzati, M.; Danadei, G. Nutrition Impact Model Study. The effect of multiple anthropometric deficits on child mortality: Meta-analysis of individual data in 10 prospective studies from developing countries. *Am. J. Clin. Nutr.* **2013**, *97*, 896–901. [CrossRef]
3. Kennedy, G.; Nantel, G.; Shetty, P. Globalization of Food Systems in Developing Countries: Impact on Food Security and Nutrition. *FAO Food Nutr. Paper. Food Control.* **2004**, *19*, 83.
4. Black, R.; Victora, C.g.; Walker, S.P.; Bhutta, Z.A.; Christian, P.; de Onis, M.; Ezzati, M.; Grantham-McGregor, S.; Katz, J.; Martorell, R.; et al. Maternal and child undernutrition and overweight in low-income and middle-income countries. *Lancet* **2013**, *382*, 427. [CrossRef]
5. Bhutta, Z.A.; Das, J.K.; Rizvi, A.; Gaffey, M.F.; Walker, N.; Horton, S.; Webb, P.; Lartey, A.; Black, R. Evidence-based interventions for improvement of maternal and child nutrition: What can be done and at what cost? *Lancet* **2013**, *382*, 452–477. [CrossRef] [PubMed]
6. Victora, C.G.; Adair, L.; Fall, C.; Hallal, P.C.; Martorell, R.; Richter, L.; Sachdev, H.S.; Maternal and Child Undernutrition Study Group. Maternal and Child Undernutrition: Consequences for Adult Health and Human Capital. *Lancet* **2008**, *371*, 340–357. [CrossRef] [PubMed]
7. Martins, V.J.B.; Florêncio, T.M.M.T.; Grillo, L.P.; Franco, M.D.C.P.; Martins, P.A.; Clemente, A.P.G.; Santos, C.D.L.; Vieira, M.d.F.A.; Sawaya, A.L. Long-Lasting Effects of Undernutrition. *Int. J. Environ. Res. Public Health* **2011**, *8*, 1817–1846. [CrossRef] [PubMed]
8. Jyoti, D.F.; Frongillo, E.A.; Jones, S.J. Food Insecurity Affects School Children's Academic Performance, Weight Gain, and Social Skills. *J. Nutr.* **2005**, *135*, 2831–2839. [CrossRef]

9. FAO; IFAD; WFP. The Multiple Dimensions of Food Security. In *The State of Food Insecurity in the World 2013*; FAO: Rome, Italy, 2013.
10. FAO; ECA; AUC. *Africa Regional Overview of Food Security and Nutrition 2020: Transforming Food Systems for Affordable Healthy Diets*; FAO: Accra, Ghana, 2021.
11. UNICEF; WHO; The World Bank Group. *Levels and Trends in Child Malnutrition*; UNICEF: New York, NY, USA, 2017.
12. UNICEF; WHO; The World Bank Group. *Levels and Trends in Child Malnutrition: Key Findings of the 2021 Edition of the Joint Child Malnutrition Estimates*; WHO: Geneva, Switzerland, 2021.
13. WHO; UNICEF. WHO child growth standards and the identification of severe acute malnutrition in infants and children. In *A Joint Statement by the World Health Organization and the United Nations Children's Fund*; WHO: Geneva, Switzerland, 2009.
14. Cornelia, M.; Chandra, J. Utilization of white oyster mushroom powder (Pleurotus ostreatus (Jacq.) P. Kumm.) in the making of biscuit as emergency food product. *EurAsian J. BioSciences* **2019**, *13*, 1859–1866.
15. Zoumas, B.L.; Armstrong, L.E.; Backstrand, J.R.; Chenoweth, W.L.; Chinachoti, P.; Klein, B.P.; Lane, H.; Marsh, K.; Tolvaven, M. *High Energy, Nutrient-Dense Emergency Relief Product*; National Academy Press: Washington, DC, USA, 2002. [CrossRef]
16. Duclercq, M.-P. Production of Ready-to-Use Food (RUF): An Overview of the Steps and Challenges Involved in the "Local" Production of RUF. CMAM Forum Technical Brief; 2014. Available online: https://www.ennonline.net/productionofreadytousefoodruf (accessed on 28 April 2024).
17. UNICEF. *UNICEF 's Approach to Scaling up Nutrition*; UNICEF: New York, NY, USA, 2012.
18. Plumpy'NUT. [Online]. Available online: https://www.nutriset.fr/products/en/plumpy-nut (accessed on 28 April 2024).
19. Scherbaum, V.; Purwestri, R.C.; Stuetz, W.; Inayati, D.A.; Suryantan, J.; Bloem, M.A.; Biesalski, H.K. Locally Produced Cereal/Nut/Legume-Based Biscuits versus Peanut/Milk-Based Spread for Treatment of Moderately to Mildly Wasted Children in Daily Programs on Nias Island, Indonesia: An Issue of Acceptance and Compliance? *Asia Pac. J. Clin. Nutr.* **2015**, *24*, 152–161. [CrossRef]
20. Okafor, J.; Okafor, G.; Ozumba, A.; Elemo, G. Quality Characteristics of Bread Made from Wheat and Nigerian Oyster Mushroom (*Pleurotus plumonarius*) Powder. *Pak. J. Nutr.* **2012**, *11*, 5–10. [CrossRef]
21. Agrahar-Murugkar, D.; Zaidi, A.; Gupta, C. Quality evaluation of composite flour laddoo containing sprouted and malted ingredients. *Nutr. Food Sci.* **2016**, *46*, 476–487. [CrossRef]
22. Ubbor, S.; Akobundu, E. Quality characteristics of cookies from composite flours of watermelon seed, cassava and wheat. *Pak. J. Nutr.* **2009**, *8*, 1097–1102. [CrossRef]
23. Fleet, A. High Energy Biscuits—Product Specification Sheet. 2017. Available online: https://www.ungm.org/Public/Notice/55262 (accessed on 28 April 2024).
24. WHO. *RUTF Biscuit (BP100). Product Specifications Sheet RUTF Biscuit (BP 100), BP100*; WHO: Geneva, Switzerland, 2017.
25. Varghese, C.; Srivastav, P.P. Formulation and sensory characterization of low-cost, high-energy, nutritious cookies for undernourished adolescents: An approach using linear programming and fuzzy logic. *Innov. Food Sci. Emerg. Technol.* **2022**, *75*, 102904. [CrossRef]
26. Kouton, S.; Vodouhè, S.; Hounkpatin, W.A.; Soumanou, M.M. Formulated Therapeutic Food Fortified of Vegetable for Nutritional Recuperation of Undernourished Infants in Benin. *J. Food Sci. Nutr. Res.* **2020**, *3*, 092–101. [CrossRef]
27. Fetriyuna, F.; Purwestri, R.C.; Susandy, M.; Köhler, R.; Jati, I.R.A.P.; Wirawan, N.N.; Biesalski, H.-K. Composite Flour from Indonesian Local Food Resources to Develop Cereal/Tuber Nut/Bean-Based Ready-to-Use Supplementary Foods for Prevention and Rehabilitation of Moderate Acute Malnutrition in Children. *Foods* **2021**, *10*, 3013. [CrossRef] [PubMed]
28. Reggi, D. Sviluppo di Biscotti Proteici per Bambini del Burundi: Prospettive di Industrializzazione. Master's Thesis, Università Cattolica del Sacro Cuore, Piacenza, Italy, 2020.
29. Minardi, A. *Food Security in Developing Countries: The Case Study of "Production of Appropriate Food: Sufficient, Safe and Sustainable" Project*; AgriSystem; Università Cattolica del Sacro Cuore: Piacenza, Italy, 2018.
30. Mourey, A. *Nutrition Manual for Humanitarian Action*; International Committee of the Red Cross Assistance Division: Geneva, Switzerland, 2008; pp. 76–96.
31. Association of Official Analytical Chemists (AOAC). *Official Methods of Analysis*; AOAC: Arlington, VA, USA, 2000.
32. *ISO 712:2009*; Cereals and Cereal Products—Determination of Moisture Content—Reference Method. ISO: Geneva, Switzerland, 2009.
33. EEC. *Commission Directive (First) No 71/250 of 15 June 1971 Establishing Community Methods of Analysis for the Official Control of Feeding-Stuffs*; EEC: Brussels, Belgium, 1971.
34. Calligaris, S.; Manzocco, L.; Kravina, G.; Nicoli, M.C. Shelf-life modeling of bakery products by using oxidation indices. *J. Agric. Food Chem.* **2007**, *55*, 2004–2009. [CrossRef]
35. EEC. *Annex III—Regulation 2568/91/EEC*; EEC: Brussels, Belgium, 1991.
36. Kaur, M.; Sandhu, K.S.; Arora, A.; Sharma, A. Gluten free biscuits prepared from buckwheat flour by incorporation of various gums: Physicochemical and sensory properties. *LWT Food Sci. Technol.* **2015**, *62*, 628–632. [CrossRef]
37. WHO. Global Panel on Agriculture and Food Systems for Nutrition. In *Food Systems and Diets: Food Systems and Diets*; WHO: London, UK, 2016.
38. Kouakou, E.; Meite, A.; Kouamé, G. Traditional process for the production of therapeutic food for the nutritional care of the malnourished. *Int. J. Curr. Res.* **2016**, *8*, 39395–39398.

39. Manary, M.J. Local Production and Provision of Ready-to-Use Therapeutic Food (RUTF) Spread for the Treatment of Severe Childhood Malnutrition. *Food Nutr. Bull.* **2006**, *27* (Suppl. S3), S83–S89. [CrossRef]
40. Arimond, M.; Hawkes, C.; Ruel, M.T.; Sifri, Z.; Berti, P.R.; Leroy, J.L.; Low, J.; Brown, L.; Frongillo, E. Agricultural Interventions and Nutrition: Lessons from the Past and New Evidence. In *CAB International FAO*; Thompson Amoroso, B., Ed.; FAO: Oxfordshire, UK, 2013; pp. 41–75.
41. Beesabathuni, K.N.; Natchu, U. Production and distribution of a therapeutic nutritional product for severe acute malnutrition in India: Opportunities and challenges. *Indian Pediatr.* **2010**, *47*, 702–706. [CrossRef]
42. Sandige, H.; Ndekha, M.J.; Briend, A.; Ashorn, P.; Manary, M.J. Home-based treatment of malnourished Malawian children with locally produced or imported ready-to-use food. *J. Pediatr. Gastroenterol. Nutr.* **2004**, *39*, 141–146. [CrossRef] [PubMed]
43. Dibari, F.; Diop, E.H.I.; Collins, S.; Seal, A. Low-cost, ready-to-use therapeutic foods can be designed using locally available commodities with the aid of linear programming. *J. Nutr.* **2012**, *142*, 955–961. [CrossRef] [PubMed]
44. Hoffmann, J.; Falvo, M. Protein—Which is Best? *J. Sports Sci. Med.* **2004**, *3*, 118–130.
45. USAID. *The United States Agency for International Development Commodity Specification for High-Energy Biscuits (HEB) for Use in International Food Assistance Programs*; USAID: Washington, DC, USA, 2016.
46. Camera di Commercio (CCIAA). *Resolution n.6—Special Regulations Vegetable Extraction Flour (18/4/2011)*; CCIAA: Milano, Italy, 2011.
47. Saini, H. Thermal stability of protease inhibitors in some cereals and legumes. *Food Chem.* **1989**, *32*, 59–67. [CrossRef]
48. Romani, S.; Tappi, S.; Balestra, F.; Estrada, M.T.R.; Siracusa, V.; Rocculi, P.; Rosa, M.D. Effect of different new packaging materials on biscuit quality during accelerated storage. *J. Sci. Food. Agric.* **2015**, *95*, 1736–1746. [CrossRef] [PubMed]
49. Beshir, E.O.; Khallaf, M.; El-Daim, Y.A.; Awad, R.; Kamel, M.A.-A. Sweet Lupin and Whey Protein Concentrate as Supplementants for Utilizing in Semi–Hard Biscuit and Its Chemical Properties. *Egypt. J. Chem.* **2021**, *64*, 7517–7527. [CrossRef]
50. Calligaris, S.; Pieve, S.; Kravina, G.; Manzocco, L.; Nicoli, C. Shelf Life Prediction of Bread Sticks Using Oxidation Indices: A Validation Study. *J. Food Sci.* **2008**, *73*, E51–E56. [CrossRef]
51. Rodrigues, L.L.; Hashimoto, J.M.; Nabeshima, E.H.; Gomes, F.d.O.; Silva, J.D.N.; Silva, K.J.D. Characteristics of biscuits containing different proportions of whole mung bean, wheat and rice brown flours. *Food Sci. Technol. Int.* **2022**, *29*, 818–830. [CrossRef]

foods

MDPI

Review

Current Trends in Food Processing By-Products as Sources of High Value-Added Compounds in Food Fortification

Helen Stephanie Ofei Darko, Lama Ismaiel, Benedetta Fanesi *, Deborah Pacetti and Paolo Lucci

Department of Agricultural, Food and Environmental Sciences, Università Politecnica delle Marche, 60131 Ancona, Italy; helenstephanieofei.darko@unito.it (H.S.O.D.); l.ismaiel@univpm.it (L.I.); d.pacetti@univpm.it (D.P.); p.lucci@univpm.it (P.L.)

* Correspondence: b.fanesi@univpm.it

Abstract: Along the food production chain of animal, fish, and vegetable products, a huge amount of by-products are generated every year. Major nutritional, financial, and environmental advantages can be achieved by transforming them into functional ingredients for food formulation and fortification. In this review, we investigated various conventional and emerging treatments recently employed to obtain functional ingredients rich in proteins, fibers, and bioactive compounds from vegetables, fish, meat, and dairy by-products. The optimal enrichment level in food as well as the nutritional, techno-functional, and sensory properties of the final food were also discussed. Novel technologies such as ultrasounds, microwaves, and high pressure have been successfully adopted to enhance the extraction of target compounds. The functional ingredients, added both in liquid or powder form, were able to improve the nutritional quality and antioxidant potential of food, although high levels of fortification may cause undesired changes in texture and flavor. This review provides important considerations for further industrial scale-up.

Keywords: animal; fish; vegetables; fruit; waste; loss; antioxidants; phenolics; functional foods

1. Introduction

One component of the Green Deal that concentrates on the food system is the Farm to Fork approach, which employs the circular economy model to increase the sustainability of food production. Improving the utilization of food waste, surplus, and by-products within the framework of food supply chains would be necessary to close the loops created [1]. Ultimately, three-quarters of edible food meant for human consumption is lost or wasted annually worldwide within the production and supply chain, amounting to around 1.3 billion tons [2]. These losses result from the entire food value chain, which starts with harvesting and continues through postharvest, industrial, and commercial processing before ending with consumer consumption. Additionally, by-products of food processing that are side streams obtained after target products are produced throughout all food production sectors also count as food losses [3]. Food processes such as vegetable oil extraction, starch, juice, and sugar production and fish and meat processing, resulting in animal waste side streams such as bones, offal, and hides, and whey protein from cheese processing generate waste or by-products at extremely intriguing volumes. A large portion of these products is typically thrown out during processing or traditionally used as compost or animal feed [4]. More recently, however, there has been investigation into their potential as alternate energy sources [5].

Currently, utilizing by-products for food fortification has emerged as a substitute strategy. Specifically, when food processing by-products contain significant amounts of favorable and valuable compounds like proteins, vitamins, minerals, fiber, and essential fatty acids, they can be applied in the food and pharmaceutical industry, thereby increasing their potential to be included in the human diet [5,6]. In addition, it is anticipated that the human population will grow by 2050, highlighting the need to create a sufficient food supply to ensure that food production will be possible in the future [7]. Efficient by-product usage has garnered significant attention in recent times due to its potential to improve sustainability, reduce waste disposal, and generate value-added goods for the food sector [8]. By-products from fruits and vegetables, for instance, are known to have a significant quantity of dietary fiber and bioactive components with significant biologically active components like phenolic and antioxidant compounds [6]. Likewise, the by-products of meat and dairy processing can provide valuable substances such as proteins, lipids, chitin, collagen, and whey proteins that can be used for food fortification [9,10]. While there has been previous discussion on the use of fish by-products in food fortification, there has not been much recent discussion about the relationship between fish by-products and new food processing techniques. This review not only emphasizes how crucial it is to pretreat fish by-products, but it also offers recommendations for the best way to use these substances in food formulation.

Many biochemical, chemical, physical, or mechanical techniques have been used to extract bioactive molecules, provided that the procedures are selective, microbiologically safe, and do not change or modify the compounds. The extraction field is expanding rapidly these days, and as a result, there is a lot of interest in finding ways to expedite the extraction process, boost quality, and protect bioactive compounds [11]. This topic is novel given that there is a growing body of research and interest in the innovative extraction techniques that can be used to successfully recover and stabilize bioactive compounds with improved functional characteristics from by-products of the food industry as well as food applications that successfully integrate high-value compounds into food systems. As a result, a more sustainable and environmentally friendly food system can be developed. In this review, we aimed to provide an overview of the most recent developments and/or technologies in the valorization of food processing by-products from vegetables, meat, fish, and dairy value chains, with a focus on the production of protein-, fiber-, and bioactive-rich functional ingredients and their incorporation into food. Comprehensive terms ('fish AND by-product' (1075), 'vegetable AND by-product' (703), brassica AND by-product' (117), 'legume AND by-product' (72), meat AND by-product' (790), and 'dairy AND by-product' (606)) were used to obtain a broad selection of articles published in the period from 2020 to 2024 from the Scopus database. Consequently, a total number of 147 articles were considered for full screening and discussion (Figure 1).

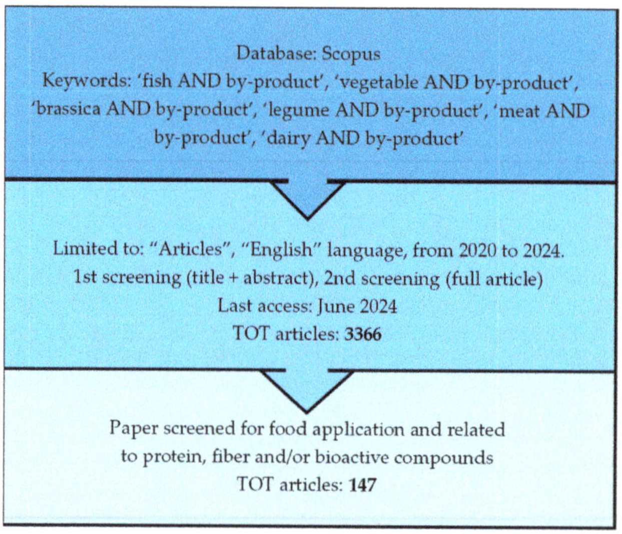

Figure 1. Flow chart of the methodology applied for screening of articles.

2. Characterization of Food Processing By-Products Based on Their Nutritional Benefit, Treatment, and Food Application

2.1. Protein

The demand for protein greatly increased both to meet consumers' needs for healthy diets and to face undernourishment in some poor countries. Moreover, life cycle assessment of meals derived from animals and plant-based, high-protein alternatives has been demonstrated to broaden the supply of sustainable high-protein food options, making these products more economically attractive [12]. Here, the importance of extracting proteins from alternative sources such as vegetable and animal by-products is discussed.

Leaves, stalks, press cakes, spent grains, and aquafaba are among the most common fruit and vegetable by-products obtained during food processing. These by-products represent an alternative source of proteins. Therefore, they can be valorized by transformation into food ingredients either in liquid form or powder. The extraction of vegetable oil from seeds, nuts, or fruits, usually obtained through screw-pressing or employing solvents, provides a large amount of residues, also called press-cake or meal, that can be ground and used as a protein-rich ingredient in quantities of ~30–40% [13–15]. The removal of lipid fraction from by-products can also be a pre-treatment for further purification processes to obtain a high protein concentration or improved functional properties. The conventional process is the wet method, consisting of alkaline extraction followed by isoelectric precipitation. Basically, ground meal is dispersed into water with a sample-to-solvent ratio of about 1:10–1:15. Later, sodium hydroxide (NaOH) is added to achieve an alkaline pH, i.e., between 9 and 12, which allows to solubilize/extract proteins while removing insoluble fibers [16]. Mild treatments (pH 9–10) have been adopted for rapeseed cake and quinoa okara to obtain stable emulsions with low particle size [17,18]. Harsh treatments (pH > 10) may lead to protein denaturation, therefore reducing recovery yield. In some cases, a stronger alkalinization (pH 12) was necessary to increase protein and polysaccharide solubilization, such as for spent brewer's yeast, which is characterized by strong and thick cell walls, and to improve emulsion stability [19]. The time–temperature relationship of alkaline treatments may vary from 30 min at 55 °C to 2–4 h at 20–25 °C [17,19]. Generally, temperatures above 60 °C may induce protein denaturation and gelation, reducing the protein concentration of the final ingredient. The time should be adequate to obtain protein dispersion without achieving protein saturation level. Afterward, protein precipitation is induced by adding hydrochloric acid (HCl) to adjust the pH towards the isoelectric point, which is usually between pH 3 and 5 for plant proteins [17]. Besides this, innovative non-thermal technologies such as ultrasound (US) and fermentation have been investigated to increase the protein content of the final food ingredient. Cavitation phenomena generated by US induce protein denaturation favoring solubilization [18,20]. Kasapoglu et al. applied US homogenization (405 W for 2.5 min) and obtained a coconut oil by-product-based drink with a protein content double the amount found in other commercial beverages [20]. On the other hand, a high-intensity US treatment (75% pulse amplitude for 10 min) applied to alkalized okara solution allowed it to have a protein concentration four times higher than the untreated control solution, demonstrating that the application of this green technology may induce conformational changes in the protein structure favoring hydrophobic interactions among proteins, which tend to aggregate and precipitate during the resting phase [18]. Additionally, fermentation was recently exploited by Castellanos-Fuentes et al. as a strategy to increase protein and fiber content while decreasing sugars as well as to improve the physicochemical and functional properties of the soy meal-based food ingredient [13]. Valorization strategies based on fermentation of vegetable by-products were deeply discussed by [21].

As an alternative to wet methods, dry fractionation allows to produce protein-rich ingredients separating the fine (protein-rich) and the coarse (starch-rich) fractions without water and chemical consumption and maintaining proteins in their native state. On the downside, the final ingredient is characterized by the presence of antinutrients and low purity. For example, Squeo et al. employed this technique with defatted durum wheat cake using an air classifier to increase the protein yield, yet the protein content did not exceed 31% [22]. Therefore, the advantages and disadvantages of introducing an additional protein concentration step should be carefully evaluated along the production process of the food ingredient.

To this end, many authors did not perform any protein concentration but employed the whole by-product flour as a protein source food ingredient. As for Brassica species [23–25] and black gram [26], by-products were stabilized (e.g., blanching and drying), milled, and sieved, with a protein concentration obtained ranging from 23 to 31%.

Emulsions, pasta, bakery goods, snacks, beverages, ice creams, and burgers are among the various food applications that have been recently investigated. Different by-product sources such as artichokes bracts, tomato seeds and skin, and broccoli leaves have been used to enrich pasta (up to 30% of enrichment). In some cases, it was possible to improve the protein content and the amino acid composition of pasta. However, the higher the enrichment level, the higher the cooking loss, an important parameter that should not exceed 10% to have an acceptable product [23,27,28]. Soy- and broccoli-based ingredients (53% and 29% protein, respectively), when used in a concentration of 5–6%, were able to improve the protein and mineral content of gluten-free bread as well as color, specific volume, and bake loss [13,24]. For muffins from cauliflower by-products [25] and protein bars from durum wheat cake [22], the optimal enrichment level was assessed at 10% both in terms of nutritional quality (higher protein and fiber content) and physicochemical parameters (e.g., color and hardness). It was possible to almost triple the enrichment level using nut by-products while maintaining good acceptability, as was tested for cashew kernel, soy okara, and peanut meal in the formulation of vegan burgers and snacks, respectively [14,29,30]. Protein-rich ingredients obtained from different by-products such as rapeseed cake [17], quinoa okara [18], aquafaba [31], and spent brewer's yeast [19] were investigated to develop emulsions. The concentrations varied from 0.1 to 6% when the food ingredient was added in the powder form and from 20 to 35% for liquid. In general, the emulsions showed good stability and low particle size. Canned chickpea aquafaba was also found as a good egg white replacer for meringue production [32].

Fish by-products provide great sources of protein (>75%), making them suitable for preparing and fortifying functional food products. Lately, they are used to cover a wide range of food applications, such as fish sausage, fish balls, coating films, emulsions, umami compounds, and ready-to-eat paste. Protein from fish by-products is subjected to hydrolysis before utilization and various concentrations (0.1–27%) have been applied to fortify food products like yoghurt and pasta. Conventionally, there are two main methods for protein extraction and hydrolysis, namely alkaline and enzymatic treatments.

Alkaline treatment can effectively solubilize protein, and it has been applied to fresh, frozen, and fermented by-products [33,34]. Samples are usually soaked with different concentrations of NaOH for various time slots. For example, alkaline treatment with 3% NaOH for 1 h was suitable to increase the concentrations of calcium and phosphorus in fortified fish sausage [33]. On the other hand, enzymatic treatment is widely applied using different types of enzymes (i.e., alcalase, protamex, flavourzyme, neutrase, trypsin, and papain). The raw material is typically minced and mixed with water. The pH and temperature are then adjusted to incubation conditions. After incubation, enzymes are inactivated. The mixture is centrifuged to separate the liquid phase. This liquid is mainly dried using a spray dryer or a freeze dryer to obtain hydrolysates.

The incubation temperature for alcalase is usually near 60 °C, and the pH is 7.6–8 for 85–180 min. Under these conditions, the protein content was increased in fortified biscuits and soup with fish by-product hydrolysates [35,36]. Additionally, in deep-fried battered squid containing brown stripe red snapper, alcalase hydrolysate showed efficiency in the reduction in oil binding capacity and increase in crispness [37]. Alternatively, a hydrolysis reaction can be achieved using a combination of enzymes depending on the desired target, such as protamex and flavourzyme at 50–55 °C for 2.5 h to improve the quality of fish sauce [38], or a mix of protamex (pH 7.0, 50 °C), alcalase (pH 8.0, 55 °C), neutrase (pH 7.0, 45 °C), and flavourzyme (pH 6.6, 50 °C) in a snack designed for physically active people [39]. Zhang et al. reported that trypsin and alcalase hydrolysates had stronger ABTS scavenging activity than papain and neutrase hydrolysates, suggesting the former as potential natural antioxidants [40]. Fortified yoghurt with 0.15% of flavourzyme hydrolysates from parrotfish by-products had considerable values for color, texture, and overall acceptance values [41]. Papain hydrolysate from yellowfin tuna red was employed as a fortifying and stabilizing agent in a mayonnaise emulsions system, replacing 50% of egg yolk. The resulting emulsions expressed a high quality with low particle size and enhanced oxidative and physicochemical stability [42].

Besides conventional hydrolysate production, some authors reported the use of cheap, eco-friendly, low-energy consumption technologies to speed up the enzymatic hydrolysis of proteins. For instance, US was used as pretreatment before the addition of the protamex enzyme, improving the extraction yield (76.34%) of umami compounds [43]. Moreover, the combined impact of natural deep eutectic solvents (NADES) and US pretreatments was studied as an extraction medium at 190 °C for the subcritical water hydrolysis of shrimp waste to be used as a biodegradable film with a high amount of protein [44].

In addition, there is increased attention on emulsions and microencapsulation prepared and stabilized with fish by-products including oil, gelatin, protein extracts, or protein hydrolysates as ready-to-use products or further application such as surimi. Gelatin has been isolated and investigated as a potential emulsifying, foaming, and thickening agent. Extracting gelatin has been subjected to various methods, including hot water-pretreated gelatin, acetic acid-pretreated gelatin, sodium hydroxide-pretreated gelatin, and enzyme-pretreated gelatin [45]. The quality and gel strength might differ depending not only on the mechanisms of extraction but also on the type of fish. Yang et al. applied pH shifting using 0.1 M NaOH and 0.05 M acetic acid on different fish skin to extract gelatin and further use it for fish oil-loaded gelatin-stabilized emulsions [46]. Gelatins and emulsions of catfish and silver carp had higher gel strengths and lower creaming index values than marine fish skin gelatin [46]. Similarly, high-value gelatins with strong gel values and significant emulsifying capacity were obtained through pepsin enzyme pretreatment after the removal of non-collagenous protein, fat, and minerals. The authors also claimed it has potential as a foaming agent for ice cream and marshmallows [45].

Pink perch surimi by-products were employed to obtain protein hydrolysate for further microencapsulation. The protein hydrolysate conditions were optimized using response surface methodology (RSM). RSM was chosen as an excellent statistical method for determining the optimal process conditions by decreasing the number of experimental trials while covering the interactions between parameters and decreasing the time and cost of tests. The protein hydrolysate produced had the highest yield (17%), good functional properties, and moderate antioxidant activity using the alcalase enzyme under the following conditions (57.9 °C, 85.8 min, with 0.15% (v/w) enzyme to substrate ratio) [47]. Furthermore, the antioxidant activity of microencapsulated, optimized protein hydrolysates was preserved; their bitterness and fishy odor were significantly decreased; and their overall acceptability was increased [47]. Optimizing protein hydrolysis conditions of surimi processing by-products proved a great potential to fortify various food formulations and provide value-added products. Correspondingly, surimi protein hydrolysate was produced with enzymatic interaction using alcalase and trypsin. The produced hydrolysate was added to silver carp surimi to prevent protein oxidation and enhance the gel properties [48]. Direct extraction of protein by thermal coagulation gave better physicochemical properties than enzymatic protein hydrolysates when preparing emulsions [49]. In another study, a simple pH shifting using 2 mol/L NaOH and HCl was applied for salmon by-products to obtain lyophilized protein powder. Subsequently, the latter was mixed with sodium alginate to encapsulate probiotics to improve its viability under pasteurization and gastrointestinal conditions [50]. Likewise, the encapsulation efficiency of corn oil was enhanced when using *Sardinella aurita* protein isolate spray-dried solution (4%) combined with maltodextrin [51]. Catfish protein concentrate extracted from catfish by-products with

0.5 N $NaHCO_3$ was used to enrich microencapsulated highly nutritional fish oil with >17% protein [52].

It is worth noting that protein content was improved in pasta enhanced with 10% seabass by-product concentrates, which were obtained simply by dipping in saline solution (8%) and drying in the oven at 60 °C for 24 h [53–55]. Moreover, high-protein cereal bars and nuggets based on fish by-products were washed, minced, and cooked at 60 °C and 180 °C, respectively [56,57]. Pressure-cooking was applied to Nile tilapia by-products followed by pressing and drying at 90 °C for 24 h to obtain fish flour with high protein content (55.41%) [58].

Meat by-products including bones, heart, kidney, liver, tongue, skin, horns, hoofs, blood, tendons, visceral organs, and gastrointestinal tract can be valorized as well considering their content of high-value protein [6,59,60]. Valorizing these by-products to produce protein for food applications is typically subjected to enzymatic hydrolysis.

As in fish by-products, flavourzyme, papain, bromelain, and alcalase are the most-used enzymes. Alcalase showed a higher degree of hydrolysis compared to other commercial ones due to its ability to break peptide bonds [61]. It was the most effective enzyme for the generation of antioxidant peptides, followed by neutrase, papain, and pepsin [62]. On the other hand, flavourzyme was the best alternative for the formation of flavor compounds [61]. The required incubation conditions for each enzyme must be respected. Da Silva Bambirra Alves et al. also used alcalase to prepare chicken blood meal hydrolysates at optimal conditions of 50 °C, pH 8.5, and enzyme/substrate (E/S) ratio of 6.5% [63]. Mattohti et al. used a mix of neutrase (100 U/mg, 45 °C, pH 7.0), alcalase (400 U/mg, 50 °C, pH 9.5), papain (1000 U/mg, 55 °C, pH 6.5), and Pepsin (300 U/mg, 37 °C, pH 2.8) in the preparation of hydrolysates with antioxidant characteristics from horse bone marrow [62].

Meat by-product pre-treatment is usually applied prior to hydrolysis. For example, de Souza Fontes et al. subjected goat viscera to a heat treatment at 90 °C for 15 min in a water bath to inactivate the endogenous enzymes to inhibit competition for the active site of the proteins during enzymatic hydrolysis [64]. Likewise, fat was removed from meat by-products either using solvents [62] or heated water [65]. This step enhances the accessibility of enzymes to the active site during the hydrolysis process and raises the protein content [66].

Recently, innovative technologies have emerged to produce high-value enzymatic hydrolysis products. For instance, high-voltage electrostatic field and US, alone or in combination, improved the extraction rate of protein, the degree of hydrolysis, and flavor sub-stances in meat and bone residue [67]. High-pressure pretreatment also increased the production of protein hydrolysates from chicken bones [61].

Collagen is another important ingredient that can be isolated from meat by-products. This can be achieved through an alkaline procedure, and then, biomodifications can increase its digestibility. Prokopová et al. isolated collagens to produce gelatins from broiler chicken stomachs using a mixture of 0.2 mol/L NaCl, 0.06 mol/L NaOH, and petroleum ether and ethanol [68]. Lukin prepared high-quality lyophilized collagen hydrolysate using protepsin. Subsequently, the produced collagen had great practical importance in the sausage formulation, obtaining the best results at 15% enrichment [69].

Meat by-products can be a source of flavoring for further food applications. Formulation and concentration of volatile compounds in flavoring supplemented with hydrolysate from goat and chicken were studied by [61,64]. The pH affected the aroma profile, resulting in a strong meaty flavor at pH 4 and sweet fatty flavor and a goat aroma at pH 6 [64]. Alternatively, protein fractions (soluble and insoluble extracts in sodium citrate buffer 0.1 M pH 5) obtained from porcine spleens were successfully employed as emulsifying (soluble protein) and thickening (insoluble) agents for sausage production [70]. Protein precipitates by pH shifting from chicken by-products were used to formulate edible coating, achieving fat uptake reduction during deep frying of chicken drumsticks [71]. The addition of dry powder of chicken feet and heads served as a feasible approach to enhance physicochemical characteristics and improve the overall quality of forcemeat formulations [72].

Whey is the primary dairy by-product composed of mainly water (94%), lactose, proteins, and fats. α-lactalbumin and δ-lactoglobulin account for 70–80% of all the proteins in whey mass. Other constituents include immunoglobulins, lactoperoxidase, bovine serum albumin, and bovine lactoferrin [73]. The two major methods used by the dairy industry to handle whey are fermentation and membrane separation (e.g., ultrafiltration). Pires et al. applied ultrafiltration (40–45 °C; 3.0–3.5 bar) to produce symbiotic kefir products from sheep's and goat's whey concentrates, obtaining a product with a high protein content (7.8–16.4%) [74]. Likewise, Pavoni et al. applied similar conditions to ovine whey [75].

Himashree et al. studied the substitution of water by different concentrations of whey (0–100%) in bread. It resulted in high protein and fat content, decreased acrylamide concentration, and improved textural properties [76]. Kusio et al. used whey protein concentrate (76.8%) to produce high-protein, fat-free dairy desserts. Rheological properties, appearance, and antioxidant activity were enhanced in the final product [73]. Similarly, Dinkçi et al. used whey protein concentrate to produce a probiotic beverage with high phenolic compounds and antioxidant activity [9]. The following (Tables 1–4) provide deep details about the ingredient preparation and their application on various food products regarding protein.

Table 1. Fruit and vegetable processing by-products with protein benefits and their food applications.

By-Products	Ingredient Preparation	Food Application	Properties	References
Artichoke bracts and tomato	Non-thermal patented treatment (Patent n°001426984) → freeze drying	Fresh pasta (3% enrichment)	Good protein network, ↑ sensory acceptability	[77]
Pepper placenta, pepper defatted seeds, tomato seed/skin	Air drying (17 h, 30–40 °C) → fat extraction (from seeds, solvent) → grinding	Pasta (10–30% enrichment)	↑ Protein, fiber	[28]
Broccoli leaf	Blanching (hot water, 1 min) → freeze drying → grinding	Pasta (2.5–5% enrichment)	↑ AA composition, ↑ cooking loss	[23]
Broccoli leaf	Blanching (hot water, 1 min) → freeze drying → grinding (0.6 mm)	Gluten-free bread (5% enrichment)	↑ Protein, minerals, ↑ antioxidant activity and anti-AGE activity, ↑ specific volume, bake loss	[24]
Cauliflower stems, leaves, stalks	Blanching (2–3 min) → drying (60 °C, 16 h) → grinding (0.8 mm)	Muffins (10–30% enrichment)	↑ Protein, fiber, TPC, TFC, antioxidant activity, ↑ dark color, ↑ hardness, ↓ specific load volume, chewiness, springiness, cohesiveness, ↓ sensory acceptability (max 10%)	[25]
Mushroom stems	Freeze drying → grinding (0.25 mm) → ergosterol removal: dispersion (3:100) in 96% ethanol (residue) → drying (30 °C, 72 h)	Spray-dried emulsions (5% enrichment)	↑ Viscosity, stability. ↑ yield, ↓ oxidation	[78]
Broken cashew kernels	Fat extraction (solvent) → grinding (0.25 mm) → water dispersion (1:15) → isoelectric precipitation (pH 3–4.5, 15 min) → drying	Vegan burgers (27% replacement)	↓ Solubility, good sensory acceptability	[30]

Table 1. *Cont.*

By-Products	Ingredient Preparation	Food Application	Properties	References
Black gram by-product	Milling → fractionation (0.4–0.8 mm)	Extrudates	↑ WAC, ↑ starch digestibility, ↓ phenolics (flavonoids, proanthocyanidins, saponins)	[26]
Quinoa okara	Water dispersion (1:10) → alkaline extraction (pH 9, 30 min, 55 °C) → US precipitation (75% amplitude, 30 s on—15 s off, 10 min) → air drying	Emulsions (2–6% enrichment)	↑ Protein, fiber, ↑ ES, ↓ particle size	[18]
Soy okara	Drying (100 °C, 4 h) → grinding (0.3–0.6 mm)	Gluten-free extruded snacks (8–40% enrichment)	↑ dark red color, ↓ expansion, ↑ nutritional/sensory quality (29–40%)	[29]
Aquafaba	Chickpea: broth, 1:2 (*w*/*v*) → cooking (100 kPa, 35 min) → freezing (liquid)	Meringue (egg white replacement)	↑ Foaming and emulsifying properties, hardness, fracturability, similar color, ↓ alveoli	[32]
Aquafaba	Pre-soaking: chickpea: water, 1:3 (*w*/*w*) (5 °C, 16 h) → Cooking: chickpea: water, 1:2 (*w*/*w*) (118 °C, 70–80 kPa, 30 min) → filtration → freezing	Emulsion (25–35% replacement)	↑ Gel-like behavior, ↓ cost	[31]
Aquafaba	Soaking (chickpea: water, 1:4, *w*/*v*, 8–10 h) → cooking (chickpea: water, 1–5, *w*/*w*, 45 min) → cooling → filtration (liquid) → freezing	Whipped cream (egg white replacement)	Good egg white replacer, ↑ foam particles, ↓ foam stability, ↓ hardness, gumminess	[79]
Walnut cake and sesame cake	Fat extraction → grinding	Low-fat mayonnaise (2–4%)	↓ Carbohydrate, ↑ protein, fiber, ↑ ES, ↓ hardness, adhesiveness, ↑ sensory properties	[15]
Soybean extruded-expelled meal	Water dispersion (1:10) → sterilization (15 min, 120 °C) → fermentation → freeze drying	Gluten-free bread (6% enrichment)	↑ Protein, fiber (cellulose lignin), ↑ specific volume, ↑ iron bio-accessibility, ↑ crumb color, ↑ aeration	[13]
Durum wheat cake	Fat extraction (solvent) → air classification (grader wheel speed 3000 rpm, airflow 50 m^3/h)	Snack protein bar (10–50% enrichment)	↑ Protein and fiber, ↓ fat, ↑ firmness, ↑ yellowness, ↑ caramel flavor	[22]
Coconut oil by-product	Fat extraction (cold press) → grinding → water dispersion (1:10) (60 °C, 1 h) → homogenization (10,000 rpm, 5 min) → US (405 W, 2.52 min) → filtrated liquid	Full-fat and low-fat vegan ice-cream (1–3% enrichment)	↑ Protein and fat, ↓ cost	[20]
Peanut meal	Fat extraction (press)	Gluten-free snacks (26% enrichment)	↑ Protein, ↑ shelf life, ↑ crispiness	[14]

Table 1. *Cont.*

By-Products	Ingredient Preparation	Food Application	Properties	References
Rapeseed press cake	Fat extraction (cold press) → water dispersion (1:9) → alkaline extraction (pH 10.5, 4 h) → isoelectric precipitation (pH 3–6.5) → freeze drying	Emulsions (0.05–5%)	↑ ES (pH 5–6), ↑ protein yield (pH 4.2), ↓ droplet size (pH 6–6.5)	[17]
Spent brewer's grain	Cooking (20 min, 100 °C) → filtration → homogenization → pasteurization (5 min, 80 °C)	Protein beverage	↑ Protein	[80]
Spent brewer's yeast	Vacuum drying (40 °C, −0.7 bar) → grinding → water dispersion (1 h) → alkaline treatment (pH 9–12, 20–25 °C, 2 h)	Emulsions (20% enrichment)	↑ Mannoprotein solubility, ↑ viscosity, ↑ structure stability, color changes	[19]

'→' followed by, '↓' low, '↑' high.

Table 2. Fish processing by-products with protein benefits and their food applications.

By-Products	Ingredient Preparation	Food Application	Properties	References
Tilapia head	Removal of non-collagenous protein and fat with 1 mol/L of NaOH → mineral removal with 0.2 mol/L of EDTA → different extraction methods: (a) Hot water-pretreated gelatin (HWG) (b) Acetic acid-pretreated gelatin (AAG) (c) Sodium hydroxide-pretreated gelatin (SHG) (d) Pepsin enzyme-pretreated gelatin (PEG)	Fish oil emulsions with gelatin solutions (2–10 mg/mL)	(1) Gel strength: PEG > AAG > HWG > SHG (2) Foaming properties, emulsion viscosity, emulsion activity, and emulsion stabilization ability: PEG > HWG ≥ AAG > SHG	[45]
Fermented fish bone residue (FFBR)	Alkaline treatment (NaOH at 3–6%) and three soaking times (0, 1, and 2 h)	Fish sausage fortification with properly treated bio-calcium (0–36 g)	↓ Salt content and intensity of fermented fish smell ↑ Calcium and phosphorus contents	[33]
Seabass (*Dicentrarchus labrax*) heads and bones	Hydrolysates production: alcalase (60 °C for 3 h/pH 7.6) → spray drying Fish meal production: steam boiling (20 min) → pressing → oven drying (60 °C for 30 h) → grinding	Biscuits (2.5–5% enrichment)	↑ Intensities of color, toasted, and fish flavors; nutritional enrichment in protein and PUFAs	[81]
Asian tiger shrimp (*Penaeus monodon*) shells	Hot air oven drying → US and NADES → subcritical water hydrolysis (150–230 °C, 3 MPa, 20 min) → filtration	Biodegradable film (2–6%) for mackerel fish samples	↑ Protein (391.96 mg BSA/g) obtained at 190 °C/UPT/NADES; average molecular size of protein molecules < 1000 Da	[44]

Table 2. *Cont.*

By-Products	Ingredient Preparation	Food Application	Properties	References
Salmon heads, fins, tails, and viscera	pH shifting: (2 mol/L NaOH) for 30 min/pH 11 → filtration → centrifugation → (2 mol/L HCl) pH 5.5 → precipitation → centrifugation → dialysis → freeze drying	Stabilized gels to encapsulate probiotic emulsions (4%, *w/v* complex protein solution; 2%, *w/v* alginate).	↑ Emulsifying ability and emulsion stability	[50]
Tuna (*Thunnus alalunga*) heads	Hydrolysates production: alcalase (60 °C for 3 h/pH 7.6) → spray drying Fish meal production: steam boiled (20 min) → pressing → oven drying (60 °C for 30 h) → grinding	Biscuits (2.5–5% enrichment)	↑ Protein and moisture	[35]
Canned sardine waste, mix of sardine meat, and by-products	Hydrolysates production: protamex, 0.50% → incubation (50 °C, 2.5 h) → addition of 0.7%; flavourzyme → adjusted to 15% salt and 14% *Aspergillus oryzae koji* → fermentation	Fermented fish sauce (50–100%)	No strong or unpleasant flavor, ↑ quality sauce	[38]
Tra fish (*Pangasius hypophthalmus*)	Washing (0.3% $NaHCO_3$, 0.3% salt) → slicing → seasoning → grinding → balls forming → blanching (40 °C, 10 min → 2 cooking methods applied for 3–6 min, (1) steaming (100 °C) (2) blanching (80–100 °C)	Fish balls	↑ Protein content, texture, color	[82]
Blue whiting (*Micromesistius poutassou*)	Protein isolate obtained commercially (92%)	Pasta (5–15% enrichment)	↓ Optimal cooking time, swelling index, and water absorption; ↑ cooking loss; ↑ protein content, firmness, and adhesiveness	[83]
Abdominal fat of striped catfish (*Pangasianodon hypophthalmus*)	Oil extraction: Abdominal fat washing and draining → chopping → heating (5 h, 70 °C → filtration → oil purification → microencapsulation by spray drying	Microencapsulated fish oil enriched with fish protein concentrates and chlorella powders	Mix of catfish oil, chlorella powder, and striped catfish protein concentrate 40%, 20%, and 40% had the best protein content (17.02%)	[52]
Fish bones and scales	Scales demineralization of gelatin: washing (10% *w/v* NaCl) → washing → treated with 2 M lactic acid solution → washed (yield of 76%) Flour: boiling bones for 30 min → drying (90 °C, 24 h) → grinding (0.6 mm)	High-protein financier and deep-fried puffed (5 g gelatin), panna cotta (6 g gelatin), whipped cream (24.6 g gelatin)	↑ Protein concentration	[84]
Nile tilapia (*Oreochromis niloticus*) and croaker (*Cynoscion virescens*)	Pressure-cooked → pressed → ground → drying → re-grinding	Instant food products and seasoned flours	↑ Protein content (55.41–39.86%) of Nile tilapia backbone and head flour	[58]

Table 2. *Cont.*

By-Products	Ingredient Preparation	Food Application	Properties	References
Parrotfish (*Chlorurus sordidus*) head	Enzymatic hydrolysis (chopped head mixed with water (1:2), neutral pH with 0.2 N NaOH → conditioning (50 °C, 5 min) → flavourzyme (0.2 AU/g) addition → incubation → inactivation/heat → centrifugation → spray drying	Yoghurt (0.1–3% enrichment)	Best enrichment (0.15%) with a viscosity value of 2.432 N·s/m^2, syneresis 49.67%, WHC 37.4%, pH 4.36, TTA 0.76%, color L* 74.55, color a* −3.76, color b* 8.2	[41]
Pink perch (*Nemipterus japonicus*) head and viscera	Enzymatic hydrolysis (alcalase, 58 °C, 85 min → inactivation/heat → centrifugation → freeze drying → microencapsulation → freeze drying	Soup fortification (15% protein + 11% protein hydrolysate, microencapsulated protein hydrolysate (27%), and sun-dried whole fish powder (16%)	↑ Overall acceptability	[36]
Pink perch (*Nemipterus japonicus*)	Enzymatic hydrolysis (alcalase, 57.9 °C, 85.8 min, 0.15% (v/w) E/S ratio) → freeze drying → microencapsulation	Microencapsulation of surimi by-product	High yield (17%), good functional properties, and moderate antioxidant activity;↑ amount of essential amino acid (35%), ↓ bitterness and odor	[47]
Pangas (*Pangasius pangasius*)	Alkaline extraction and precipitation	Fish sausages (2.5–10% enrichment)	↑ Crude protein and lipid content, ↓ gel strength	[34]
Roe, milt, and liver of plaice (*Pleuronectes platessa*), herring (*Clupea harengus*), and cod (*Gadus morhua*)	Washing → mixing →cooking	Fish pate	↑ Acceptability, ↑ protein content (21%) in roe pate	[85]
Sea bass (*Dicentrarchus labrax*)	Dipped in saline solution 8% → dried (60 °C/24 h)	Pasta (10% enrichment)	↑ Protein, fat, fiber, EPA, DHA; ↑ nutritional availability after cooking	[54]
Sea bass (*Dicentrarchus labrax*)	Dipped in saline solution 8% → dried (60 °C/24 h)	Pasta (10% enrichment)	↓ Texture properties except for adhesiveness, ↑ Nutritional profile, low impact on the technological quality	[86]
Chub mackerel (*Scomber japonicus*) tail	Minced fresh tail offcuts	Ready-to-eat fish paste (3.5–61.1%)	Paste with 61.1%: ↑ overall acceptability, flavor, color, texture, crude protein; ↓ crude fat percentage	[87]

Table 2. *Cont.*

By-Products	Ingredient Preparation	Food Application	Properties	References
Carp skins (*Cyprinus carpio*)	Gelatin extraction: 2.6% NaCl → filtration → freeze dying Hydrolysate production: protamex, alcalase, neutrase and flavourzyme → enzyme inactivation → centrifugation → SDS-PAGE separation → freeze drying	Snack	↑ Nutritional value and antioxidant power, good sensory qualities	[39]
Sea bass (*Dicentrarchus labrax*)	Dipped in saline solution 8% → drying (60 °C/24 h)	Fresh pasta (10% enrichment)	↑ Nutritional values, ↑ protection against oxidation, ↑ ω-3 and ω-6, adequate sensory profiles	[55]
Atlantic salmon (*Salmo salar*) and cod (*Gadus morhua*) backbones	Enzymatic hydrolysis, direct protein extracts by thermal coagulation without enzyme	Emulsions	↑ Emulsion activity in the products based on direct protein extraction	[49]
Abeo rohita head	US enzymatic extraction: mincing → pH adjusted (2 M NaOH) → US → pH 7.5 (2 M NaOH or 2 M HCl) → protamex incubation → inactivation → centrifugation → freeze drying	Umami compounds	↑ Efficiency of extraction best results at degree hydrolysis (22.73%) and extraction yield (76.34%) of umami extract achieved with a liquid/solid ratio of 3 (v/w), US time 19.20 min, hydrolysis time 140.45 min	[43]
Sardinella aurita	Protein isolates by alkaline process: homogenization → adjusting pH 11.0 (2 N NaOH) → centrifugation → precipitation (2 N HCl) → centrifugation → adjusting pH 7 → spray drying	Stabilized microencapsulated corn oil	↑ Encapsulation efficiency, ↑ protection of corn oil, ↑ best ratio protein isolates and maltodextrin 1:4	[51]
Brown stripe red snapper (*Lutjanus vitta*) protein hydrolysate	Enzymatic hydrolysis by alcalase → freeze drying	Deep-fried battered squid with hydrolysate (2–6% and 8%)	↑ Hardness and crispness, ↓ fat content at 8%, ↓ oil binding capacity, maximum acceptability scores at 4%	[37]
Surimi by-products	Lyophilized by-product (25 g protein) → alcalase 2.4 L (270,000 U/g) and trypsin (200,000 U/g) → inactivate protease → centrifuge → freeze drying	Gel for silver carb surimi	↑ Delaying protein oxidation, ↑ initial gelation properties, ↓ loss in gelation and water-holding capacity	[48]
Silver carp (*Hypophthalmichthys molitrix*) fin	Hydrolysate by four enzymes (trypsin, alcalase, papain, and neutrase) → freeze drying	Bighead carp (*Hypophthalmichthys nobilis*) fillets fortified	Trypsin and alcalase hydrolysates had strong ABTS scavenging activity and inhibition of protein oxidation, ↓ lipid oxidation	[40]

Table 2. *Cont.*

By-Products	Ingredient Preparation	Food Application	Properties	References
Yellow fin tuna red (*Thunnus albacares*)	Hydrolysate by papain enzyme: E/S ratio and hydrolysis time as 0.34% and 30 min, respectively at 60 °C and pH of 6.5 → spray drying	Mayonnaise	(50%) Replacement of egg yolk with protein hydrolysate: ↑ desirability, ↑ quality of emulsion, ↓ particle size, ↑ oxidative and physicochemical stability	[42]
Tilapia, salmon, and tuna	Thawing → washing → cooking (BHT at 0.5 mg/kg) → draining → pressing → milling → oven (60 °C, 24 h)	Cereal bars (90% tilapia protein concentrate and 10% salmon or tuna protein concentrate)	No change in the nutrients, sensory and microbiological profile ↑ mineral matter	[57]
Priacanthus arenatus	1st Heating (180 °C, 100 min) → milling → 2nd heating (200 °C, 40 min) → milling	Fish nuggets (10–40%)	40% fish waste flour: ↑ protein, lipid, and ash ↓ carbohydrate ↑ hue angle, a*, and b* values ↑ overall liking and preference	[56]
Chinese long snout catfish skin, silver carp skin, salmon skin, and Alaska pollack skin	Soaking with 0.1 M NaOH (pH 7) → soaking in 0.05 M acetic acid (pH 7) → incubation (55 °C) → filtration → freeze drying	Fish oil-loaded gelatin-stabilized emulsions (2–10 mg/mL gelatin)	↑ Gel strength of catfish and silver carp gelatins than marine fish skin gelatins, ↑ amino acids, ↑ contents of hydrophobic amino acids, ↑ molecular weights	[46]
Blue shark (*Prionace glauca*) skin	Pepsin soluble collagen extraction: shark skin mixed with 0.1 M NaOH (remove non-collagenous proteins) → fat removal → extraction (0.5 M acetic acid + 0.2% pepsin) → centrifugation → dialysation → freeze drying	Chitosan–collagen composite coating for red porgy (*Pagrus major*) meat	↑ Deterioration indexes at a ratio of 1:0.8 chitosan/pepsin, soluble collagen, ↑ physicochemical indexes	[88]

'→' followed by, '↓' low, '↑' high.

Table 3. Meat processing by-products with protein benefits and their food applications.

By-Products	Ingredient Preparation	Food Application	Properties	References
Porcine spleens	Soluble proteins extraction (sodium citrate buffer 0.1 M, pH 5) → centrifuging→ spray drying → washing of insoluble fraction → Pasteurization (80 °C, 45 min)	Sausages	Good functional properties	[70]
Broiler chicken stomachs	Separation of organic matter (0.2 mol/L NaCl, 0.06 mol/L NaOH, 1:1 petroleum ether/ethanol) → conditioning of collagen (protamex, pH 6.5, 0.1–0.2 wt%) → hot water	Gelatins	↑ Gelatin yield (65%), gel strength (25–439 Bloom), digestibility (100%), melting point, ↓ ash	[68]

Table 3. *Cont.*

By-Products	Ingredient Preparation	Food Application	Properties	References
Trimmings, skin, meat-left-on-bone of chicken drumstick	Homogenization → alkaline extraction (pH 11.5, 10 min) → centrifugation → isoelectric precipitation (pH 5.5, 10 min) → centrifugation	Edible coating	↑ Moisture content, ↓ fat uptake, ↑frying yield in coated samples	[71]
Chicken bones	High pressure pretreatment (121 °C, 4 h) → oven drying (55 °C, 5 h) → enzymatic hydrolysis → Maillard reaction → autoclaving (113 °C, 10 min).	Flavoring agents	↑ Protein and lipid content	[61]
Chicken feet and head	Grinding → dispersion (1:1) in 10% solution of ascorbic acid → drying	Forcemeat	↑ Protein, ash content	[72]
Cattle lips and ears	Cleaning → crushing → enzymatic hydrolysis (protepsin, 1:2 E/S, 0.5–1.5 h, 17–22 °C) → filtration → freeze drying	Forcemeats (5–20% enrichment)	↑ Moisture-retaining capacity	[69]
Goat viscera	Cleaning → crushing → homogenization → enzymes inactivation (90 °C, 15 min) → cooling → enzymatic hydrolysis (alcalase, 0.8% E/S), 240 min) → inactivation (90 °C, 15 min) → cooling → Maillard reaction → homogenization → autoclaving (121 °C, 60 min)	Meat flavorings	↑ Meat aroma	[64]

'→' followed by, '↓' low, '↑' high.

Table 4. Dairy processing by-products with protein benefits and their food applications.

By-Products	Ingredient Preparation	Food Application	Properties	References
Whey	Heating of milk (70 °C, 20 min) → coagulation (1% citric acid) → filtration	Bread (0–100% water substitution)	↑ Protein, fat	[76]
Whey	Whey-protein concentrate (76.8%) obtained commercially	High-protein fat-free dairy desserts	↑ Texture and appearance, ↓ water activity	[73]
Sheep and goat cheese whey	Ultrafiltration (40–45 °C, 3.0–3.5 bar → pasteurization (65 °C, 30 min) → homogenization → freezing	Symbiotic kefir	↑ Protein concentration	[74]
Cow, sheep, and goat whey	Whey-protein concentrate (>60%) obtained commercially	Probiotic beverages	↑ Protein, free amino acids, TPC, antioxidant activity	[9]

'→' followed by, '↓' low, '↑' high.

2.2. Fiber

The importance of dietary fiber (DF) has been related to its numerous health benefits, such as constipation prevention, cholesterol and blood glucose reduction, and improvement of gut microflora [89]. The World Health Organization has also established their recommended daily intake as 25 g/day, but many people do not achieve such amounts [90]. Thus, researchers and companies have focused their attention on the formulation of fiber-rich food exploiting by-products as alternative and sustainable sources of DF.

Fruits and vegetables including legumes and cereals are the main sources of DF. DF can be distinguished into two groups based on their solubility in hot water: soluble and insoluble dietary fibers. Soluble fibers consist of non-cellulosic polysaccharides such as β-glucan, inulin, pectin, and resistant starch, whereas insoluble fibers include cellulose, hemicellulose, chitosan, and lignin [91]. Peel, seeds, and pomace from different fruits (e.g., banana, apple, grape, tomato, and pepper) and vegetables (e.g., carrot, onion, and lettuce) are commonly used by-products, followed by meals from nuts, cereals, and oilseeds obtained after fat extraction and spent brewer's yeast as a by-product of beer production (Table 5). DF is a very heterogenous group; therefore, their processing and food origin may influence their functionalities both in terms of physicochemical and fermentable properties [90]. Most by-products are minimally processed to obtain a shelf stable powder; thus, freeze or air drying are applied to stabilize the by-product and further grinding and sieving to achieve a homogeneous powder [29,92–96]. In some cases, disinfection and/or blanching have been necessary to ensure microbial safety and enzyme inactivation prior to drying [25,97–99]. Instead, defatted meals from hempseeds [100], walnut and sesame [15,101], durum wheat [22], and pepper seeds [28] were directly ground. A few simple steps are sufficient to convert a by-product into a fiber-rich food ingredient with a fiber content varying from about 10% to 65% depending on the type of plant source.

Additionally, fiber fractions (e.g., pectin, cellulose, and β-glucans) may be extracted, obtaining a purer ingredient with a high commercial value [102]. Acid extraction is commonly used to extract pectin from agro-industrial by-products. The powdered by-product is dispersed into water, and then, pH is adjusted to 1.5–2 and heated to 80–90 °C for 45 min up to 3 h. Insoluble material is removed by filtration, and the rest is dispersed into 70–95% ethanol with a sample-to-solvent ratio of 1:2–1:4 to induce pectin precipitation. Afterward, the precipitate is collected, dried, ground, and sieved [103–105]. As a green alternative to the use of solvents, US has been employed to extract pectin from celery root peel [102]. The by-product dispersed into water was subjected to US treatment (40–60 °C for 10–30 min) with a power level between 40% and 60% to improve extraction efficiency in a shorter time. After that, water-soluble pectin was extracted by distillation, with 75 °C as the chosen optimal temperature to ensure enzyme inactivation and the bioavailability of nutritional compounds.

Cellulose was obtained from oil palm empty fruit bunch after several acid treatments (pH 4) with acetic acid addition to the by-product dispersed in water and sodium chloride, for about 2 h at 70–80 °C. The solution was rinsed, dried, and extracted with alkali for 2 h at room temperature [106]. Another study used a harsh alkaline treatment (pH 12) to degrade the strong cell wall of spent brewer's yeast and solubilize its components, which are mainly β-glucans and mannoproteins, often covalently linked together [19].

Fiber-rich ingredients and fiber fractions are often used in the food industry not only to improve the nutritional quality of food but also because of their potential functional properties, such as water-holding capacity and emulsifying and gelling capacities. These ingredients have been mostly employed for the development of pasta, bread, and other bakery goods with or without gluten. Pasta was enriched with 5% to 30% of fiber-rich ingredients, and it also fulfilled the claim "source of fiber" when ≥10% was replaced with hempseed flour containing 46% of dietary fibers [100]. The addition of ≤10% fiber-rich ingredients allowed to improve nutritional and technological quality, while exceeding this amount may negatively affect sensory quality and promote cooking loss [94,100]. In gluten and gluten-free bread, an enrichment level of 3 to 12% provided a higher amount of nutrients and fermentable sugars to yeasts, thus resulting in higher CO_2 production and dough development [96,103,107]. In turn, the higher the substitution level, the weaker the gluten network [108]. Other bakery goods like biscuits, muffins, and macarons were enriched up to 20% [98], 30% [25], and 50% [101], respectively. However, they were mostly considered acceptable when the substitution level did not exceed 10% [25,98,101,109]. It is noteworthy that the higher the fiber content, the darker and harder the final products. The substitution of starch with fibers (or proteins) determines a reduction in the expansion volume and the porosity, thus resulting in a harder texture. This was highlighted also for the formulation of extruded and freeze-dried snacks [29,99,110,111].

The gelling, emulsifying, and water-binding capacities of fibers are suitable also for the development of emulsions and semi-solid foods. Stable and viscous products were obtained by adding both powder ingredients at 2–6% enrichment [15,18] and liquid ones at 20–35% [19,31] in emulsions or 15–62% in a yogurt-based beverage [102]. By contrast, a high enrichment level may strongly affect the odor, thus reducing consumer acceptability.

Fiber derived from fish by-products is becoming increasingly popular due to its potential benefits, environmental sustainability, and various applications. This type of fiber is mostly derived from fish skin and bones [45,46] or from shellfish shells [44], where chitin and chitosan (chitin-derived fiber) are obtained. Recently, the application of chitosan in the food industry has been focused on the synthesis of biodegradable film or coating material with valuable properties such as biodegradability, low toxicity, and strong antimicrobial and antioxidant potential [44,88]. In oenology, chitosan originating from shrimp and crab shells was tested for wine clarification as a potential replacer of fungus-derived chitosan [112].

The three primary stages of chitin extraction to obtain chitosan from crustaceans are pre-treatment, deacetylation, and posttreatment. The pre-treatment includes raw material washing and milling. This step is targeting the removal of mineral using acid washing and the removal of protein, glycoprotein, and branching polysaccharide using alkaline washing. In the deacetylation, the acetyl group from the chitin chains is removed using a high concentration of alkaline at high temperatures for a short time. In the posttreatment, recovery of deacetylated chitin (i.e., chitosan) takes place using low-concentration acid, and the chitosan is further precipitated at pH 10, washed, and dried [112]. Nonetheless, there are some variations in the literature regarding alkaline or acid concentrations, type of organic acids used, temperature, and time. The most challenging step is to avoid the use of environmentally unfriendly deacetylation procedures such as the enzymatic approach since the reaction of deacetylase requires long time. Different cases are detailed in the literature, with a description of each extraction step and its impact [113].

For example, four different organic acids (malic, acetic, succinic, and hydrochloric acid) were used to dissolve (1% v/v) commercially available chitosan from shellfish by-products for further use in wine sedimentation. However, all dissolved chitosan solutions showed great and fast wine clarification compared to the conventional chitosan from *Aspergillus niger* culture [112]. In another study, Asian tiger shrimp (*Penaeus monodon*) shells were dried in a hot air oven and milled before chitosan extraction. In detail, the hydrolysate powder was demineralized for 24 h at room temperature while being continuously stirred using a solid-to-solvent ratio of 1:10 (w/v) and 1 M HCl. A paste was produced after centrifugation at 5000 rpm, neutralization, and overnight drying at 45 °C. Afterward, chitosan was obtained by a deacetylation step using 60% NaOH (w/v) [44]. One gram of the extracted chitosan was employed to create a biodegradable packaging film, which was then applied to mackerel fish samples and showed a strong ability to postpone the fish samples' oxidation. A high-quality edible coating material was obtained by a composite of chitosan–collagen, and it extended red porgy fillet shelf life, as reported by [88]. As noticed, the main use of fiber components from fish by-products concerns are non-dietary (i.e., chitosan). Thus, their main application in the food sector is in the field of edible films and food packaging materials [88]. The following (Tables 5 and 6) provide deep details about the ingredient preparation and their application on various food products regarding fiber.

Table 5. Fruit and vegetable processing by-products with fiber benefits and their food applications.

By-Products	Ingredient Preparation	Food Application	Properties	References
Grape pomace	Addition (or not) of lyoprotectors → freeze drying → grinding	Gluten-free cookies (10–20% enrichment) Filling (17.2 g pomace flour)	↑ Fiber, protein, fat; ↑ sensory properties	[114]
Carrot pomace	Air drying (60 °C, 12 h) → grinding (250 mm)	Muffins (5–20% enrichment)	↑ Crude fiber, ↑ firmness, ↑ taste, ↑ appearance (10% the best sensory results)	[109]
Coconut residue	Drying → grinding	Spread (16% enrichment)	↑ Protein, fiber, good spreadability, ↑ sensory acceptability	[92]
Banana peel	Disinfection → drying (63 °C, 24 h) → grinding (0.8 mm)	Biscuits (10–20% enrichment)	↑ Fiber (+20% at 10%), ↑ dark color, good sensory acceptability (10% maximum)	[98]
		Bread (10% enrichment)	↑ Fiber (+12.5%)	
		Pasta (5–10% enrichment)	↑ Fiber (+71.4% at 10%)	
Hemp seed meal	Fat extraction (cold press) → grinding	Gnocchi (5–20% enrichment)	↑ Fiber (source of fiber claim, >10%), ↑ cooking loss, ↓ cooking resistance, ↓ sensory acceptability (bitter taste, odor)	[100]
Pineapple peel, banana peel, and pumpkin seed	Banana, pineapple: cutting → sanitation → drying (60 °C, 8 h) → grinding Pumpkin seed: roasting (15 min) → grinding	Gluten-free muffins (17% enrichment)	↑ Fiber, good sensory acceptability (100% pineapple peel; 50% pineapple peel; 50% pumpkin seed)	[97]
Grape pomace and skin	Grape by-products obtained commercially	Frankfurt and Spanish sausage (0.5% grape skin, 3–6% inulin, 0.5–1% β-glucan)	↓ Fat reduction, ↓ boar taint, ↓ brightness, ↓ hardness	[115]
Male date palm flowers	Drying (40 °C, 24 h) → grinding	Biscuits (3–9% enrichment)	↑ Fiber (source of fiber claim, >6%), darkness, hardness, good sensory acceptability (6% maximum)	[93]
Walnut oil cake	Fat extraction → grinding (0.8 mm)	Macarons (10–50% enrichment)	↑ Fiber, TPC, antioxidant capacity; ↓ sensory acceptability (10% maximum) ↓ costs	[101]
Brewery bagasse	Cooking (20 min, 100 °C) → filtration (residue) → drying (60 °C) → grinding (1 mm)	Bread (6% enrichment)	↑ Fiber	[80]

Table 5. *Cont.*

By-Products	Ingredient Preparation	Food Application	Properties	References
Jackfruit core	Pectin extraction: drying (60 °C, 20–24 h) → grinding → water dispersion (1:29) → acid extraction (pH 2, 80 °C, 105 min) → filtration → ethanol precipitation → drying (45 °C, overnight) → grinding	Vegetable soup	↑ Sensory properties	[105]
Tomato seed and skin, peppers, placenta	Drying (30–40 °C, 17 h) → grinding (0.25 mm)	Bread (3–12% enrichment)	↑ Fiber, WAC, dough development, softening; ↓ stability time, ↑ redness and yellowness	[96]
Celery root peel	Grinding → water dispersion (1:5) → homogenization → US extraction (25 kHz frequency, 40–60 °C, 40–60% power level, 10–30 min) → distillation (75–85 °C, 1–6 h) → refrigeration (4 °C)	Ayran Turkish beverage (15–62% enrichment)	↑ Pectin, sensory and structural properties; ↑ flavor, WAC, viscosity; ↓ color, phase separation; ↑ shelf life	[102]
Sugar beet pulp	Sucrose extraction in water (75 °C, 45 min) → filtration → cooling → bleaching (pH 11, 120 min) → fiber extraction (pH 6.5) → saponins removal (water: ethanol dispersion 1:4, overnight) → pectin removal (pH 1.6, 90 °C, 3 h) → filtration → ethanol dispersion (1:4, pH 3, 24 h) → filtration, rinsing → boiling (20 min, pH 6.5) → air drying → grinding (1 mm)	Gluten-free muffins (0.2–0.4% enrichment) depectinized or not-depectinized ingredient	↑ Viscosity, hardness (with pectin); ↑ fiber, swelling capacity, water and oil binding capacity (w/o pectin)	[104]
Paprika, pitted pepper, tomato pomace	Not specified	Bread (3–12% enrichment)	↓ Gluten network strength	[108]
Carrot pomace and beetroot–apple (7:3) pomace	Drying (40 °C) → grinding (0.4 mm)	Pasta (10–30% enrichment)	↑ Fiber, ↑ WAC (carrot pomace), ↑ technological and sensory (color, texture) properties	[94]
Apple and blackcurrant pomace	Obtained commercially	Freeze-dried snacks (2% enrichment)	↓ Porosity, hygroscopicity; ↑ brittle and fragile texture, color changes	[110,111]
Non-compliant broccoli and carrots and pomace	Washing → cutting → steam-blanching → (juice extraction) → freeze drying → grinding (0.8 mm)	Extruded snacks (3% enrichment with pomace, 20–100% enrichment with whole vegetable powder)	↓ Starch content, ↑ fiber, ↓ phenols, ↓ expansion volume	[99]

Table 5. *Cont.*

By-Products	Ingredient Preparation	Food Application	Properties	References
Orange and apple pomace, tomato peel, pepper peel, prickly pear peel, and prickly pear seed peel	Drying → grinding (0.5 mm) → pectin extraction: acid extraction (2 g sample +40 mL HCl, 90 °C, 45 min) → filtration (removal of insoluble material) → pectin precipitation: 95% ethanol dispersion (1:2), overnight → pectin purification: 70% ethanol dispersion (x2) → collection of the precipitate → drying (65 °C)	Gluten-free bread (2.5–7.5% enrichment)	↑ Dough height, ↑ CO_2 production and retention, ↑ specific volume	[103]
Pea	Pea fiber obtained commercially	Bread (3–7% enrichment)	↑ Fermentation, ↑ CO_2 production	[107]
Oil palm empty fruit bunch	Cleaning → drying (60 °C, 24 h) → grinding (0.2 mm) → α-cellulose extraction: NaCl dispersion + acetic acid (1:50, pH 4, 70–80 °C, 2 h, x 5) → rinsing→ drying (60 °C, overnight) → NaOH dispersion (1:25, 2 h, room T °C) → filtration→ washing→ drying (60 °C, 24 h) → carboxymethyl cellulose extraction: aqueous-alkaline dispersion (1:2), isopropyl alcohol addition (1 h, room T °C) → monochloroacetate addition (45 °C, 3 h) → pH adjustment with glacial acetic acid (neutral) → soaking in methanol (to remove impurities) → purification in ethanol	Low-fat ice cream (0.3% enrichment)	↓ Viscosity, hardness; ↓ overrun	[106]

'→' followed by, '↓' low, '↑' high.

Table 6. Fish processing by-products with fiber benefits and their food applications.

By-Products	Ingredient Preparation	Food Application	Properties	References
Shrimp and crab shells	Chitosan provided commercially with a deacetylation degree of 80–95%	Wine clarification	↑ Faster and greater sedimentation	[112]
Asian tiger shrimp (*Penaeus monodon*) shells	Shells drying → milling (710 μm) → chitosan extraction → demineralization (1 M HCl) → centrifugation → deacetylation: chitin conversion to chitosan using 60% NaOH (*w*/*v*) Chitosan film: 1 g chitosan + water (containing 1% acetic acid) → glycerol (plasticizer) → protein hydrolysates → drying	Biodegradable film	Excellent lipid oxidation-delaying capacity	[44]

'→' followed by, '↓' low, '↑' high.

2.3. Bioactive Compounds

A bioactive compound is a substance with a positive or negative biological activity in the organism depending on the type, the dose, and the bioavailability of the substance [116]. Plants (e.g., fruit, vegetables, cereals, and legumes), animals, fish, and microorganisms contain various bioactive compounds with proven biological activities such as antioxidant, anti-inflammatory, anti-diabetic, and antitumor ones; therefore, a sufficient intake may help to prevent or treat diseases [117].

Phenolics (e.g., phenolic acids, flavonoids, tannins, carotenoids, stilbenes, and lignans), terpenoids (e.g., tocotrienol, tocopherols, carotenoids, limonoids, and phytosterols), alkaloids, nitrogen-containing compounds, and organosulfur compounds are among the bio-active compounds mostly found in plants, including fruit, vegetables, cereals, and legumes [118]. Many researchers have focused their attention on the recovery of bioactive compounds, especially phenolics, from fruit and vegetable by-products for food enrichment both to face environmental issues and to meet consumers' needs for healthy choices. For this purpose, various by-products have been already exploited, such as peel, stems, leaves, pomace, pods, press-cakes, as well as non-compliant fruit and vegetables.

Many studies added the by-product directly in the liquid or powder form into the food formulation without applying any isolation and purification of bioactive compounds. In general, by-products are washed, disinfected, cut, and blanched. Blanching is needed to inactivate enzymes that may degrade and oxidize bioactive compounds, such as polyphenol oxidase, peroxidase, and myrosinase. Hot water (98–100 °C for 1–5 min) or steam blanching are commonly used [24,25,99,119–121]; however, some authors demonstrated the potential of microwaves [122,123]. Later, by-products are stabilized by air drying or freeze drying or even sun drying, as reported by Traffano-Schiffo et al., and then ground to a fine powder [124]. The particle size of powders may affect not only the techno-functional properties of the ingredient when it is wholly included in the food formulation but also the extraction efficiency of bioactive compounds [119]. Depending on the extraction method, it is important to consider different parameters including but not limited to sample-to-solvent ratio, type of solvent, food matrix, treatment time, and temperature [118,121]. Novel extraction methods such as ultrasounds, microwaves, and high pressure (HP) have gained popularity in this field because of higher extraction efficiency and their lesser use of hazardous solvents and energy than conventional ones, which include, for example, Soxhlet, magnetic stirring, shaking, or maceration [118].

No toxic solvents such as ethanol and water, at different ratios from 0% [121,125] to 96% ethanol [126,127], are suitable for the extraction of hydrophilic compounds such as phenolics and glucosinolates, whereas deep eutectic solvents are a potential alternative to harmful solvents (e.g., hexane, acetone, and petroleum ether) for carotenoids extraction [128]. The sample-to-solvent ratio must be adequate to avoid saturation effects as well as the waste of solvents. Most sample dispersions were prepared in a ratio of 1:10–30, with few exceptions [102,126,129–131]. By contrast, 1:4 was found to be the optimal liquid/liquid ratio to obtain a phenol-rich extract from olive mill wastewater, using ethyl acetate as solvent [132].

As regards time and temperature, conventional methods usually take longer (up to 5 days) [133] and apply high temperatures, near 80 °C [131]. In general, the higher the temperature, the shorter the time. However, long extraction times and high temperatures are deleterious for thermolabile bioactive compounds [134]. On the other hand, novel technologies allow to reduce time and temperature to a maximum of 2 h and 60 °C [102,128]. In the case of ultrasounds, low frequencies (20–40 kHz) are commonly used to improve extractability. Traffano-Schiffo et al. showed that a high-intensity ultrasound (500 W) increased the extraction of phenolics from cowpea pods [124]. High-pressure processing (300–600 MPa) was also successfully applied for the extraction of lycopene from tomato by-products to enrich extra-virgin olive oil, directly used as a green solvent [135]. The results were very promising, and the pressure seemed to not affect carotenoid recovery. After treatment, the extract rich in bioactive compounds is separated from insoluble material by filtration or centrifugation and then stabilized either by drying or freezing.

A wide range of food products have been enriched with bioactive compounds, from bakery and pasta products to snacks, beverages, and emulsions. Most of the authors developing bakery and pasta products used the whole by-product powder with an enrichment level varying from 2.5% [136] to 15% [137] for pasta and from 5% [24,138] to 50% [101] for bakery goods. However, 10% is usually the maximum to obtain a product with good technological properties and sensory acceptability. The food matrix helps to prevent the thermal degradation of bioactive compounds during cooking [121]. Another strategy to protect these compounds is encapsulation. Kaderides et al. encapsulated pomegranate peel polyphenol using orange bagasse fibers as a carrier agent for spray drying and obtained cookies with a higher phenolic content and antioxidant activity despite bitter flavor and astringent taste [139]. Moreover, encapsulated systems, e.g., microemulsions [140], nanoparticles [141], and hydrogels [124,142], allowed the increase in bio-accessibility of bioactive compounds. Similarly, emulsion-based foods such as yogurt [102,143], butter [144], dressing [145], ice cream [146], and mayonnaise [133] registered an increase in bioactive content, bio-accessibility, and antioxidant activity, in some cases improving also color and flavor.

Bioactive-rich extracts have been widely applied for liquid food formulations such as instant drinks [120] and soups [95], fermented soy milk [147], kombucha [79,148], smoothies [131], distilled spirits [149], wine [150], and beer [151], with the aim to improve the nutritional quality and antioxidant activity of the final product. The antioxidant potential of bioactive compounds, mainly phenolics, has been exploited to enhance the oxidative stability of vegetable oils from rapeseed [130], soybean [129,152], olive [135], sesame [153], and sunflower [132].

Valorization of fish by-products for their bioactive compounds focuses mainly on extracting high-quality oil rich in nutrients, especially polyunsaturated fatty acids such as EPA and DHA, which have been recommended to be included in a routine diet to maintain good health. These bioactives from fish by-products are characterized by antioxidant, anti-obesity, antitumor, anti-inflammatory, and anti-allergenic activities as well as a protective action against cardiovascular disease. Bioactive oil is usually extracted from the side streams of fish by-products, including backbones, heads, gills, and viscera. These by-products may be subjected to freeze drying prior to oil extraction [154–157]. Differently, several by-products are processed wet [156,158,159]. Occasionally, by-products are submitted to a cooking step at 95 °C for 12 min before pressing [160].

Food-grade solvents (i.e., hexane and ethanol) are widely used for oil extraction. Some traditional methods are applied, such as Soxhlet [158,159] or the Bligh and Dyer method [156]. Alternatively, some mild processes include US, microwave-assisted extraction (MAE), pressure extraction, and super-critical fluid extraction [155–157,161].

Different studies have reported the effects of raw material treatment and extraction methods on the oil yield and its nutritional characteristics. For example, Kalogianni et al. reported the highest antioxidant activity measured by DPPH and ABTS assays in oil extracted using supercritical fluid extraction starting from freeze-dried samples [156]. The authors performed the extraction at 37 °C and 20 MPa pressurized with CO_2 for 30 min with a flow rate of 3 mL/min. Their findings also highlighted significant amounts of total PUFA (20.7%) represented mainly by EPA and DHA with a considerable oil yield [156]. The use of MAE for freeze-dried by-products of salmon allowed the recovery of >60% of oil in <19 min [155,157]. MAE oil was characterized by a healthy lipid profile due to the high content of EPA and DHA as well as high antimicrobial and anti-inflammatory properties [155]. High-quality crude oils were produced by pressing cooked by-products of tuna, seabass, seabream, and wild sardines. The oil derived from tuna, seabass, and seabream had low levels of anisidine value, peroxide, and free fatty acid content. Seabass and sea-bream oil had the highest α-tocopherol content, whereas sardine oil had the highest concentration of EPA and DHA (32.8%) [160].

Furthermore, fortifying food products with omega-3-rich oil offers the potential to scale up and valorize the use of fish by-products. For instance, a functional flavored yogurt was incorporated with fish by-product oil derived from Nile perch, containing 20.5% omega-3 fatty acids (consisting of 3% EPA, 6.2% DPA, and 9.0% DHA) [162]. Consequently, the target daily consumption of omega-3 fatty acids was met, and there was no observable impact of the fish oil on flavor [162]. Similarly, pork sausage was fortified with oil extracted from skipjack tuna heads and viscera. The overall acceptance, ABTS+, DPPH radical scavenging activity, and antimicrobial properties of the sausage were improved, with significant increases in EPA (4.15%) and DHA (21.96%) [154]. Deep-fried chips enriched with fish by-product oil and artichoke bract powder had a booster in their nutritional quality. They showed reduced triglycerides, cholesterol, and LDL levels, along with increased average lipid content and essential fatty acids [158]. The incorporation of fish oil and artichoke powder also effectively improved their vulnerability to oxidation. These findings are likely due to the capacity of these enriched chips to block lipid peroxidation, enhance antioxidant status, and obstruct lipolysis and the subsequent insulin-induced inhibition of free fatty acid release. Thus, reducing the associated risk of cardiovascular diseases [158].

Regarding dairy by-products, whey is a source of protein with high biological value as well as branched chain amino acids, essential amino acids, and high-sulfur-containing amino acids, which support antioxidant activity. They promote glutathione synthesis, enhance probiotic bacteria growth, and protect from oxidative stress due to their antioxidant effect [9]. The application of whey for antioxidant purposes was studied by Alwohaibi et al. by valorizing different dairy by-products. The study was focused on using sweet whey, sweet buttermilk, and skimmed milk to produce a functional dairy–millet beverage fermented with *Lactobacillus paracasei* as an adjunct culture [163]. Four fermented milk beverages (water-based, whey-based, buttermilk-based, and skimmed-milk based) were developed and analyzed for several characteristics, including their antioxidant properties. All fermented milk beverages produced using the dairy by-products exhibited significant antioxidant properties in comparison to the water-based fermented beverage. Additionally, all samples had a low glycemic index and glycemic load [163]. The following Tables 7–9 provide deep details about the ingredient preparation and their application on various food products regarding bioactive compounds.

Table 7. Fruit and vegetable processing by-products with bioactive benefits and their food applications.

By-Products	Ingredient Preparation	Food Application	Properties	References
Longan peel	Fat extraction → freeze drying → 80% ethanolic extraction → solvent evaporation	Soybean oil	↑ Oxidative stability during frying, tocopherols; ↓ peroxides, conjugated dienes, TBARS	[152]
Black currant pomace	Obtained commercially	Freeze-dried vegetable snacks (1–5% enrichment)	↑ TPC, antioxidant activity; ↑ dark color, ↓ water activity, ↓ porosity, ↑ hardness, crispiness	[164]
Pomegranate peels, olive oil pomace, broccoli root and leaves	Broccoli: drying (35 °C, 48 h) → grinding (0.5 mm) pomegranate: washing and disinfection → drying (50 °C, 48 h) → grinding (0.5 mm) olive: drying (35 °C, 72 h) → grinding (0.5 mm)	Fresh pasta (10% broccoli and 3–6% olive and pomegranate by-products enrichment)	↑ Phenols, flavonoids; ↓ microbial growth; ↑ shelf life	[165]

Table 7. *Cont.*

By-Products	Ingredient Preparation	Food Application	Properties	References
Pomegranate peels, olive oil pomace, broccoli root and leaves	Pomegranate, broccoli: washing → disinfection → cutting → drying (45 °C) → grinding (0.5 mm) Olive: drying (45 °C) → grinding (0.5 mm)	Fresh pasta (3–8% enrichment each)	↑ Phenols, flavonoids, antioxidant activity, shelf life (especially pomegranates)	[166]
Papaya peel	Washing → cutting → freeze-drying → grinding (<2 mm) → carotenoid extraction: dispersion (1:20) in THF+0.01% BHT+ magnesium carbonate →homogenization → US treatment (30 min) → centrifugation → re-extraction with acetone (pellet) → supernatant collection → diethyl ether addition → organic phase collection → filtration → vacuum drying	HPP-treated microemulsions (7 mg/g soy-bean/sunflower/coconut oil)	↓ Particle size, ↑ carotenoid bio-accessibility (sunflower, soybean oil)	[140]
Coconut husk	Washing → cutting → grinding → MW blanching (2450 MHz–900 W, 40–160 s) → cooling → grinding	Kheer dairy-based dessert (5–15% enrichment)	↑ Flavor, color (10% maximum)	[122]
Apple pomace	Drying → milling	Gluten-free bread (5–15% enrichment)	↑ Phenols, flavonoids, chlorogenic acid, phlorizin; ↑ antioxidant activity; ↑ sensory acceptability	[138]
Apple pomace	Freeze drying → UV sterilization	Fermented soy milk	↑ Phenols, ↑ antioxidant activity, ↑ viable cell count	[147]
Lemon peel	Grinding	Flavoring gel	↑ thermal stability ↑ safety, quality	[167]
Mango peel	Cutting → blanching (98 °C, 1 min) → drying (50 °C, 4 h)	Instant drink (2–6% enrichment)	↑ TPC, vitamin C, ↑ antioxidant activity, ↑ acidity	[120]
Pomegranate peel, orange juice residue	Pomegranate: drying (40 °C, 48 h) → grinding (0.1 mm) → water dispersion (1:32) → US treatment (50 W, 35 °C, pulse 7 s on—6 s off, 10 min) → homogenization with hydrated orange fibers (1:9) → encapsulation (spray drying)	Cookies (enrichment at 5000 ppm of phenolic concentration)	↑ Retention of encapsulated phenolics, ↑ antioxidant activity, ↑ bitterness, astringency, ↑ color, odor, ↓ texture	[139]
Banana, pineapple, squash, mango, calamansi, potato peels	Washing → cutting → drying	Distilled spirit (sample-to-solvent, 1:10, solvent: 40% ethanol/gin/vodka/tequila)	↑ TPC, DPPH (mango), gin/vodka good substitute for food-compatible solvents	[149]
Star fruit peel	Drying (40 °C, 24 h) → grinding → phenol extraction (sample: acetone, 1:20, 4–24 h, 30 °C, x2) > solvent evaporation	Sesame oil (200–1000 ppm)	↑ Oxidative stability during storage and frying (1000 ppm), good alternative to synthetic antioxidants	[153]
Non-compliant melon	Washing → peeling → deseeding → cutting → juice extraction → decantation (1 °C, 24 h, SO_2 addition) → must	Melon-based wine	↑ TPC, antioxidants, ↑ fruity, floral aroma	[150]

Table 7. *Cont.*

By-Products	Ingredient Preparation	Food Application	Properties	References
Grape seed, olive stone	Flour obtained commercially	Crackers (10–30% enrichment)	↑ TPC, ↑ dark color, ↓ sensory acceptance, ↓ hardness (olive stone)	[168]
Olive mill wastewater	Filtration → liquid-liquid extraction with ethyl acetate (1:1–5, 4–70 °C, 0.25–24 h) → organic phase separation → solvent evaporation	Olive and sunflower oil (0.3% enrichment)	↑ lipophilicity and bioactivity of phenols, ↑ thermal stability, ↓ oxidation	[132]
Olive wastewater (oil + brine)	Centrifugation	Stuffing or paste for olives (63% enrichment oil fraction) and pickled vegetable (brine)	↑ Phenols, ↑ sensory properties	[169]
Olive leaf	Drying (T °C room, 15 d) → dispersion (1:10) in vinegar → homogenization → maceration (T °C room, 5 d) → filtration	Low-fat vegan mayonnaise	↑ Phenols (oleuropein), ↑ oxidative stability, ↑ shelf life, ↑ color, taste	[133]
Olive leaf	Washing → drying (120 °C, 8 min) → grinding → water dispersion (1:20→ US extraction → filtration (liquid extract) → freeze drying	Ripened sausage (0.02–0.08%)	↑ Weight loss, ↑ oxidative stability, ↓ hardness, ↑ microbial safety, natural additives (nitrite substitute)	[125]
Red onion peel	Washing → drying (40 °C) → grinding (1 mm) → UV sterilization → dispersion (1:14) in 70% acidified ethanol → US treatment (40 kHz, 25 °C, 40 min, 100 W) → centrifugation (liquid extract)	Ricotta cheese (1–3% enrichment)	↑ Phytochemical content, ↑ antioxidant activity ↑ texture, color	[170]
Onion skin	Washing → drying (50 °C, 12 h) → grinding (0.5 mm)	Pasta (2.5–7.5% enrichment)	↑ Phenol (quercetin, phenolic acids), ↑ antioxidant, anti-inflammatory activity	[136]
Onion skin	Washing → drying (T °C room, 14 d) → grinding (0.2 mm)	Extrudates (3–9% enrichment)	↑ Quercetin, ↑ TPC, antioxidant activity	[171]
Tomato by-products	Dispersion (1:10) in DES-LA (deep eutectic solvents/lactic acid, 1:2) → US (40 kHz, 30 °C, 1–2 h) → centrifugation (liquid extract) → extract: chitosan solution (40:60, *w*/*w*)	Edible films (40% liquid extract)	↑ Thermal stability, hydrophilicity; ↑ free radical scavenging rate, ↓ oxidation, ↑ shelf life	[128]
Tomato seeds and skin	Blanching → freezing → HPP extraction (by-product+ oil, 300–600 Mpa, 10 min) → agitation (45 °C, 45 min)	Extra virgin olive oil (10–20% enrichment)	↑ Lycopene recovery, good sensory acceptability (10%)	[135]
Tomato pomace, linseed meal	Tomato: grinding → drying (45 °C, 10 h) Linseed: cold pressing	Gluten-free pasta (10–15% enrichment)	↑ Fiber, lipid, tocols, phenols, ↑ antioxidant capacity	[137]
Artichoke bracts and tomato	Non-thermal patented treatment (Patent n°001426984) → freeze drying	Fresh pasta (3% enrichment)	↑ Phenols, antioxidant activity	[77]

Table 7. *Cont.*

By-Products	Ingredient Preparation	Food Application	Properties	References
Black carrot pomace	Freeze drying → grinding → UV sterilization → dispersion (1:10) in 70% acidified ethanol → US treatment (40 kHz, 25 °C, 40 min, 100 W) → centrifugation (liquid extract)	Yoghurt (1–2% enrichment)	↑ Nutritional quality, ↑ phytochemicals, antioxidant activity, ↑ color, appealing (2% best formulation)	[143]
Carrot pomace	Drying (60 °C, 5 h) → grinding (0.4 mm) → water-in-oil emulsion (addition of other ingredients + homogenization 5 min, 20 °C)	Butter (21% enrichment)	↑ Carbohydrates, proteins, vitamins, minerals, ↑ oxidative stability, ↓ microbial growth, ↓ spreadability, good acceptability	[144]
Red chili by-products	Freeze drying → grinding (0.5 mm) → carotenoid extraction: dispersion (1:100) in 60% acetone-40% petroleum ether (20 min stirring, protect from light) → centrifugation → solvent evaporation	Soybean oil (0.05–0.5 ppm enrichment)	↓ Oxidation, ↑ thermal stability	[129]
Bell pepper residue	Drying (60 °C, 16 h) → grinding (0.25 mm)	Sausage (50–270 mg GAE/kg meat)	↑ Ash, ↓ fat, ↑ lipid oxidative stability (180–2700 mg GAE/kg), natural additives (nitrite substitute)	[172]
Coffee husk 'cascara'	Obtained commercially	Kombucha (10 g/L)	↑ Phenols, flavonoids, ↑ antioxidant, antibacterial activity, ↓ pH, balanced aroma and taste	[173]
Outer leaves of lettuce, onion peels, banana peels, whey protein, and brewery yeast	Freeze drying → grinding	Instant soups	↑ Antioxidant activity, ↑ phenols, flavonoids, hyperglycemia and hyperlipidemia, ↑ bone health, ↑ feces weight	[95]
Beet leaves	Washing → cutting → dispersion (1:50) in 80% ethanol (pH 6, 80 °C, 40 min) → centrifugation → solvent evaporation (40 °C) → resuspension in water	Vegetable smoothie (30% enrichment)	↑ TPC (+50%), ↑ antioxidant capacity, ↑ shelf life (+1 week), ↑ nutritional retention	[131]
Green tea leaves and branches, hops	MW hydrodiffusion and gravity/US/MW/ autohydrolysis (ethanol/water, 55–200 °C, 2–120 min)	Gluten-free hydrogel	Wide range of texture properties, ↑ thermal stability	[142]
Cowpea pod	Sun drying → grinding (0.25 mm) → water dispersion (1:15) → US treatment (20 kHz, 500 W, 10–20 min, 20–80% amplitude) → centrifugation → filtration (liquid) → freeze drying	Hydrogels	↑ TPC, ↑ antioxidant activity, ↑ beads size, ↓ compactness (in the presence of proteins)	[124]

Table 7. *Cont.*

By-Products	Ingredient Preparation	Food Application	Properties	References
Sesame, pomegranate, and grape seed	Grinding (1 mm) → sterilization (110 °C, 10 min, 1.5 atm)	Ice cream (2% enrichment)	↑ TPC, antioxidant activity (grape seed), ↑ viable probiotics	[146]
Peanut skin	Grinding (0.4 mm) → dispersion (1:10–30) in 95% ethanol → US treatment (10–60 min) → filtration → solvent evaporation (40 °C) → re-suspension in 95% ethanol	Dipped-bottom mushroom (0.005–0.05 mg/mL) and sacha inchi oil (0.01–0.05–1%)	↓ Oxidation, ↑ antioxidant activity	[127]
Rapeseed meal	Grinding (0.32 mm) → dispersion (1:4) in 75% ethanol (T °C room, 30 min, x4) → filtration → vacuum-concentration (50 °C) → freeze drying → phenols extraction: water dispersion (1:50, 20 min) → US treatment (20 kHz, 10–35% amplitude, 2–8 min) → centrifugation → filtration	Rapeseed oil (30% enrichment)	↓ Oxidation	[130]
Broccoli stalks, non-compliant carrots	Washing → disinfection → cutting → MW blanching (700 W, 2 min, x2) → cooling in ice → juice extraction	Beverage (82.5% broccoli, 17.5% carrots) (pasteurization, HPP, US treatment)	↑ Carotenoids, sulfur compounds, ↑ shelf life (HPP, US treatment)	[123]
Kimchi cabbage leaves and cortex	Steaming (120 °C, 11 min) → drying (80 °C, 48 h) → 0–100% methanol extraction → US treatment (0–30 min, 0–40% amplitude) → centrifugation (liquid) → freeze drying	Nanoparticles	↑ GLS protection (gastric phase), ↑ GLS release in intestine	[141]
Broccoli by-product	Obtained commercially	Beer (0.1% enrichment)	↑ Sulforaphane, ↑ alcohol content, ↓ bitterness, ↑ color, ↑ off-flavors, ↓ sensory properties	[151]
Non-compliant and broccoli stems	Blanching → freeze drying → grinding(powder) → dispersion (1:15) in 0–50% ethanol (40 °C, 1 h) → centrifugation (liquid extract)	Biscuit (10% flour, 30 mL extract/kg dough)	↑ Glucosinolates, TPC, carotenoids, color (flour), ↑ stickiness, ↓ workability (liquid extracts)	[121]
Brassica leaves	Washing → drying (60 °C, overnight)	Kombucha (7 g/L)	↓ TPC, antioxidant activity (than black tea), ↓ aroma, bitterness, ↑ color, appearance	[148]
Cauliflower stalks and leaves	Cutting → blanching (100 °C, 5 min) → freeze drying → grinding	Pizza (10–30% enrichment)	↑ Glucosinolates, carotenoids, phytosterols, good rheological properties (10%)	[16]
Broccoli stalks	Washing → cutting → brine	Fermented broccoli stalks	Source of phenols and GLS, ↓ broccoli flavor (fermentation), ↓ pH good sensory acceptability	[174]

Table 7. *Cont.*

By-Products	Ingredient Preparation	Food Application	Properties	References
Broccoli stem and leaves	Grinding (liquid nitrogen) → freeze drying	Salad dressing (1% enrichment)	↑ phenols bio-accessibility (modulated the presence of oil)	[145]
Fruit and vegetable waste	Quercetin extract powder obtained commercially	Bars (0.02–0.08% enrichment)	↑ Quercetin, TPC, ↑ antioxidant capacity, ↓ water activity, ↑ lightness, hardness, good sensory acceptability (0.06%)	[175]
Fruit and vegetable by-products	Washing → drying (60 °C, 5 h) → grinding → dispersion (1:3) in 96% ethanol (65 °C, 3 h) → filtration → concentration	Fish meal (300 ppm enrichment coffee sediment extract)	↑ Antioxidant activity, inhibits lipid oxidation	[126]

'→' followed by, '↓' low, '↑' high.

Table 8. Fish processing by-products with bioactive benefits and their food applications.

By-Products	Ingredient Preparation	Food Application	Properties	References
Skipjack tuna (*Katsuwonus pelamis*) by-products	Freeze drying → fat extraction (oil)	Pork sausage	↑ EPA (4.2%), DHA (22%), ↑ ABTS, DPPH, ↑ overall acceptance, ↑ antimicrobial properties	[154]
Mullet roe by-products	Three mild processes: pressure (PE) for fresh by-products, supercritical fluid extraction (SFE), and solvent extraction (SE) for freeze-dried by-products	Fish oil	↑ Antioxidant activity, DPPH, ABTS (SFE oil), ↑ oil yield (SE oil), ↑ EPA (20.7%), DHA (24.3%)	[156]
Fish by-products	Freeze drying → powder → microwave-assisted extraction (MAE)	Bioactive fish oil	↑ Oil yield (<19 min), no effect on DHA, linoleic, EPA, ↑ antimicrobial, antioxidant, anti-inflammatory activities	[157]
Catfish by-product	Fat extraction (US treatment)	Nanoemulsions (2 wt% oil)	↑ Antioxidant, antibacterial, anti-inflammatory properties, no toxicity against normal skin cells	[161]
Salmon (*Salmo salar*) backbones, heads, viscera	Freeze drying → powder → MAE: 14.6 min, 291.9 W, 80.1 g/L for backbones; 10.8 min, 50.0 W, 80.0 g/L for heads; and 14.3 min, 960.6 W, 99.5 g/L for viscera	Bioactive fish oil	↑ Oil recovery (69% heads, backbone; 92% viscera), ↑ cytotoxic, antioxidant, anti-inflammatory, and antimicrobial properties	[155]
Tuna (*Thunnus obesus*) and sea bass (*Dicentrarchus labrax*) by-product	4% Saline solution (5 min) → drying (50 °C) → grinding	Pasta (3% enrichment)	↑ DHA, EPA, cohesiveness (tuna), ↑ brightening (sea bass), ↓ hardness, fracturability, ↓ weight gain, swelling index	[176]

Table 8. *Cont.*

By-Products	Ingredient Preparation	Food Application	Properties	References
Northern shrimp (*Pandalus borealis*) by-product	Wet by-products → oil extraction (Soxhlet, hexane/acetone, 2:3)	Bioactive shrimp oil	↑ Phospholipids, n-3 PUFA and astaxanthin-esters, ↓ fat accumulation in 3T3-L1 cells, regulated adipogenesis and lipogenesis	[159]
Sardine (*Sardina pilchardus*) gill and viscera	Oil extraction (Soxhlet)	Chips (25 mL fish oil to 0.5 kg wheat)	↑ linoleic, linolenic, EPA, DHA, ↓ reducing triglycerides, LDL cholesterol, ↑ nutritional, antidiabetic, antihyperlipidemic, histoprotective effects	[158]
Nile perch (*Lates niloticus*) by-products	Oil extraction → centrifuging (recover of omega-3 FA)	Yoghurt (3.5 g oil)	Omega-3 daily intake was achieved through a single serving (150 g) of fortified yoghurt	[162]
Tuna, seabass, seabream, wild sardines by-products	Grinding → cooking (95 °C, 12 min → pressing → centrifuged (crude oil)	Crude oils	↓ Free fatty acid, peroxide, TOTOX, anisidine value (tuna, seabass, seabream) ↑ α-tocopherol (seabass, seabream), ↑ EPA, DHA (32.8% sardine oil)	[160]

'→' followed by, '↓' low, '↑' high.

Table 9. Dairy processing by-products with bioactive benefits and their food applications.

By-Products	Ingredient Preparation	Food Application	Properties	References
Sweet whey, sweet buttermilk, and skimmed milk	Pasteurization (72 °C, 15 s) → cooling → rennet addition → coagulation → draining → aging (4 °C, 24 h) → churning	Sweet whey fermented milk beverage, sweet buttermilk FMB, and skimmed milk FMB	↑ Antioxidant properties, ↓ glycemic index, glycemic load	[163]

'→' followed by, '↓' low, '↑' high.

3. Conclusions

The increasing demand for more nutritious, healthy, and sustainable food products has moved the attention of researchers and companies to find alternative sources of macro- and micro-nutrients, especially proteins, fibers, and bioactive compounds. Food processing by-products from animal, fish, and vegetable origin represent a valuable alternative. The effective use of by-products can help to increase resource efficiency, lower the burden of waste disposal, and produce useful inputs for the food industry. Besides conventional extraction methods, novel technologies may be adopted to produce functional high-quality ingredients in the form of powders or liquid extracts. Protein-, fiber-, and bioactive-rich ingredients allow the improvement of the nutritional quality of fortified food. Among them, bioactive compounds (e.g., phenols, carotenoids, omega-3 fatty acids, etc.) can both provide health benefits for the organism and increase the oxidative stability and the shelf life of the final food. Depending on the type of food, proper enrichment levels should be carefully evaluated to obtain the optimal physicochemical properties.

Author Contributions: Conceptualization, B.F., D.P. and P.L.; methodology, B.F., L.I. and H.S.O.D.; investigation, B.F., L.I. and H.S.O.D.; data curation, B.F., L.I. and H.S.O.D.; writing—original draft preparation, B.F., L.I. and H.S.O.D.; writing—review and editing, P.L. and D.P.; visualization, L.I.; supervision, P.L. All authors have read and agreed to the published version of the manuscript.

Funding: This work was supported by the European project Vitality (Next Generation EU, Project Code ECS00000041, CUP I33C22001330007) funded under the National Recovery and Resilience Plan (NRRP) and by the project WINfood (PSR Marche 2014/2020, ID 59459, Sottoprodotti della lavorazione industriale dei vegetali: da scarti ad ingredienti funzionali per la formulazione di alimenti innovativi).

Institutional Review Board Statement: Not applicable.

Informed Consent Statement: Not applicable.

Data Availability Statement: No new data were created or analyzed in this study. Data sharing is not applicable to this article.

Conflicts of Interest: The authors declare no conflicts of interest.

References

1. Rao, M.; Bast, A.; de Boer, A. Valorized Food Processing By-Products in the EU: Finding the Balance between Safety, Nutrition, and Sustainability. *Sustainability* **2021**, *13*, 4428. [CrossRef]
2. Orozco-Angelino, X.; Espinosa-Ramírez, J.; Serna-Saldívar, S.O. Extrusion as a Tool to Enhance the Nutritional and Bioactive Potential of Cereal and Legume By-Products. *Food Res. Int.* **2023**, *169*, 112889. [CrossRef]
3. Delgado, L.; Schuster, M.; Torero, M. Quantity and Quality Food Losses across the Value Chain: A Comparative Analysis. *Food Policy* **2021**, *98*, 101958. [CrossRef]
4. Chandrasekaran, M. *Valorization of Food Processing By-Products: Fermented Foods and Beverages Series*; Chandrasekaran, M., Ed.; CRC Press (Imprint of the Taylor and Francis Group): Boca Raton, FL, USA, 2012.
5. Reguengo, L.M.; Salgaço, M.K.; Sivieri, K.; Maróstica Júnior, M.R. Agro-Industrial by-Products: Valuable Sources of Bioactive Compounds. *Food Res. Int.* **2022**, *152*, 110871. [CrossRef]
6. Galali, Y.; Omar, Z.A.; Sajadi, S.M. Biologically Active Components in By-Products of Food Processing. *Food Sci. Nutr.* **2020**, *8*, 3004–3022. [CrossRef]
7. Food Wastage Footprint (Project) Food Wastage Footprint: Impacts on Natural Resources. 2013. Available online: https://scholar.google.com/scholar_lookup?title=Food+Wastage+Footprint.+Impact+on+Natural+Resources&author=FAO&publication_year=2013 (accessed on 17 August 2024).
8. Rațu, R.N.; Veleșcu, I.D.; Stoica, F.; Usturoi, A.; Arsenoaia, V.N.; Crivei, I.C.; Postolache, A.N.; Lipșa, F.D.; Filipov, F.; Florea, A.M.; et al. Application of Agri-Food By-Products in the Food Industry. *Agriculture* **2023**, *13*, 1559. [CrossRef]
9. Dinkçi, N.; Akdeniz, V.; Akalın, A.S. Probiotic Whey-Based Beverages from Cow, Sheep and Goat Milk: Antioxidant Activity, Culture Viability, Amino Acid Contents. *Foods* **2023**, *12*, 610. [CrossRef]
10. Toniasso, D.P.W.; Giacomelli da Silva, C.; de Souza Brum Junior, B.; Somacal, S.; Emanuelli, T.; Hashime Kubota, E.; Cristina Prestes Dornelles, R.; Mello, R. Collagen Extracted from Rabbit: Meat and by-Products: Isolation and Physicochemical Assessment. *Int. Food Res.* **2022**, *162*, 111967. [CrossRef]
11. Castro-Muñoz, R.; Díaz-Montes, E.; Gontarek-Castro, E.; Boczkaj, G.; Galanakis, C.M. A Comprehensive Review on Current and Emerging Technologies toward the Valorization of Bio-Based Wastes and by Products from Foods. *Compr. Rev. Food Sci. Food Saf.* **2022**, *21*, 46–105. [CrossRef]
12. Varela-Ortega, C.; Blanco-Gutiérrez, I.; Manners, R.; Detzel, A. Life Cycle Assessment of Animal-Based Foods and Plant-Based Protein-Rich Alternatives: A Socio-Economic Perspective. *J. Sci. Food Agric.* **2022**, *102*, 5111–5120. [CrossRef]
13. Castellanos-Fuentes, A.P.; Genevois, C.E.; Flores, S.K.; de Escalada Pla, M.F. Novel Application of a Food Ingredient Based on Soybean Extruded-Expelled Meal Containing Probiotics for Improving Gluten-Free Bread Quality. *Int. J. Food Sci.* **2023**, *58*, 6855–6861. [CrossRef]
14. Kahlon, T.S.; Avena-Bustillos, R.J.; Kahlon, A.K.; Brichta, J.L. Consumer Sensory Evaluation and Quality of Sorghum-Peanut Meal-Okra Snacks. *Heliyon* **2021**, *7*, e06874. [CrossRef]
15. Leahu, A.; Ropciuc, S.; Ghinea, C.; Damian, C. Physico-Chemical, Textural and Sensory Evaluation of Emulsion Gel Formulated with By-Products from the Vegetable Oil Industry. *Gels* **2023**, *9*, 964. [CrossRef]
16. Nartea, A.; Fanesi, B.; Pacetti, D.; Lenti, L.; Fiorini, D.; Lucci, P.; Frega, N.G.; Falcone, P.M. Cauliflower By-Products as Functional Ingredient in Bakery Foods: Fortification of Pizza with Glucosinolates, Carotenoids and Phytosterols. *Curr. Res. Food Sci.* **2023**, *6*, 100437. [CrossRef]
17. Ahlström, C.; Thuvander, J.; Rayner, M.; Matos, M.; Gutiérrez, G.; Östbring, K. The Effect of Precipitation PH on Protein Recovery Yield and Emulsifying Properties in the Extraction of Protein from Cold-Pressed Rapeseed Press Cake. *Molecules* **2022**, *27*, 2957. [CrossRef]

18. Ragonese, V.E.; Moscoso Ospina, Y.A.; Cabezas, D.M.; Kakisu, E.J. Effect of Ultrasound Treatment on the Composition and Emulsifying Properties of Quinoa Okara. *Int. J. Food Sci. Technol.* **2024**, *59*, 1481–1489. [CrossRef]
19. López-García, A.; Moraga, G.; Hernando, I.; Quiles, A. Providing Stability to High Internal Phase Emulsion Gels Using Brewery Industry By-Products as Stabilizers. *Gels* **2021**, *7*, 245. [CrossRef]
20. Kasapoglu, M.Z.; Sagdic, O.; Avci, E.; Tekin-Cakmak, Z.H.; Karasu, S.; Turker, R.S. The Potential Use of Cold-Pressed Coconut Oil By-Product as an Alternative Source in the Production of Plant-Based Drink and Plant-Based Low-Fat Ice Cream: The Rheological, Thermal, and Sensory Properties of Plant-Based Ice Cream. *Foods* **2023**, *12*, 650. [CrossRef]
21. Sabater, C.; Ruiz, L.; Delgado, S.; Ruas-Madiedo, P.; Margolles, A. Valorization of Vegetable Food Waste and By-Products Through Fermentation Processes. *Front. Microbiol.* **2020**, *11*, 581997. [CrossRef]
22. Squeo, G.; Latrofa, V.; Vurro, F.; De Angelis, D.; Caponio, F.; Summo, C.; Pasqualone, A. Developing a Clean Labelled Snack Bar Rich in Protein and Fibre with Dry-Fractionated Defatted Durum Wheat Cake. *Foods* **2023**, *12*, 2547. [CrossRef]
23. Drabińska, N.; Nogueira, M.; Ciska, E.; Jeleń, H. Effect of Drying and Broccoli Leaves Incorporation on the Nutritional Quality of Durum Wheat Pasta. *Pol. J. Food Nutr. Sci.* **2022**, *72*, 273–285. [CrossRef]
24. Krupa-Kozak, U.; Drabińska, N.; Baczek, N.; Šimková, K.; Starowicz, M.; Jeliński, T. Application of Broccoli Leaf Powder in Gluten-Free Bread: An Innovative Approach to Improve Its Bioactive Potential and Technological Quality. *Foods* **2021**, *10*, 819. [CrossRef]
25. Tukassar, A.; Shukat, R.; Iahtisham-Ul-Haq, M.S.; Butt, M.S.; Nayik, G.A.; Ramniwas, S.; Al Obaid, S.; Ali Alharbi, S.; Ansari, M.J.; Konstantinos Karabagias, I.; et al. Compositional Profiling and Sensory Analysis of Cauliflower By-Products-Enriched Muffins. *Food Sci. Nutr.* **2023**, *11*, 6020–6031. [CrossRef] [PubMed]
26. Kamani, M.H.; Luithui, Y.; Meera, M.S. Upgrading Black Gram By-Product to a New Texturized Vegetable Source by Extrusion: Evaluation of Functionality, Antinutrients and In Vitro Digestibility. *Waste Biomass Valorization* **2021**, *12*, 4319–4330. [CrossRef]
27. la Gatta, B.; Rutigliano, M.; Dilucia, F.; Liberatore, M.T.; Viscecchia, R.; Bimbo, F.; Di Luccia, A. Protein Network Assessment and Sensory Characteristics of Durum Wheat Fresh Pasta Fortified with Artichoke Bracts and Tomato Powders. *Food Biosci.* **2023**, *54*, 102858. [CrossRef]
28. Teterycz, D.; Sobota, A. Use of High-Protein and High-Dietary-Fibre Vegetable Processing Waste from Bell Pepper and Tomato for Pasta Fortification. *Foods* **2023**, *12*, 2567. [CrossRef]
29. Aussanasuwannakul, A.; Teangpook, C.; Treesuwan, W.; Puntaburt, K.; Butsuwan, P. Effect of the Addition of Soybean Residue (Okara) on the Physicochemical, Tribological, Instrumental, and Sensory Texture Properties of Extruded Snacks. *Foods* **2022**, *11*, 2967. [CrossRef] [PubMed]
30. Lima, J.R.; Araújo, Í.M.D.S.; Pinto, C.O.; Goiana, M.L.; Passos Rodrigues, M.D.C.; De Lima, L.V. Obtaining Cashew Kernel Protein Concentrate from Nut Processing By-Product and Its Use to Formulate Vegetal Burger. *Braz. J. Food Technol.* **2021**, *24*, e2020232. [CrossRef]
31. Bovi Karatay, G.; Theóphilo Galvão, M.; Bhunia, A.K.; Amaral, J.S.; Byrne, D.V.; Varzakas, T.; Sendra, E.; Adhikari, B.P.; Grossi Bovi Karatay, G.; Maria Medeiros Theóphilo Galvão, A.; et al. Storage Stability of Conventional and High Internal Phase Emulsions Stabilized Solely by Chickpea Aquafaba. *Foods* **2022**, *11*, 1588. [CrossRef] [PubMed]
32. Choya, F.; Combarros-Fuertes, P.; Camino, A.; Renes Bañuelos, D.; Prieto Gutiérrez, E.; Rodríguez, T.; Fresno Baro, M.E.; Fuentes Choya, P.; Combarros-Fuertes, P.; Camino, D.A.; et al. Study of the Technological Properties of Pedrosillano Chickpea Aquafaba and Its Application in the Production of Egg-Free Baked Meringues. *Foods* **2023**, *12*, 902. [CrossRef]
33. Gawborisut, S.; Ketkaew, C.; Buasook, T. Production of Biocalcium from Fermented Fish Bone Residue for Fish Emulsion Sausage Fortification. *Foods* **2024**, *13*, 882. [CrossRef]
34. Surasani, V.K.R.; Mandal, A.; Sofi, F.R.; Joshi, S. Utilization of Pangas Protein Isolates and By-Products in Fish Sausages: Effect on Quality Attributes and Acceptability. *J. Food Process. Preserv.* **2022**, *46*, e16798. [CrossRef]
35. Honrado, A.; Ardila, P.; Leciñena, P.; Beltrán, J.A.; Calanche, J.B. Transforming 'Bonito Del Norte' Tuna By-Products into Functional Ingredients for Nutritional Enhancement of Cereal-Based Foods. *Foods* **2023**, *12*, 4437. [CrossRef] [PubMed]
36. Kumari, A.; Kaushik, N.; Slizyte, R. Formulation of a Ready-to-Cook Soup Mix Using Microencapsulated Protein Hydrolysate Obtained from Pink Perch By-Product. *Pol. J. Food Nutr. Sci.* **2023**, *73*, 253–264. [CrossRef]
37. Chew, R.M.; Mohd Zin, Z.; Ahmad, A.; Mohtar, N.F.; Rusli, N.D.; Zainol, M.K. Physicochemical and Sensory Properties of Deep Fried Battered Squid Containing Brownstripe Red Snapper (Lutjanus Vitta) Protein Hydrolysate. *Food Res.* **2020**, *4*, 1245–1253. [CrossRef] [PubMed]
38. Zhang, Y.; Yang, Z.; Zhao, S.; Gai, J.; Wang, L.; Ning, X.; Tao, N. Quality Characteristics of Sardine (Sardina Pilchardus) Fish Sauce Produced Using Five Kinds of Material. *Int. J. Gastron. Food Sci.* **2023**, *34*, 100801. [CrossRef]
39. Tkaczewska, J.; Kulawik, P.; Morawska-Tota, M.; Zając, M.; Guzik, P.; Tota, Ł.; Pająk, P.; Duliński, R.; Florkiewicz, A.; Migdał, W. Protocol for Designing New Functional Food with the Addition of Food Industry By-Products, Using Design Thinking Techniques—A Case Study of a Snack with Antioxidant Properties for Physically Active People. *Foods* **2021**, *10*, 694. [CrossRef]
40. Zhang, L.; Shan, Y.; Hong, H.; Luo, Y.; Hong, X.; Ye, W. Prevention of Protein and Lipid Oxidation in Freeze-Thawed Bighead Carp (Hypophthalmichthys Nobilis) Fillets Using Silver Carp (Hypophthalmichthys Molitrix) Fin Hydrolysates. *LWT* **2020**, *123*, 109050. [CrossRef]
41. Nurdiani, R.; Zahira, A.; Ainnayah, Q.; Rat, F.R.H.; Firdaus, M.; Astuti, R.T. Characteristics Of Yoghurt Enriched With Protein Hydrolysate From Parrotfish (Chlorurus Sordidus) Head. *Carpathian J. Food Sci. Technol.* **2023**, *15*, 184–196. [CrossRef]

42. Unnikrishnan, P.; Puthenveetil Kizhakkethil, B.; Anant Jadhav, M.; Sivam, V.; Ashraf, P.M.; Ninan, G.; Aliyamveetil Abubacker, Z. Protein Hydrolysate from Yellowfin Tuna Red Meat as Fortifying and Stabilizing Agent in Mayonnaise. *J. Food Sci. Technol.* **2020**, *57*, 413–425. [CrossRef]
43. Bruno Siewe, F.; Kudre, T.G.; Narayan, B. Optimisation of Ultrasound-Assisted Enzymatic Extraction Conditions of Umami Compounds from Fish by-Products Using the Combination of Fractional Factorial Design and Central Composite Design. *Food Chem.* **2021**, *334*, 127498. [CrossRef] [PubMed]
44. Ali, M.S.; Roy, V.C.; Park, J.S.; Haque, A.R.; Mok, J.H.; Zhang, W.; Chun, B.S. Protein and Polysaccharide Recovery from Shrimp Wastes by Natural Deep Eutectic Solvent Mediated Subcritical Water Hydrolysis for Biodegradable Film. *Mar. Biotechnol.* **2024**. [CrossRef] [PubMed]
45. Li, L.; Kan, G.; Peng, J.; Gong, H.; Zi, Y.; Shi, C.; Wang, X.; Zhong, J. Tilapia Head Gelatins to Stabilize Fish Oil Emulsions and the Effect of Extraction Methods. *Int. J. Biol. Macromol.* **2024**, *269*, 132137. [CrossRef]
46. Yang, L.; Yang, M.; Xu, J.; Nie, Y.; Wu, W.; Zhang, T.; Wang, X.; Zhong, J. Structural and Emulsion Stabilization Comparison of Four Gelatins from Two Freshwater and Two Marine Fish Skins. *Food Chem.* **2022**, *371*, 131129. [CrossRef]
47. Kumari, A.; Kaushik, N.; Slizyte, R.; Khushboo. Production and Microencapsulation of Protein Hydrolysate of Pink Perch (Nemipterus Japonicus) By-Products Obtained from Surimi Industry for Its Sustainable Utilization. *Waste Biomass Valorization* **2023**, *14*, 209–226. [CrossRef]
48. Zhang, L.; Li, Q.; Hong, H.; Luo, Y. Prevention of Protein Oxidation and Enhancement of Gel Properties of Silver Carp (Hypophthalmichthys Molitrix) Surimi by Addition of Protein Hydrolysates Derived from Surimi Processing by-Products. *Food Chem.* **2020**, *316*, 126343. [CrossRef] [PubMed]
49. Steinsholm, S.; Oterhals, Å.; Underhaug, J.; Aspevik, T. Emulsion and Surface-Active Properties of Fish Solubles Based on Direct Extraction and after Hydrolysis of Atlantic Cod and Atlantic Salmon Backbones. *Foods* **2021**, *10*, 38. [CrossRef]
50. Wang, K.; Ni, J.; Tian, X.; Xiang, S.; Li, H.; Shang, W.; Liu, B.; Tan, M.; Su, W. Survivability of Probiotics in Pickering Emulsion Gels Stabilized by Salmon By-Product Protein/Sodium Alginate Soluble Complexes at Neutral PH. *Int. J. Biol. Macromol.* **2024**, *255*, 128190. [CrossRef]
51. Nasri, R.; Taktak, W.; Hamdi, M.; Ben Amor, N.; Kabadou, A.; Li, S.; Nasri, M. Sardinelle Protein Isolate as a Novel Material for Oil Microencapsulation: Novel Alternative for Fish by-Products Valorisation. *Mater. Sci. Eng. C* **2020**, *116*, 111164. [CrossRef]
52. Syahrul, S.; Dewita, D.; Effendi, I.; Suwondo, S. The Use of Striped Catfish (Pangasianodon Hypophthalmus) Abdominal Fat as a Raw Material for Highly Nutritional Fish Oil. *AACL Bioflux* **2023**, *16*, 979–988.
53. Ainsa, A.; Iranzo, L.; Honrado, A.; Marquina, P.; Roncalés, P.; Beltrán, J.A.; Calanche, J. Effects of Cooking over the Stability of Fatty Acids as Bioactive Compounds in Enriched Pasta with a Fish By-Product. *Cereal Chem.* **2022**, *99*, 286–294. [CrossRef]
54. Ainsa, A.; Roldan, S.; Marquina, P.L.; Roncalés, P.; Beltrán, J.A.; Calanche Morales, J.B. Quality Parameters and Technological Properties of Pasta Enriched with a Fish By-Product: A Healthy Novel Food. *J. Food Process. Preserv.* **2022**, *46*, e16261. [CrossRef]
55. Ainsa, A.; Marquina, P.L.; Roncalés, P.; Beltrán, J.A.; Calanche, M.J.B. Enriched Fresh Pasta with a Sea Bass By-Product, a Novel Food: Fatty Acid Stability and Sensory Properties throughout Shelf Life. *Foods* **2021**, *10*, 255. [CrossRef] [PubMed]
56. Bonfim, B.d.C.; Monteiro, M.L.G.; Santos, A.F.G.N.d.; Vilar, J.d.S.; Conte-Junior, C.A. Nutritional Improvement and Consumer Perspective of Fish Nuggets with Partial Substitution of Wheat Flour Coating by Fish (Priacanthus Arenatus, Cuvier, 1829) Waste Flour. *J. Aquat. Food Prod.* **2020**, *29*, 28–42. [CrossRef]
57. Vitorino, K.C.; Chambo, A.P.S.; Coradini, M.F.; Matiucci, M.A.; Graton Michka, J.M.; Goes, E.S.d.R.; Gonçalves, A.A.; de Souza, M.L.R. Cereal Bars Flavored with Fish Protein Concentrate from Different Species. *J. Aquat. Food Prod.* **2020**, *29*, 65–72. [CrossRef]
58. Santos Corrêa, S.; Gonçalves Oliveira, G.; Coradini Franco, M.; Gasparino, E.; Feihrmann, A.C.; Siemer, S.; Vieira Dantas Filho, J.; Cavali, J.; de Vargas Schons, S.; Rodrigues de Souza, M.L. Quality of Oreochromis Niloticus and Cynoscion Virescens Fillets and Their By-Products in Flours Make for Inclusion in Instant Food Products. *PLoS ONE* **2023**, *18*, e0279351. [CrossRef] [PubMed]
59. Long, J.M.; Mohan, A. Development of Meat Powder from Beef Byproduct as Value-Added Food Ingredient. *LWT* **2021**, *146*, 111460. [CrossRef]
60. Usman, M.; Ishaq, A.; Regenstein, J.M.; Sahar, A.; Aadil, R.M.; Sameen, A.; Khan, M.I.; Alam, A. Valorization of Animal By-Products for Gelatin Extraction Using Conventional and Green Technologies: A Comprehensive Review. *Biomass Convers. Biorefin.* **2023**, *1*, 1–13. [CrossRef]
61. de Souza Cunha, R.C.; de Carvalho, L.M.; de Sousa Fontes, V.M.; de Sousa Galvão, M.; Olegário, L.S.; de Medeiros, L.L.; Morgano, M.A.; Pacheco, M.T.B.; Madruga, M.S.; Bezerra, T.K.A. Volatilomic Evaluation of Protein Hydrolysates from Free-Range Chicken Bones Treated with Hot-Pressure Process. *LWT* **2023**, *174*, 114368. [CrossRef]
62. Mattohti, W.; Rozi, P.; Ababakri, G.; Esmayil, Z.; Zhou, C.; Cao, B.; Yadikar, N.; Yang, X. Isolation and Structural Characterization of Antioxidant Peptides from Horse Bone Marrow Protein Hydrolysates. *J. Food Meas. Charact.* **2023**, *17*, 732–750. [CrossRef]
63. da Silva Bambirra Alves, F.E.; Carpiné, D.; Teixeira, G.L.; Goedert, A.C.; de Paula Scheer, A.; Ribani, R.H. Valorization of an Abundant Slaughterhouse By-Product as a Source of Highly Technofunctional and Antioxidant Protein Hydrolysates. *Waste Biomass Valorization* **2021**, *12*, 263–279. [CrossRef]
64. de Sousa Fontes, V.M.; de Sousa Galvão, M.; Moreira de Carvalho, L.; do Nascimento Guedes, F.L.; dos Santos Lima, M.; Alencar Bezerra, T.K.; Madruga, M.S. Thiamine, Cysteine and Xylose Added to the Maillard Reaction of Goat Protein Hydrolysate Potentiates the Formation of Meat Flavoring Compounds. *Food Chem.* **2024**, *445*, 138398. [CrossRef]

65. Londoño-Zapata, L.; Franco-Cardona, S.; Restrepo-Manotas, S.; Gomez-Narvaez, F.; Suarez-Restrepo, L.; Nuñez-Andrade, H.; Valencia-Araya, P.; Simpson, R.; Vega-Castro, O. Valorization of the By-Products of Poultry Industry (Bones) by Enzymatic Hydrolysis and Glycation to Obtain Antioxidants Compounds. *Waste Biomass Valorization* **2022**, *13*, 4469–4480. [CrossRef]
66. Song, Y.; Gu, H.; Jo, J.M.; Shin, M.; Kim, S.Y.; Gam, D.H.; Imamura, S.; Kim, J.W. Production of Functional Peptide with Anti-Obesity Effect from Defatted Tenebrio Molitor Larvae Using Proteolytic Enzyme. *Biotechnol. Bioprocess Eng.* **2020**, *25*, 374–383. [CrossRef]
67. Tang, T.; Zhang, M.; Liu, Y. Valorization of Meat and Bone Residue by Ultrasound and High Voltage Electrostatic Field Assisted Two-Stage Enzymatic Hydrolysis: Nutritional Characteristics and Flavor Analysis. *Food Biosci.* **2023**, *56*, 103203. [CrossRef]
68. Prokopová, A.; Gál, R.; Mokrejš, P.; Pavlačková, J. Preparation of Gelatin from Broiler Chicken Stomach Collagen. *Foods* **2022**, *12*, 127. [CrossRef]
69. Lukin, A. On the Possibility of Obtaining High-Quality Lyophilized Collagen Hydrolysate and Its Applicability in the Sausage Production. *Food Sci. Technol.* **2019**, *40*, 721–728. [CrossRef]
70. Toldrà, M.; Parés, D.; Saguer, E.; Carretero, C. Utilisation of Protein Fractions from Porcine Spleen as Technofunctional Ingredients in Emulsified Cooked Meat Sausages. *Int. J. Food Sci. Technol.* **2020**, *55*, 871–877. [CrossRef]
71. Ananey-Obiri, D.; Matthews, L.; Tahergorabi, R. Chicken Processing By-Product: A Source of Protein for Fat Uptake Reduction in Deep-Fried Chicken. *Food Hydrocoll.* **2020**, *101*, 105500. [CrossRef]
72. Akimova, D.; Suychinov, A.; Kakimov, A.; Kabdylzhar, B.; Zharykbasov, Y.; Yessimbekov, Z. Effect of Chicken By-Products on the Physicochemical Properties of Forcemeat Formulations. *Future Foods* **2023**, *7*, 100238. [CrossRef]
73. Kusio, K.; Szafrańska, J.O.; Radzki, W.; Sołowiej, B.G. Effect of Whey Protein Concentrate on Physicochemical, Sensory and Antioxidative Properties of High-Protein Fat-Free Dairy Desserts. *Appl. Sci.* **2020**, *10*, 7064. [CrossRef]
74. Pires, A.; Tan, G.; Gomes, D.; Pereira-Dias, S.; Díaz, O.; Cobos, A.; Pereira, C. Application of Ultrafiltration to Produce Sheep's and Goat's Whey-Based Synbiotic Kefir Products. *Membranes* **2023**, *13*, 473. [CrossRef] [PubMed]
75. Pavoni, J.M.F.; Leidens, N.; Luchese, C.L.; Baldasso, C.; Tessaro, I.C. In Natura Ovine Whey Proteins Concentration by Ultrafiltration Combining Batch and Diafiltration Operating Modes. *J. Food Process. Eng.* **2020**, *43*, e13554. [CrossRef]
76. Himashree, P.; Gokul Devi, E.; Hariharan, K.C.; Sunil, C.K. Impact of Whey Incorporation on Acrylamide Formation and the Quality of Bread. *J. Food Meas. Charact.* **2023**, *17*, 5748–5758. [CrossRef]
77. la Gatta, B.; Rutigliano, M.; Liberatore, M.T.; Dilucia, F.; Spadaccino, G.; Quinto, M.; Di Luccia, A. Preservation of Bioactive Compounds Occurring in Fresh Pasta Fortified with Artichoke Bracts and Tomato Powders Obtained with a Novel Pre-Treatment. *LWT* **2023**, *187*, 115298. [CrossRef]
78. Gunel, Z. Microencapsulation of Apricot Kernel Oil: Utilization of Mushroom by-Product as an Emulsifier in Oil-in-Water Emulsion. *Acad Bras. Cienc.* **2024**, *96*, e20220448. [CrossRef] [PubMed]
79. Thi Minh Nguyet, N.; Pham Tan Quoc, L.; Gia Buu, T. Evaluation of Textural and Microstructural Properties of Vegan Aquafaba Whipped Cream from Chickpeas. *Chem. Eng. Trans.* **2021**, *83*, 421–426. [CrossRef]
80. Lamas, D.L.; Gende, L.B. Valorisation of Brewers' Spent Grain for the Development of Novel Beverage and Food Products. *Appl. Food Res.* **2023**, *3*, 100314. [CrossRef]
81. Honrado, A.; Ardila, P.; Leciñena, P.; Beltrán, J.A.; Calanche, J. Nutritional Enrichment of Cereal Foods: Feasibility Study and Characterisation of Biscuits Fortified with Seabass by-Products. *Int. J. Food Sci. Technol.* **2024**, *59*, 5376–5388. [CrossRef]
82. Vu, N.D.; Tran, T.T.; Nguyen, V.M. Effect of Cooking and Preservation Time on Fish Balls Quality Produced from Pangasius Hypophthalmus Meat By-Product. *Squalen Bull. Mar. Fish. Postharvest Biotechnol.* **2023**, *18*, 42–51. [CrossRef]
83. Khodaei, D.; Forde, A.; Noci, F.; Ryan, L. Physicochemical and Sensory Characteristics of Pasta Enriched with Blue Whiting (Micromesistius Poutassou) Fish Protein Hydrolysate. *Int. J. Food Sci.* **2023**, *58*, 2782–2789. [CrossRef]
84. Boronat, Ò.; Sintes, P.; Celis, F.; Díez, M.; Ortiz, J.; Aguiló-Aguayo, I.; Martin-Gómez, H. Development of Added-Value Culinary Ingredients from Fish Waste: Fish Bones and Fish Scales. *Int. J. Gastron. Food Sci.* **2023**, *31*, 100657. [CrossRef]
85. Furey, A.E.; Hoeche, U.; McLaughlin, C.; Noci, F. Incorporation of Roe, Milt and Liver from Plaice (Pleuronectes Platessa), Herring (Clupea Harengus) and Cod (Gadus Morhua) in Newly Developed Seafood Pâtés: Sensory Evaluation by Teenage Consumers in Ireland and Their Attitudes to Seafood. *Int. J. Gastron. Food Sci.* **2022**, *28*, 100524. [CrossRef]
86. Aínsa, A.; Vega, A.; Honrado, A.; Marquina, P.; Roncales, P.; Gracia, J.A.B.; Morales, J.B.C. Gluten-Free Pasta Enriched with Fish By-Product for Special Dietary Uses: Technological Quality and Sensory Properties. *Foods* **2021**, *10*, 3049. [CrossRef]
87. Samarakoon, H.M.M.C.; Patabandi, P.P.S.K.; Walpita, C.N.; Priyadarshana, M.K.C.; Udayathilake, E. Development of Ready to Eat Fish Paste Using Chub Mackerel (Scomber Japonicus) Tail Offcuts. *J. Agric. Sci.* **2021**, *16*, 247–259. [CrossRef]
88. Liu, J.; Shibata, M.; Ma, Q.; Liu, F.; Lu, Q.; Shan, Q.; Hagiwara, T.; Bao, J. Characterization of Fish Collagen from Blue Shark Skin and Its Application for Chitosan- Collagen Composite Coating to Preserve Red Porgy (Pagrus Major) Meat. *J. Food Biochem.* **2020**, *44*, e13265. [CrossRef]
89. Jahan, K.; Qadri, O.S.; Younis, K. Dietary Fiber as a Functional Food. In *Functional Food Products and Sustainable Health*; Springer: Singapore, 2020; pp. 155–167. [CrossRef]
90. Ye, S.; Shah, B.R.; Li, J.; Liang, H.; Zhan, F.; Geng, F.; Li, B. A Critical Review on Interplay between Dietary Fibers and Gut Microbiota. *Trends Food Sci.* **2022**, *124*, 237–249. [CrossRef]
91. He, Y.; Wang, B.; Wen, L.; Wang, F.; Yu, H.; Chen, D.; Su, X.; Zhang, C. Effects of Dietary Fiber on Human Health. *Food Sci. Hum. Wellness* **2022**, *11*, 1–10. [CrossRef]

92. Hitlamani, V.; Huded, P.; Kumar, G.S.; Chetana, R. Development of High-Fiber and High-Protein Virgin Coconut Oil-Based Spread and Its Physico-Chemical, and Sensory Qualities. *J. Food Sci. Technol.* **2024**, *1*, 1–9. [CrossRef]
93. Karra, S.; Sebii, H.; Bouaziz, M.A.; Blecker, C.; Attia, H.; Besbes, S. Male Date Palm Flower Powder: Effect of Incorporation on Physico-Chemical, Textural, and Sensory Quality of Biscuits. *J. Food Process. Preserv.* **2020**, *44*, e14687. [CrossRef]
94. Kultys, E.; Moczkowska-Wyrwisz, M. Effect of Using Carrot Pomace and Beetroot-Apple Pomace on Physicochemical and Sensory Properties of Pasta. *LWT* **2022**, *168*, 113858. [CrossRef]
95. Mohamed, R.S.; Abozed, S.S.; El-Damhougy, S.; Salama, M.F.; Hussein, M.M. Efficiency of Newly Formulated Functional Instant Soup Mixtures as Dietary Supplements for Elderly. *Heliyon* **2020**, *6*, e03197. [CrossRef] [PubMed]
96. Wirkijowska, A.; Zarzycki, P.; Teterycz, D.; Nawrocka, A.; Blicharz-Kania, A.; Łysakowska, P. The Influence of Tomato and Pepper Processing Waste on Bread Quality. *Appl. Sci.* **2023**, *13*, 9312. [CrossRef]
97. Brigagão, T.C.S.; Fortes, R.R.; Lourenço, C.O.; Carvalho, E.E.N.; Cirillo, M.Â.; Nachtigall, A.M.; Vilas Boas, B.M. Optimization of Gluten-Free Muffins Made with Pineapple Peel, Banana Peel, and Pumpkin Seed Flours. *J. Food Process. Preserv.* **2021**, *45*, e16037. [CrossRef]
98. Segura-Badilla, O.; Kammar-García, A.; Mosso-Vázquez, J.; Ávila-Sosa Sánchez, R.; Ochoa-Velasco, C.; Hernández-Carranza, P.; Navarro-Cruz, A.R. Potential Use of Banana Peel (Musa Cavendish) as Ingredient for Pasta and Bakery Products. *Heliyon* **2022**, *8*, e11044. [CrossRef]
99. Ying, D.; Sanguansri, L.; Cheng, L.; Augustin, M.A. Nutrient-Dense Shelf-Stable Vegetable Powders and Extruded Snacks Made from Carrots and Broccoli. *Foods* **2021**, *10*, 2298. [CrossRef]
100. Merlino, M.; Tripodi, G.; Cincotta, F.; Prestia, O.; Miller, A.; Gattuso, A.; Verzera, A.; Condurso, C. Technological, Nutritional, and Sensory Characteristics of Gnocchi Enriched with Hemp Seed Flour. *Foods* **2022**, *11*, 2783. [CrossRef]
101. Pop, A.; Paucean, A.; Socaci, S.A.; Alexa, E.; Man, S.M.; Muresan, V.; Chis, M.S.; Salanta, L.; Popescu, I.; Berbecea, A.; et al. Quality Characteristics and Volatile Profile of Macarons Modified with Walnut Oilcake By-Product. *Molecules* **2020**, *25*, 2214. [CrossRef]
102. Aşkın Uzel, R. Sustainable Green Technology for Adaptation of Circular Economy to Valorize Agri-Food Waste: Celery Root Peel as a Case Study. *Manag. Environ. Qual. Int. J.* **2023**, *34*, 1018–1034. [CrossRef]
103. Djeghim, F.; Bourekoua, H.; Różyło, R.; Bieńczak, A.; Tanaś, W.; Zidoune, M.N. Effect of By-Products from Selected Fruits and Vegetables on Gluten-Free Dough Rheology and Bread Properties. *Appl. Sci.* **2021**, *11*, 4605. [CrossRef]
104. Saeidy, S.; Nasirpour, A.; Barekat, S. Effect of Sugar Beet Fiber and Different Hydrocolloids on Rheological Properties and Quality of Gluten-Free Muffins. *J. Sci. Food Agric.* **2023**, *103*, 1404–1411. [CrossRef] [PubMed]
105. Islam, M.R.; Biswas, M.M.H.; Esham, M.K.H.; Roy, P.; Khan, M.R.; Hasan, S.M.K. Jackfruit (Artocarpus Heterophyllus) by-Products a Novel Source of Pectin: Studies on Physicochemical Characterization and Its Application in Soup Formulation as a Thickener. *Food Chem. Adv.* **2023**, *2*, 100273. [CrossRef]
106. Parid, D.M.; Rahman, N.A.A.; Baharuddin, A.S.; Kadir Basha, R.P.; Mohammed, M.A.; Mat Johari, A.; Abd Razak, S.Z. Effects of Carboxymethyl Cellulose Extracted from Oil Palm Empty Fruit Bunch Stalk Fibres on the Physical Properties of Low-Fat Ice Cream. *Food Res.* **2021**, *5*, 1–7. [CrossRef]
107. Silchuk, T.; Zuiko (Kulinich), V.; Nazar, M.; Tsirulnikova, V.; Tyshchenko (Usatiuk), O.; Pushka, O.; Kyrpichenkova, O.; Bortnichuk, O. Investigation of the Influence of Peas Dietary Fibers on the Process of Formation of Wheat Dough. *J. Hyg. Eng. Des.* **2021**, *35*, 56–60.
108. Nawrocka, A.; Zarzycki, P.; Kłosok, K.; Welc, R.; Wirkijowska, A.; Teterycz, D. Effect of Dietary Fibre Waste Originating from Food Production on the Gluten Structure in Common Wheat Dough. *Int. Agrophys.* **2023**, *37*, 101–109. [CrossRef]
109. Bamal, P.; Dhull, S.B. Development of Functional Muffins from Wheat Flour-Carrot Pomace Powder Using Fenugreek Gum as Fat Replacer. *Curr. Res. Nutr.* **2024**, *12*, 306–319. [CrossRef]
110. Karwacka, M.; Ciurzyńska, A.; Galus, S.; Janowicz, M. Freeze-Dried Snacks Obtained from Frozen Vegetable by-Products and Apple Pomace—Selected Properties, Energy Consumption and Carbon Footprint. *Innov. Food Sci. Emerg. Technol.* **2022**, *77*, 102949. [CrossRef]
111. Karwacka, M.; Rybak, K.; Smetana, S.; Galus, S.; Janowicz, M. Analysis of Selected Functional Properties, Resource Demands, and Energy Consumption of Freeze-Dried Vegetable Snacks. *J. Food Process. Preserv.* **2022**, *46*, e16721. [CrossRef]
112. Vendramin, V.; Spinato, G.; Vincenzi, S.; Castro Marín, A.; Chinnici, F. Shellfish Chitosan Potential in Wine Clarification. *Appl. Sci.* **2021**, *11*, 4417. [CrossRef]
113. Bastiaens, L.; Soetemans, L.; D'Hondt, E.; Elst, K. Sources of Chitin and Chitosan and Their Isolation. In *Chitin and Chitosan*; Broek, L.A.M., Boeriu, C.G., Eds.; John Wiley & Sons Ltd.: Hoboken, NJ, USA, 2019; pp. 1–34. [CrossRef]
114. Jofre, C.M.; Campderrós, M.E.; Rinaldoni, A.N. Integral Use of Grape: Clarified Juice Production by Microfiltration and Pomace Flour by Freeze-Drying. Development of Gluten-Free Filled Cookies. *Food Chem. Adv.* **2024**, *4*, 100583. [CrossRef]
115. Egea, M.; Álvarez, D.; Peñaranda, I.; Panella-Riera, N.; Linares, M.B.; Garrido, M.D. Fat Replacement by Vegetal Fibres to Improve the Quality of Sausages Elaborated with Non-Castrated Male Pork. *Animals* **2020**, *10*, 1872. [CrossRef]
116. Mahfoudhi, N.; Ksouri, R.; Hamdi, S. Nanoemulsions as Potential Delivery Systems for Bioactive Compounds in Food Systems: Preparation, Characterization, and Applications in Food Industry. *Emulsions* **2016**, *3*, 365–403. [CrossRef]
117. Banwo, K.; Olojede, A.O.; Adesulu-Dahunsi, A.T.; Verma, D.K.; Thakur, M.; Tripathy, S.; Singh, S.; Patel, A.R.; Gupta, A.K.; Aguilar, C.N.; et al. Functional Importance of Bioactive Compounds of Foods with Potential Health Benefits: A Review on Recent Trends. *Food Biosci.* **2021**, *43*, 101320. [CrossRef]

118. Jha, A.K.; Sit, N. Extraction of Bioactive Compounds from Plant Materials Using Combination of Various Novel Methods: A Review. *Trends Food Sci.* **2022**, *119*, 579–591. [CrossRef]
119. Nartea, A.; Kuhalskaya, A.; Fanesi, B.; Orhotohwo, O.L.; Susek, K.; Rocchetti, L.; Di Vittori, V.; Bitocchi, E.; Pacetti, D.; Papa, R. Legume Byproducts as Ingredients for Food Applications: Preparation, Nutrition, Bioactivity, and Techno-Functional Properties. *Compr. Rev. Food Sci. Food Saf.* **2023**, *22*, 1953–1985. [CrossRef] [PubMed]
120. Ahmed, A.; Abid, H.M.R.; Ahmad, A.; Khalid, N.; Shibli, S.; Amir, R.M.; Malik, A.M.; Asghar, M. Utilization of Mango Peel in Development of Instant Drink. *Asian J. Agric. Biol.* **2020**, *8*, 260–267. [CrossRef]
121. Fanesi, B.; Ismaiel, L.; Nartea, A.; Orhotohwo, O.L.; Kuhalskaya, A.; Pacetti, D.; Lucci, P.; Falcone, P.M. Bioactives and Technological Quality of Functional Biscuits Containing Flour and Liquid Extracts from Broccoli By-Products. *Antioxidants* **2023**, *12*, 2115. [CrossRef] [PubMed]
122. Naik, B.; Kumar, V.; Gupta, A.K. Valorization of Tender Coconut Mesocarp for the Formulation of Ready-to-Eat Dairy-Based Dessert (Kheer): Utilization of Industrial by-Product. *J. Agric. Res.* **2023**, *12*, 100572. [CrossRef]
123. Pérez, P.; Hashemi, S.; Cano-Lamadrid, M.; Martínez-Zamora, L.; Gómez, P.A.; Artés-Hernández, F. Effect of Ultrasound and High Hydrostatic Pressure Processing on Quality and Bioactive Compounds during the Shelf Life of a Broccoli and Carrot By-Products Beverage. *Foods* **2023**, *12*, 3808. [CrossRef]
124. Traffano-Schiffo, M.V.; Aguirre Calvo, T.R.; Avanza, M.V.; Santagapita, P.R. High-Intensity Ultrasound-Assisted Extraction of Phenolic Compounds from Cowpea Pods and Its Encapsulation in Hydrogels. *Heliyon* **2020**, *6*, e04410. [CrossRef]
125. Difonzo, G.; Totaro, M.P.; Caponio, F.; Pasqualone, A.; Summo, C. Olive Leaf Extract (OLE) Addition as Tool to Reduce Nitrate and Nitrite in Ripened Sausages. *Foods* **2022**, *11*, 451. [CrossRef]
126. Hue, H.T.; Tinh, H.T.; Van Bao, N.; Dao, P.T.A. Screening for Antioxidant Activity of Vegetable and Fruit By-Products and Evaluating the Ability of Coffee Sediment to Preserve Fish Meal. *SN Appl. Sci.* **2020**, *2*, 1282. [CrossRef]
127. Tsai, S.Y.; Sun, W.J.; Wu, Y.C.; Lin, C.P. Peanut Skin Extracts Applied on Shelf Life of Button Mushrooms and Antioxidant Effect of Sacha Inchi Oil via Isothermal Calorimetry. *J. Therm. Anal. Calorim.* **2023**, *148*, 1629–1638. [CrossRef]
128. Yu, J.; Xu, S.; Chen, R.; Shao, P. A Promising Bioactive Chitosan Film in Strawberry Fresh-Keeping: Plasticized with Tomato Processing by-Product Extract of Deep Eutectic Solvent. *Food Hydrocoll.* **2024**, *151*, 109859. [CrossRef]
129. Benbouriche, A.; Haddadi-Guemghar, H.; Mehidi-Terki, D.; Boulekbache-Makhlouf, L.; Bachir-bey, M. Enrichment of Soybean Oil with Carotenoids from Chili Paste By-Product: Enhancement of Quality, Oxidative Stability, and Thermo-Resistance. *J. Food Meas. Charact.* **2024**, *18*, 1107–1120. [CrossRef]
130. Cisneros-Yupanqui, M.; Chalova, V.I.; Kalaydzhiev, H.R.; Mihaylova, D.; Krastanov, A.I.; Lante, A. Ultrasound-Assisted Extraction of Antioxidant Bioactive Compounds from Wastes of Rapeseed Industry and Their Application in Delaying Rapeseed Oil Oxidation. *Environ. Technol. Innov.* **2023**, *30*, 103081. [CrossRef]
131. Fernandez, M.V.; Bengardino, M.; Jagus, R.J.; Agüero, M.V. Enrichment and Preservation of a Vegetable Smoothie with an Antioxidant and Antimicrobial Extract Obtained from Beet By-Products. *LWT* **2020**, *117*, 108622. [CrossRef]
132. Fotiadou, R.; Vougiouklaki, D.; Houhoula, D.; Stamatis, H. Improvement of the Oxidative Stability of Edible Oils through Enzymatic Esterification with Hydroxytyrosol-Rich Extract. *Eur. J. Lipid Sci. Technol* **2024**, *126*, 2400014. [CrossRef]
133. De Leonardis, A.; Macciola, V.; Iftikhar, A.; Lopez, F. Characterization, Sensory and Oxidative Stability Analysis of Vegetable Mayonnaise Formulated with Olive Leaf Vinegar as an Active Ingredient. *Foods* **2022**, *11*, 4006. [CrossRef]
134. Dzah, C.S.; Duan, Y.; Zhang, H.; Wen, C.; Zhang, J.; Chen, G.; Ma, H. The Effects of Ultrasound Assisted Extraction on Yield, Antioxidant, Anticancer and Antimicrobial Activity of Polyphenol Extracts: A Review. *Food Biosci.* **2020**, *35*, 100547. [CrossRef]
135. Fernandez-Pan, I.; Horvitz, S.; Ibañez, F.C.; Arroqui, C.; Beriain, M.J.; Virseda, P. Extra-Virgin Olive Oil Enriched with Lycopene: From Industrial Tomato by-Products to Consumer. *Food Sci. Nutr.* **2024**, *12*, 5815–5823. [CrossRef] [PubMed]
136. Michalak-Majewska, M.; Złotek, U.; Szymanowska, U.; Szwajgier, D.; Stanikowski, P.; Matysek, M.; Sobota, A. Antioxidant and Potentially Anti-Inflammatory Properties in Pasta Fortified with Onion Skin. *Appl. Sci.* **2020**, *10*, 8164. [CrossRef]
137. Betrouche, A.; Estivi, L.; Colombo, D.; Pasini, G.; Benatallah, L.; Brandolini, A.; Hidalgo, A. Antioxidant Properties of Gluten-Free Pasta Enriched with Vegetable By-Products. *Molecules* **2022**, *27*, 8993. [CrossRef] [PubMed]
138. Gumul, D.; Ziobro, R.; Korus, J.; Kruczek, M. Apple Pomace as a Source of Bioactive Polyphenol Compounds in Gluten-Free Breads. *Antioxidants* **2021**, *10*, 807. [CrossRef]
139. Kaderides, K.; Mourtzinos, I.; Goula, A.M. Stability of Pomegranate Peel Polyphenols Encapsulated in Orange Juice Industry By-Product and Their Incorporation in Cookies. *Food Chem.* **2020**, *310*, 125849. [CrossRef]
140. Lara-Abia, S.; Lobo, G.; Pérez-Pascual, N.; Welti-Chanes, J.; Cano, M.P. Improvement in the Stability and Bioaccessibility of Carotenoid and Carotenoid Esters from a Papaya By-Product Using O/W Emulsions. *Foods* **2023**, *12*, 2654. [CrossRef] [PubMed]
141. Park, S.J.; Lee, M.J.; Choi, Y.J.; Yun, Y.R.; Lee, M.A.; Min, S.G.; Seo, H.Y.; Park, D.H.; Park, S.H. Optimization of Extraction and Nanoencapsulation of Kimchi Cabbage By-Products to Enhance the Simulated in Vitro Digestion of Glucosinolates. *Heliyon* **2023**, *9*, e16525. [CrossRef]
142. Sanz, V.; Domínguez, H.; Torres, M.D. Formulation and Thermomechanical Characterization of Functional Hydrogels Based on Gluten Free Matrices Enriched with Antioxidant Compounds. *Appl. Sci.* **2021**, *11*, 1962. [CrossRef]
143. Stoica, F.; Rațu, R.N.; Motrescu, I.; Cara, I.G.; Filip, M.; Țopa, D.; Jităreanu, G. Application of Pomace Powder of Black Carrot as a Natural Food Ingredient in Yoghurt. *Foods* **2024**, *13*, 1130. [CrossRef]

144. Naurzbayeva, G.; Smolnikova, F.; Baytukenova, S.; Kapshakbayeva, Z.; Mustafayeva, A.; Baytukenova, S.; Smailova, Z.; Dikhanbayeva, F.; Temerbayeva, M.; Baybalinova, G. Incorporating Carrot Pomace-Based Emulsion to Enhance the Nutritional Value and Shelf Life of Butter. *Int. J. Food Prop.* **2023**, *26*, 2455–2475. [CrossRef]
145. Kamiloglu, S.; Ozdal, T.; Tomas, M.; Capanoglu, E. Oil Matrix Modulates the Bioaccessibility of Polyphenols: A Study of Salad Dressing Formulation with Industrial Broccoli by-Products and Lemon Juice. *J. Sci. Food Agric.* **2022**, *102*, 5368–5377. [CrossRef]
146. Akca, S.; Akpinar, A. The Effects of Grape, Pomegranate, Sesame Seed Powder and Their Oils on Probiotic Ice Cream: Total Phenolic Contents, Antioxidant Activity and Probiotic Viability. *Food Biosci.* **2021**, *42*, 101203. [CrossRef]
147. Vlad, C.C.; Păcularu-Burada, B.; Vasile, A.M.; Milea, Ș.A.; Bahrim, G.E.; Râpeanu, G.; Stănciuc, N. Upgrading the Functional Potential of Apple Pomace in Value-Added Ingredients with Probiotics. *Antioxidants* **2022**, *11*, 2028. [CrossRef] [PubMed]
148. Martin-Gómez, H.; Diez, M.; Abadias, M.; Rivera, A.; Aguiló-Aguayo, I. Promoting a Circular Economy by Developing New Gastronomic Products from Brassica Non-Edible Leaves. *Int. J. Food Sci. Technol.* **2024**, *59*, 1071–1079. [CrossRef]
149. Ortinero, C.V.; Rafael, R.R.; Rayos, C.E.R.; Bautista, K.D.A.; Feliciano, M.A.M.; Natividad, L.R.; Natividad, G.M. Distilled Spirit Extraction of Phenolic Antioxidants from Fruit and Vegetable Residues. *J. Ecol. Eng.* **2021**, *22*, 125–131. [CrossRef]
150. Salas-Millán, J.Á.; Aznar, A.; Conesa, E.; Conesa-Bueno, A.; Aguayo, E. Fruit Wine Obtained from Melon By-Products: Physico-Chemical and Sensory Analysis, and Characterization of Key Aromas by GC-MS. *Foods* **2022**, *11*, 3619. [CrossRef] [PubMed]
151. Sánchez-Bravo, P.; Abellán, Á.; Zapata, P.J.; García-Viguera, C.; Domínguez-Perles, R.; Giménez, M.J. Broccoli Products Supplemented Beers Provide a Sustainable Source of Dietary Sulforaphane. *Food Biosci.* **2023**, *51*, 102259. [CrossRef]
152. Rakariyatham, K.; Teerawutgulrag, A.; Laokuldilok, T.; Osiriphun, S.; Ackcharoensuk, N.; Tanbamrung, W. Combinatorial Effects of Longan (Dimocarpus Longan) Peel Extract and Lecithin on Stability of Soybean Oil and the Oxidative Stability of Fried Shrimp Crackers during Storage. *Lebensm. Wiss. Und-Technol.* **2024**, *198*, 116065. [CrossRef]
153. Pereira, J.C.; Sivakanthan, S.; Vasantharuba, S. Effect of Star Fruit (*Averrhoa Carambola* L.) By-Product on Oxidative Stability of Sesame (Sesamum Indicum) Oil under Accelerated Oven Storage and during Frying. *J. Oleo Sci.* **2020**, *69*, 837–849. [CrossRef]
154. Abeysinghe, A.M.A.K.; Ho, T.C.; Surendhiran, D.; Roy, V.C.; Park, J.S.; Han, J.M.; Chun, B.S. Oil Extraction and Methyl Esterification of Fatty Acids Obtained from Skipjack Tuna By-Products: Antimicrobial Property and Preservative Capability on Pork Sausage. *Biomass Convers. Biorefin.* **2024**, *14*, 1–11. [CrossRef]
155. de la Fuente, B.; Pinela, J.; Mandim, F.; Heleno, S.A.; Ferreira, I.C.F.R.; Barba, F.J.; Berrada, H.; Caleja, C.; Barros, L. Nutritional and Bioactive Oils from Salmon (Salmo Salar) Side Streams Obtained by Soxhlet and Optimized Microwave-Assisted Extraction. *Food Chem.* **2022**, *386*, 132778. [CrossRef]
156. Kalogianni, E.P.; Georgiou, D.; Charisis, A.; Exarhopoulos, S.; Tzika, P. Valorization of Mullet Roe By-Products for the Production of Polyunsaturated Fatty Acids Rich Oils. *J. Am. Oil Chem. Soc.* **2023**, *100*, 317–327. [CrossRef]
157. Pinela, J.; de la Fuente, B.; Rodrigues, M.; Pires, T.C.S.P.; Mandim, F.; Almeida, A.; Dias, M.I.; Caleja, C.; Barros, L. Upcycling Fish By-Products into Bioactive Fish Oil: The Suitability of Microwave-Assisted Extraction. *Biomolecules* **2023**, *13*, 1. [CrossRef] [PubMed]
158. Benkhoud, H.; Baâti, T.; Njim, L.; Selmi, S.; Hosni, K. Antioxidant, Antidiabetic, and Antihyperlipidemic Activities of Wheat Flour-Based Chips Incorporated with Omega-3-Rich Fish Oil and Artichoke Powder. *J. Food Biochem.* **2021**, *45*, e13297. [CrossRef]
159. Phadtare, I.; Vaidya, H.; Hawboldt, K.; Cheema, S.K. Shrimp Oil Extracted from Shrimp Processing By-Product Is a Rich Source of Omega-3 Fatty Acids and Astaxanthin-Esters, and Reveals Potential Anti-Adipogenic Effects in 3T3-L1 Adipocytes. *Mar. Drugs* **2021**, *19*, 259. [CrossRef]
160. Šimat, V.; Vlahović, J.; Soldo, B.; Generalić Mekinić, I.; Čagalj, M.; Hamed, I.; Skroza, D. Production and Characterization of Crude Oils from Seafood Processing By-Products. *Food Biosci.* **2020**, *33*, 100484. [CrossRef]
161. Azmi, N.A.N.; Elgharbawy, A.A.M.; Salleh, H.M.; Moniruzzaman, M. Preparation, Characterization and Biological Activities of an Oil-in-Water Nanoemulsion from Fish By-Products and Lemon Oil by Ultrasonication Method. *Molecules* **2022**, *27*, 6725. [CrossRef]
162. Murage, M.W.; Muge, E.K.; Mbatia, B.N.; Mwaniki, M.W. Development and Sensory Evaluation of Omega-3-Rich Nile Perch Fish Oil-Fortified Yogurt. *Int. J. Food Sci.* **2021**, *2021*, 8838043. [CrossRef]
163. Alwohaibi, A.A.A.; Ali, A.A.; Sakr, S.S.; Mohamed Ahmed, I.A.; Alhomaid, R.M.; Alsaleem, K.A.; Aladhadh, M.; Barakat, H.; Hassan, M.F.Y. Valorization of Different Dairy By-Products to Produce a Functional Dairy–Millet Beverage Fermented with Lactobacillus Paracasei as an Adjunct Culture. *Fermentation* **2023**, *9*, 927. [CrossRef]
164. Karwacka, M.; Galus, S.; Janowicz, M. Development and Physicochemical Characteristics of Multicomponent Freeze-Dried Snacks Obtained with Blackcurrant Pomace Powder and Calcium Ions as Structuring Agents. *J. Food Sci. Technol.* **2024**, *61*, 1363–1373. [CrossRef]
165. Lordi, A.; Panza, O.; Conte, A.; Del Nobile, M.A. Best Combination of Vegetable By-Products for the Shelf-Life Extension of Fresh Pasta. *Foods* **2023**, *13*, 44. [CrossRef] [PubMed]
166. Panza, O.; Lacivita, V.; Tarantino, F.; Manzi, A.; Conte, A.; Del Nobile, M.A. Fruit and Vegetable By-Products as Source of Bioactive Compounds to Preserve Handmade Fresh Pasta. *LWT* **2023**, *190*, 115584. [CrossRef]
167. Long, J.M.; Mohan, A. Food Flavoring Prepared with Lemon By-Product. *J. Food Process. Preserv.* **2021**, *45*, e15462. [CrossRef]
168. Giannoutsos, K.; Zalidis, A.P.; Koukoumaki, D.I.; Menexes, G.; Mourtzinos, I.; Sarris, D.; Gkatzionis, K. Production of Functional Crackers Based on Non-Conventional Flours. Study of the Physicochemical and Sensory Properties. *Food Chem. Adv.* **2023**, *2*, 100194. [CrossRef]

169. García-Pastor, M.E.; Ródenas-Soriano, M.; Dobón-Suárez, A.; Zapata, P.J.; Giménez, M.J. Use of Olive Industry By-Products for Value-Added Food Development. *Agronomy* **2023**, *13*, 718. [CrossRef]
170. Lipșa, F.D.; Stoica, F.; Rațu, R.N.; Veleșcu, I.D.; Cârlescu, P.M.; Motrescu, I.; Usturoi, M.G.; Râpeanu, G. Red Onion Peel Powder as a Functional Ingredient for Manufacturing Ricotta Cheese. *Foods* **2024**, *13*, 182. [CrossRef] [PubMed]
171. Tonyali, B.; Sensoy, I.; Karakaya, S. Effects of Processing on Onion Skin Powder Added Extrudates. *J. Food Sci. Technol.* **2020**, *57*, 3426. [CrossRef]
172. Cocan, I.; Cadariu, A.I.; Negrea, M.; Alexa, E.; Obistioiu, D.; Radulov, I.; Poiana, M.A. Investigating the Antioxidant Potential of Bell Pepper Processing By-Products for the Development of Value-Added Sausage Formulations. *Appl. Sci.* **2022**, *12*, 12421. [CrossRef]
173. Nguyen, L.A.D.; Huynh, T.K.D.; Nguyen, T.N.H.; Nguyen, Q.T.; Tran, M.P. Effect of Rosemary (Rosmarinus Officinalis) Extract on the Protection of the Fishballs from Knife Fish (Chitala Chitala) and Striped Catfish by-Product (Pangasianodon Hypophthalmus) against Spoilage during Frozen Storage. *Food Res.* **2023**, *7*, 85–95. [CrossRef]
174. Salas-Millán, J.Á.; Aznar, A.; Conesa, E.; Conesa-Bueno, A.; Aguayo, E. Functional Food Obtained from Fermentation of Broccoli By-Products (Stalk): Metagenomics Profile and Glucosinolate and Phenolic Compounds Characterization by LC-ESI-QqQ-MS/MS. *LWT* **2022**, *169*, 113915. [CrossRef]
175. Bansal, U.; Bhardwaj, A.; Singh, S.N.; Khubber, S.; Sharma, N.; Bansal, V. Effect of Incorporating Plant-Based Quercetin on Physicochemical Properties, Consumer Acceptability and Sensory Profiling of Nutrition Bars. *Funct. Foods Health Dis.* **2022**, *12*, 116–127. [CrossRef]
176. Ainsa, A.; Honrado, A.; Marquina, P.L.; Roncalés, P.; Beltrán, J.A.; Calanche, M.J.B. Innovative Development of Pasta with the Addition of Fish By-Products from Two Species. *Foods* **2021**, *10*, 1889. [CrossRef] [PubMed]

foods

MDPI

Article

Optimization of Heat–Moisture Treatment Conditions for High-Amylose Starch and Its Application in High-Resistant Starch Triticale Noodles

Hua Li [1,2,*], Hua Li [1], Yu Liu [1], Ruixin Liu [1,2] and Sirithon Siriamornpun [3,4,*]

[1] Department of Cuisine and Nutrition, Yangzhou University, Yangzhou 225127, China
[2] Key Laboratory of Chinese Cuisine Intangible Cultural Heritage Technology Inheritance, Ministry of Culture and Tourism, Yangzhou 225127, China
[3] Research Unit of Thai Food Innovation (TFI), Mahasarakham University, Kantarawichai 44150, Thailand
[4] Department of Food Technology and Nutrition, Faculty of Technology, Mahasarakham University, Kantarawichai 44150, Thailand
* Correspondence: lihua216@yzu.edu.cn (H.L.); sirithon.s@msu.ac.th (S.S.); Tel.: +86-13852577568 (H.L.); +66-88-747-4136 (S.S.)

Abstract: Heat–moisture treatment (HMT) is a widely used method for modifying starch properties with the potential to reduce the digestibility of high-amylose starch (HAS). This study aimed to optimize the HMT conditions for HAS and apply the resulting HMT-HAS to triticale noodles to develop low-glycemic-index products. HMT significantly increased the resistant starch (RS) content and decreased the rapidly digestible starch (RDS) content of HAS. The treatment conditions—temperature, heating time, and moisture content—were found to significantly influence the starch composition. Optimal HMT conditions were determined using response surface methodology: a temperature of 108 °C, a heating time of 5.8 h, and a moisture content of 25.50%. Under these conditions, the RS content of HMT-HAS was 60.23%, nearly double that of the untreated sample. Increasing the level of HMT-HAS in triticale noodles led to significant decreases in short-range order, relative crystallinity, and viscosities, while the RS content increased from 12.08% to 34.41%. These findings suggest that incorporating HMT-HAS into triticale noodles effectively enhances starch digestive resistance, supporting the development of functional, low-glycemic-index triticale-based foods.

Keywords: starch digestibility; resistant starch; starch structure; heating time; relative crystallinity; pasting property; infrared spectroscopy

1. Introduction

High-amylose maize starch, abbreviated as high-amylose starch (HAS), could decrease blood glucose levels, increase satiety, improve the gut microbiota, and reduce the risk of obesity and cardiovascular disease [1]. HAS contains a much higher content of resistant starch (RS) than low- or normal-amylose starch, which makes it digested and absorbed by the body at a slower rate, thereby lowering the glycemic index (GI) of food and improving the postprandial glucose and insulin responses [2]. Therefore, HAS has been considered a good raw material for the development of low-GI foods.

Heat–moisture treatment (HMT) is a physical modification technique that involves treating starch at high temperatures (90–130 °C) with low moisture content (15–35%) for a certain period. During HMT, the movement of water molecules is intensified by thermal energy to modify the internal structure of starch [3]. HMT may affect the sensitivity of digested enzymes to starch, depending on the source and variety of the starch as well as the treatment conditions [4]. For example, starch, particularly HAS, is apt to be converted into forms with low GI by HMT [5]. Animal studies have indicated that HMT-HAS exhibits stronger resistance to enzyme digestion in the mouse digestive tract [6] and could effectively reduce plasma cholesterol levels in ovariectomized rats [7]. Therefore, optimizing the HMT conditions for HAS is essential for regulating the GI, promoting the application of HMT technology, and enhancing the utilization of HAS.

Noodles are a popular staple in the world, especially in China, and are generally regarded as a high-GI food. To the best of our knowledge, we are the first to report on the development of low-GI triticale noodles as an alternative product. Triticale (× *Triticosecale* Wittmack), a hybrid grain from wheat (*Triticum* sp.) and rye (*Secale cereale* L.), has been increasingly incorporated into various foods such as bread, biscuits, tortillas, pasta, and noodles due to its high nutritional value and health-promoting properties [8]. Previous research, such as Lorenz et al. [9], has demonstrated the suitability of triticale for noodle production based on its cooking and sensory properties. Additionally, studies have explored the nutrients, polyphenols, and sensory qualities of triticale noodles [10,11], with Liu et al. [12] advancing the development of the gluten network in triticale whole-wheat noodles using enzymes like glucose oxidase, xylanase, and laccase. Compared to traditional wheat noodles, triticale noodles offer superior nutritional value and bioactivity, attributed to higher contents of protein, minerals, vitamins, and phenolics [13]. Despite the growing interest in low-GI foods for managing postprandial blood glucose, limited research has focused on reducing the GI of triticale noodles. Although numerous studies have investigated the effects and mechanisms of HMT on the physicochemical properties and starch digestibility of various starches, the results have been inconsistent. Therefore, optimizing HMT conditions for HAS, and considering factors like heating time, temperature, and moisture content, remains crucial to enhancing the RS content.

Therefore, in this study, we optimized the HMT conditions for HAS by response surface methodology using RS content as the response variable and then incorporated the HAS produced by HMT at optimal conditions into the triticale noodle preparation to attenuate the starch digestibility of the noodle. Meanwhile, some physical characteristics, including microstructure, crystalline characteristics, pasting properties, and short-range order, were also explored to better understand the change in the noodle's starch characteristics because of the addition of HMT-HAS. This investigation may be helpful in developing low-GI triticale products to meet the needs of diabetes patients and increase the added value of triticale as well.

2. Materials and Methods

2.1. Materials and Reagents

The triticale cultivar (ND2201) was obtained from Huinong Fumin Technology Co., Ltd. (Beijing, China). High-amylose maize starch (1945, 71% amylose) was purchased from Kehe Biotechnology Co., Ltd. (Shanghai, China). Amyloglucosidase and α-amylase were purchased from Aladdin Corp. (Shanghai, China). Phenol and 3,5-dinitrosalicylic acid were acquired from Suyi Chemical Reagent Co., Ltd. (Shanghai, China) and Lanji Technology Development Co., Ltd. (Shanghai, China), respectively. All other reagents of analytical grade were supplied by Sinopharm Chemical Reagent Co., Ltd. (Shanghai, China).

2.2. HMT of HAS

HAS was heat–moisture treated as follows: The moisture content of HAS in a reaction vessel was adjusted to 15–35% by adding distilled water, followed by standing at room temperature for 24 h, and then the sealed vessel was heated in a drying oven (DHG-9055A, Yiheng Technology, Shanghai, China) at a certain temperature (90–130 °C) for a certain period of time (2–6 h). Subsequently, the vessel was cooled for 30 min, and then the samples treated with heat and moisture were dried at 45 °C for 12 h. Finally, the dried samples were ground (Q-250A, Bingdu Electrical Appliance, Shanghai, China) and passed through an 80-mesh sieve to obtain the HMT-HAS samples.

2.3. Optimization of HMT Conditions for HAS

2.3.1. Single-Factor Experimental Design

Three independent variables, i.e., moisture content, treatment temperature, and heating time, were investigated for their effects on the starch digestibility of HAS. The experiments were designed as follows: (1) The temperature and time were 110 °C and 4 h, respectively; the moisture content was 15%, 20%, 25%, 30%, or 35%. (2) The moisture content and time were 25% and 4 h, respectively; the temperature was 90 °C, 100 °C, 110 °C, 120 °C, or 130 °C. (3) The moisture content and temperature were 25% and 110 °C, respectively; the time was 2, 3, 4, 5, or 6 h.

2.3.2. Response Surface Experimental Design

To obtain the optimal HMT conditions for HAS, on the basis of the preliminary single-factor experiments, moisture content, treatment temperature, and heating time were set as the independent variables, with RS content as the response variable. The Box–Behnken design was conducted using Design Expert 13 (Stat-Ease, Minneapolis, MN, USA). The factors and levels of the response surface experiment are shown in Table 1.

Table 1. Factors and levels of response surface experiment.

Levels	Factors		
	A: Moisture Content (%)	**B: Treatment Temperature (°C)**	**C: Heating Time (h)**
−1	20	100	4
0	25	110	5
1	30	120	6

2.4. Preparation of Triticale Noodle Samples

The triticale noodles were prepared as follows: Briefly, 100 g of the flour (consisting of triticale and HMT-HAS) containing 0%, 10%, 20%, 30%, 40%, or 50% HMT-HAS was blended with 60 mL of water and kneaded into a uniform dough. Subsequently, the dough was set to rest for 30 min at 30 °C, followed by being cut into noodles that were 1 mm thick, 5 mm wide, and 20 cm long. After being freeze-dried, some of the noodles were ground into a powder, passing through an 80-mesh sieve. Both the ground and unground noodles were sealed and preserved in a desiccator at room temperature until used.

2.5. Scanning Electron Microscopy Analysis

As per the method supplied by Lv [14], the microstructural characteristics of freeze-dried noodle samples were explored using a scanning electron microscope (S-4800II, Hitachi, Tokyo, Japan) with a magnification of 800× at an accelerating voltage of 20 kV.

2.6. Analysis of Short-Range Order

The short-range order of the freeze-dried noodle powders was investigated using Fourier transform infrared spectroscopy (670-IR+610-IR, Varian, Palo Alto, CA, USA), as reported by Zhang [15]. The scan was performed in the range of 400–4000 cm^{-1} with a resolution of 4 cm^{-1}. The spectra were obtained by scanning 32 times with KBr as the background and analyzed using OMNIC 9.7 (Thermo Fisher Scientific Inc., Waltham, MA USA). The spectra were deconvoluted and normalized in the range of 900 to 1250 cm^{-1}, with a peak width of 40 cm^{-1} and an enhancement factor of 1.9.

2.7. X-ray Diffraction Analysis

Following the method of Li et al. [16], the X-ray diffraction patterns of the freeze-dried noodle powders were determined using an X-ray diffractometer (D8 Advance, Bruker AXS, Karlsruhe, Baden-Württemberg, Germany) at 40 kV and 40 mA with Cu-Kα radiation. The scanning region ranged from 5° to 40° (2θ) at a scanning speed of 2°/min. The relative crystallinity was calculated using the X-ray diffractometer software (Topas 4.0).

2.8. Determination of Pasting Properties

The pasting properties of the freeze-dried noodle powders were analyzed with a rapid viscosity analyzer (TechMaster RVA, Perten, Segeltorp, Sweden). A mixture of 1.5 g of the noodle sample and 25 g of distilled water was put into an aluminum box for RVA. The program parameters were set as follows: equilibration at 50 °C for 1 min, heating to 95 °C at a rate of 12 °C/min, holding at 95 °C for 2.5 min, and then cooling to 50 °C at a rate of 12 °C/min and holding for 2 min. The rotational speed of the helical rotor was 960 rpm for the first 10 s and then reduced to 160 rpm.

2.9. Analysis of Total Starch Content

The total starch content was analyzed using the acid hydrolysis method outlined in the "National Food Safety Standard—Determination of Starch in Foods (GB5009.9-2016)" [17].

2.10. Analysis of In Vitro Starch Digestibility

The RDS, SDS, and RS contents were analyzed following the method reported by Englyst et al. [18] with minor modifications. In brief, 200 mg of the sample was thoroughly mixed with 15 mL of sodium acetate buffer (0.2 mol/L, pH 5.2) and then incubated in a boiling water bath for 15 min. Following a water bath at 37 °C for 5 min, the sample was hydrolyzed with 5 mL of α-amylase (290 U/mL) and 5 mL of amyloglucosidase (15 U/mL) at 160 rpm and 37 °C. After hydrolysis for 20 and 120 min, 1 mL of the hydrolysate was immediately blended with 4 mL of a 66% anhydrous ethanol solution. Following centrifugation (Allegra X-30R, Beckman Coulter, Brea, CA, USA) at 6000 rpm for 10 min, the level of glucose in the supernatant was measured using the DNS method.

2.11. Statistical Analysis

The data were presented as the mean ± standard deviation for three replications. Significance analysis was carried out using SPSS 19.0 (IBM, Armonk, NY, USA) at the $p < 0.05$ level. Graphs were constructed by Origin 2018 (OriginLab, Northampton, MA, USA).

3. Results and Discussion

3.1. Single Factor Analysis for HMT

3.1.1. Effect of Moisture Content

Starch digestibility has been demonstrated to be dependent on the parameters in HMT, including moisture content, treatment temperature, and heating time [19]. Table 2 shows the effect of moisture content during HMT on the starch composition of HAS. HMT led to an evident change in the contents of RDS, SDS, and RS. Moreover, the contents of those different types of starches varied significantly with the moisture content. Compared to the untreated counterpart, the RDS content decreased by 21–50% and the RS content increased by 31–80%, with the highest RS and lowest RDS contents observed at 25% of moisture content. The impact of HMT on starch digestibility may be attributed to starch structural and morphological changes under high temperature and low moisture content, such as crystallite disruption, crystallite reorientation, amylose–lipid interaction, starch chain interaction, and formation of fissures and cracks on the granule surface [19]. In our study, the reasons for native starch showing the highest RDS content were likely as follows: HMT can cause the formation of ordered molecular aggregation architecture, which increases crystalline perfection and improves resistance to enzymes [20]. In addition, it has been reported that there is up to 1.0% lipid contents in typical washed cereal starch [21]. Therefore, the starch–lipid complex, which does not fit into the active sites of α-amylases, may be formed during HMT, as reported by Liu et al. [22], who also found the formation of the starch–lipid complex after HMT despite the low lipid content (0.5%) in buckwheat starch. At high temperatures, the intensified movement of water molecules disrupts the ordered structure of starch, causing migration and rearrangement. This leads to enhanced interaction between the amorphous regions and the starch molecular chains, thereby obstructing the entry of amylase into the interior of the starch granules and inhibiting starch digestion [23]. When higher moisture levels were used, the disruption of starch structure by HMT might be intensified rather than rearranged [24]. When the moisture content exceeded 25%, with more water molecules at high temperatures, starch gelatinization intensified, leading to structural disruption of starch and increased attachment sites for amylase, thus promoting starch digestion. Therefore, a moisture content of 20% to 30% was selected for the response surface experiment.

Table 2. Starch digestibility of HAS under HMT with different moisture contents.

Moisture Content (%)	RDS (%)	SDS (%)	RS (%)
untreated	45.76 ± 0.15 [a]	23.26 ± 0.61 [c]	30.98 ± 0.67 [f]
15	32.47 ± 0.14 [c]	27.09 ± 0.09 [a]	40.44 ± 0.14 [e]
20	30.57 ± 0.08 [d]	24.55 ± 0.48 [b]	44.87 ± 0.41 [c]
25	22.79 ± 0.85 [f]	21.30 ± 0.72 [d]	55.91 ± 0.13 [a]
30	29.58 ± 0.21 [e]	20.62 ± 0.35 [d]	49.80 ± 0.47 [b]
35	36.01 ± 0.69 [b]	20.78 ± 0.82 [d]	43.21 ± 0.44 [d]

Values are presented as mean ± standard deviation (n = 3), and values with different lowercases in the same column are considered significantly different ($p < 0.05$). HMT: heat–moisture treatment. HAS: high-amylose starch. RDS: rapidly digestible starch. SDS: slowly digestible starch. RS: resistant starch.

3.1.2. Effect of Treatment Temperature

The effect of temperature during HMT on the starch composition of HAS is shown in Table 3. Similarly, the temperature also significantly affected the contents of RDS, SDS, and RS. The sample by HMT at 100 °C exhibited the lowest RDS content, about half of that in the untreated counterpart. Meanwhile, HMT at 100 °C induced an increase in RS content, 1.8 times higher than the untreated one. It may be due to the intensified movement of water molecules under thermal energy, promoting the perfection of crystalline structure and the formation of more stable starch-lipid complexes. When the temperature was higher than 90 °C, the RDS content as well as the RS content at different temperatures was comparable, although significant. It may be attributed to the stable starch structure, in accordance with the findings of Wang et al. [25]. Increasing the temperature leads to a more stable starch structure through the processes of gelatinization and retrogradation. The stability arises from the disruption of the crystalline regions in the starch granules and the formation of a more amorphous, gel-like structure that is less prone to reverting to its original state [26]. Considering the reduction in cost and starch discoloration at excessively high temperatures, temperatures ranging from 90 °C to 110 °C were selected for the response surface experiment.

Table 3. Starch digestibility of HAS under HMT at different treatment temperatures.

Treatment Temperature (°C)	RDS (%)	SDS (%)	RS (%)
untreated	45.76 ± 0.15 [a]	23.26 ± 0.61 [ab]	30.98 ± 0.67 [d]
90	33.81 ± 0.22 [b]	23.96 ± 0.07 [a]	42.23 ± 0.24 [c]
100	22.36 ± 0.87 [d]	22.34 ± 0.91 [b]	55.30 ± 0.12 [b]
110	23.62 ± 0.15 [c]	20.73 ± 0.40 [c]	55.65 ± 0.51 [ab]
120	23.18 ± 0.97 [cd]	20.91 ± 0.91 [c]	55.91 ± 0.13 [ab]
130	23.02 ± 0.23 [cd]	20.76 ± 0.10 [c]	56.22 ± 0.14 [a]

[a–d]: values with different lowercases in the same column are considered significantly different ($p < 0.05$).

3.1.3. Effect of Heating Time

Table 4 displays the effect of heating time on the starch composition of the HMT-HAS. The contents of RDS, SDS, and RS varied significantly with time. As the time increased, the RS content greatly increased, reaching a maximum of 62.06% at 6 h, approximately two times that in the untreated sample. On the contrary, the RDS content sharply decreased from 45.76% (the untreated sample) to 19.31% (the sample with HMT for 6 h). This phenomenon may be attributed to the enhanced interaction between the amorphous regions and molecular chains under the influence of water and heat with prolonged HMT time, therefore reinforcing the stability of the internal structure and hindering the entry of digestive enzymes [27]. In the preliminary study, we observed that HMT for more than 6 h could cause starch discoloration and a burnt smell, thereby limiting the treatment time to 4 to 6 h for the response surface experiment.

Table 4. Starch digestibility of HAS under HMT at different heating times.

Heating Time (h)	RDS (%)	SDS (%)	RS (%)
untreated	45.76 ± 0.15 [a]	23.26 ± 0.61 [a]	30.98 ± 0.67 [e]
2	25.90 ± 0.15 [b]	19.98 ± 0.85 [b]	54.12 ± 0.73 [d]
3	24.60 ± 0.28 [c]	20.94 ± 0.44 [b]	54.46 ± 0.26 [d]
4	21.84 ± 0.20 [d]	22.25 ± 0.27 [a]	55.91 ± 0.13 [c]
5	20.62 ± 0.88 [e]	20.49 ± 0.78 [b]	58.89 ± 0.90 [b]
6	19.31 ± 0.48 [f]	18.63 ± 0.73 [c]	62.06 ± 0.25 [a]

[a–f]: values with different lowercases in the same column are considered significantly different ($p < 0.05$).

3.2. Response Surface Analysis of HMT Conditions

The HMT conditions were optimized by response surface methodology. The results are shown in Table 5. Through regression analysis, the relationship between the RS content and the independent variables (treatment temperature, heating time, and moisture content) was presented by a quadratic polynomial equation as follows:

$$RS = 55.67 + 6.05A + 3.56B + 2.20C + 0.58AB + 0.45AC + 0.96BC - 3.33A^2 - 2.43B^2 - 3.86C^2$$

Table 5. Box–Behnken design arrangement and results of response surface experiment.

Run	A: Treatment Temperature (°C)	B: Heating Time (h)	C: Moisture Content (%)	RS (%)
1	100	4	20	44.15
2	90	5	30	44.14
3	90	5	20	40.94
4	110	4	25	52.21
5	110	5	30	56.91
6	100	5	25	56.37
7	100	5	25	55.45
8	100	6	20	49.92
9	100	5	25	56.39
10	100	5	25	55.76
11	90	6	25	46.44
12	110	5	20	51.91
13	100	6	30	56.51
14	100	4	30	46.92
15	110	6	25	59.93
16	90	4	25	41.03
17	100	5	25	54.36

The variance analysis for the regression equation is displayed in Table 6. The regression model is highly significant ($p < 0.01$), with a non-significant lack of fit, indicating the reasonableness of the regression model and its good fit. Thus, this model could be used to predict the relationship between HMT conditions and RS content. The determination coefficient (R^2) and adjusted coefficient (R^2_{Adj}) of the equation were 0.9938 and 0.9858, respectively, indicating a high degree of fit between the predicted and actual values. Among these factors, A, B, C, A^2, B^2, and C^2 were highly significant ($p < 0.01$), BC was significant

($p < 0.05$), and the rest were not significant. According to the F-values, the influence of HMT conditions on the RS content of HAS was treatment temperature > heating time > moisture content.

Table 6. Variance analysis for regression equation.

Variation Source	Sum of Squares	Degree of Freedom	Mean Square	F-Value	p-Value	Significance
Model	588.16	9	65.35	124.35	<0.0001	**
A	292.94	1	292.94	557.40	<0.0001	**
B	101.46	1	101.46	193.06	<0.0001	**
C	38.54	1	38.54	73.34	<0.0001	**
AB	1.33	1	1.33	2.54	0.1551	
AC	0.81	1	0.81	1.54	0.2544	
BC	3.65	1	3.65	6.94	0.0337	*
A^2	46.74	1	46.74	88.93	<0.0001	**
B^2	24.90	1	24.90	47.38	0.0002	**
C^2	62.71	1	62.71	119.32	<0.0001	**
Residual	3.68	7	0.5255			
Lack of Fit	0.90	3	0.2993	0.4305	0.7427	
Pure Error	2.78	4	0.6952			
Total Deviation	591.83	16		$R^2 = 0.9938$	$R^2_{Adj} = 0.9858$	$C.V.\% = 1.42$

* significant ($p < 0.05$); ** highly significant ($p < 0.01$).

Based on the response surface analysis, the optimal HMT conditions for the highest RS content in HAS were as follows: treatment temperature of 108.25 °C, heating time of 5.82 h, and moisture content of 25.54%, with a predicted RS content of 60.43%. Considering operability, the treatment conditions were adjusted to a temperature of 108 °C, a time of 5.80 h, and a moisture content of 25.50%. Validation experiments in triplicate were conducted on the optimal treatment conditions, and the average RS content was 60.23%, close to the predicted value, indicating the reliability and practicality of the model.

3.3. Macroscopic and Microscopic Structures

The macroscopic and microscopic structural images of triticale noodles with different levels of HMT-HAS are depicted in Figure 1. At the macroscopic level, pure triticale noodles, due to the presence of bran, exhibited a relatively rough surface, a dark appearance, and a brownish color. The addition of HMT-HAS lightened the color of the noodles, gradually transitioning towards white, and the surface became smooth and elastic, making it more readily accepted sensorially. At the microscopic level, the addition of HMT-HAS led to a smoother surface for the noodles. However, as the protein content decreases, gluten networks may have difficulty forming, and the gel structure gradually becomes loose, even showing tiny pores, in agreement with the findings by Zhang et al. [28]. This microscopic change inevitably affects the texture and mouthfeel of the noodles.

3.4. Short-Range Order

The Fourier transform infrared spectra of the triticale noodles with different levels of HMT-HAS are illustrated in Figure 2, revealing similar peak distributions among the spectra with varying addition levels and HMT-HAS. The broad band at 3381 cm^{-1} was related to the stretching vibration of O-H, while the characteristic peaks at 2929 and 1652 cm^{-1} corresponded to the C-H deformation vibration of the glucose element and the bending vibration of O-H in water, respectively, which is associated with the amorphous region of starch [29]. The peaks at 1157, 1109, and 1082 cm^{-1} were attributed to the asymmetric C-O and C-C skeleton stretching vibrations [29,30]. However, the triticale noodle samples showed a peak at 1540 cm^{-1}, corresponding to the stretching vibration of protein amide II [31], which did not exist in HMT-HAS. The IR bands at ca. 1047 and ca. 1022 cm^{-1} are sensitive to the amount of ordered and amorphous structures of starch, respectively, so the absorbance ratio of the two bands could be used to evaluate the short-range order of double helices [29]. Our preliminary data have shown that the short-range order of HMT-HAS (0.860) was significantly lower than that of triticale flour (0.964). Therefore, we tried to determine how the short-range order changes when adding HMT-HAS to the triticale noodle. As shown in Table 7, with the increasing level of HMT-HAS, the absorbance ratio at $R_{1051\ cm^{-1}}$ and $R_{1021\ cm^{-1}}$ decreased from 0.936 to 0.883, suggesting a notable weakening of the short-range order of starch in the noodles. In other words, there was an inverse relationship between the amylose content and the content of the short-range ordered structure, consistent with the findings of Shi et al. [32] and Karwasra et al. [33]. Starch with a high content of amylose tends to have relatively low short-range order, possibly because amylose interferes with the formation of double-helix structures between amylopectin molecules.

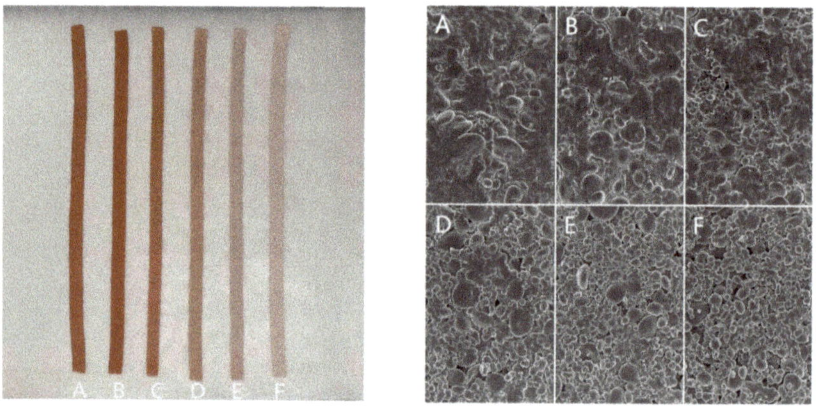

(**a**) (**b**)

Figure 1. Photographs of triticale noodles with different levels of HMT-HAS: (**a**) macro images; (**b**) SEM micrographs. The levels of HMT-HAS are 0% (**A**), 10% (**B**), 20% (**C**), 30% (**D**), 40% (**E**), and 50% (**F**).

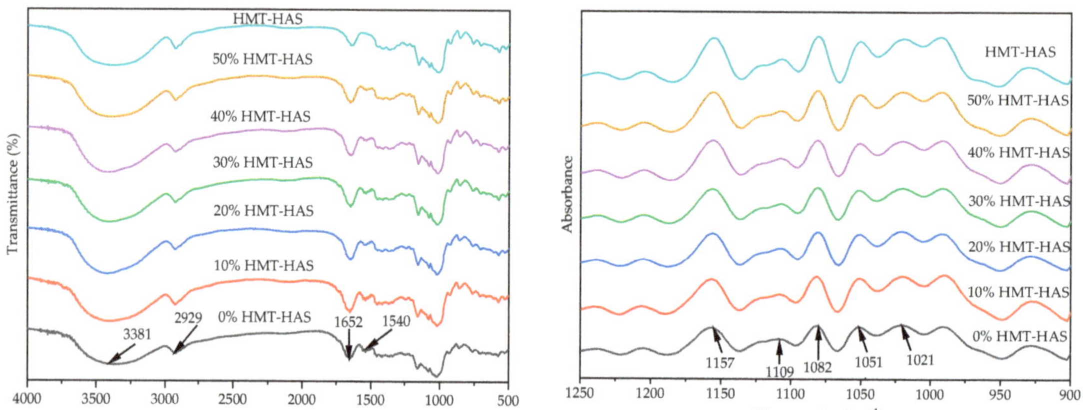

(**a**) (**b**)

Figure 2. Fourier transform infrared spectra of triticale noodles with different levels of HMT-HAS: (**a**) transmittance spectra; (**b**) deconvoluted spectra.

Table 7. Absorbance ratios at $R_{1051\ cm^{-1}}$ and $R_{1021\ cm^{-1}}$ in Fourier transform infrared spectroscopy.

Samples	$R_{1051\ cm^{-1}}/R_{1021\ cm^{-1}}$
0% HMT-HAS	0.936 ± 0.001 [a]
10% HMT-HAS	0.909 ± 0.002 [b]
20% HMT-HAS	0.894 ± 0.002 [c]
30% HMT-HAS	0.892 ± 0.000 [c]
40% HMT-HAS	0.889 ± 0.001 [c]
50% HMT-HAS	0.883 ± 0.005 [d]
HMT-HAS	0.860 ± 0.002 [e]

[a–e]: values with different lowercases in the same column are considered significantly different ($p < 0.05$).

3.5. X-ray Diffraction and Relative Crystallinity

The X-ray diffraction patterns and relative crystallinity of the triticale noodles with different levels of HMT-HAS are displayed in Figure 3. The crystalline structure of starch generally includes three main types (A-, B-, and C-type) on the basis of their characteristics and X-ray diffraction patterns [34]. Triticale starch exhibited a typical A-type crystalline structure, with diffraction peaks at 15°, 17°, 18°, and 23° (2θ), whereas HMT-HAS showed single peaks at 5.6°, 15°, and 17° and a merging of doublet at 22° and 24°, still exhibiting a B-type structure, although partly changed by HMT. Additionally, HMT-HAS showed a peak at 20°, indicating the presence of a V-type crystalline structure. Undoubtedly, after the addition of HMT-HAS, a prominent V-type diffraction peak appeared at 20°. Moreover, with the level of HMT-HAS increasing, the peak became more pronounced, indicating an increase in the content of amylose–lipid complexes [35]. Due to the increasing level of HMT-HAS, the content of amylose in the noodles increased, and the amorphous region expanded. Consequently, the relative crystallinity decreased from 29.22% to 21.63%.

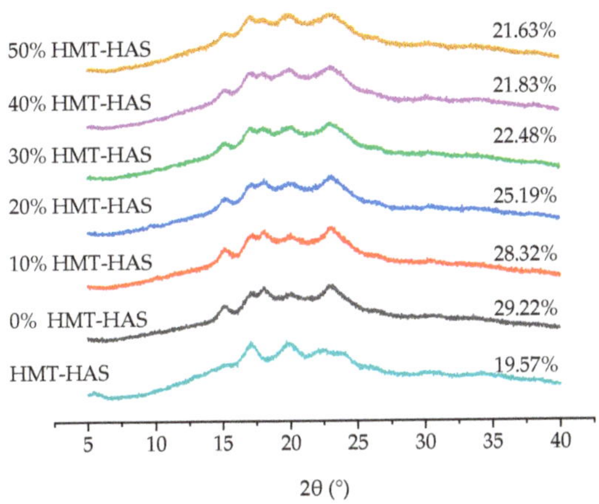

Figure 3. X-ray diffraction patterns of triticale noodles with different levels of HMT-HAS.

3.6. Pasting Properties

The pasting properties of a food reflect the changes in viscosity during the heating-cooling process in the presence of water, and viscosity development is closely related to the formation of a tightly packed array of swollen and deformable granules and the leaching of amylose [36]. These changes, mainly induced by the alteration of starch structure, may affect the sensory characteristics and starch digestibility of the food product.

As shown in Table 8, the addition of HMT-HAS significantly reduced the peak viscosity, trough viscosity, final viscosity, breakdown, setback, and peak time of triticale noodles. The pasting properties of starch were influenced by some factors such as amylose content and starch molecular structure [37]. The addition of HMT-HAS led to an increase in the ratio of amylose to amylopectin in triticale noodles. Amylopectin can enhance the cohesion and network structure of starch granules, thus increasing viscosity during starch gelatinization. Therefore, there was a significant negative correlation between these indicators of viscosity and the amylose/amylopectin ratio [38]. The phenomenon was also observed in wheat starch: as the amylose content increased, these values for pasting properties—peak and final viscosities and breakdown—all had a tendency to decrease [39]. This implied that with the increase in amylose content, the viscosity characteristics of starch showed a gradual weakening trend. Additionally, HMT resulted in a stable internal structure of starch, increased heat stability, and a low water-binding capacity, thus reducing viscosity and decreasing setback value [40].

The pasting properties of starch are closely related to noodle quality. Hormdok and Noomhorm [36] investigated the effects of HMT-rice starch substitution on the cooking and textural properties of rice noodles and found that cooking loss correlated positively with breakdown, while tensile strength and extensibility of the noodles correlated negatively with setback. Whereas, for Chinese dry white noodles, peak viscosity and breakdown were positively associated with the elasticity, stickiness, and smoothness of the noodles [41]. Similarly, for Chinese fresh white noodles, breakdown exhibited a significant positive correlation with the noodles' viscoelasticity and smoothness [42]. Consequently, the decrease in sensory quality of triticale noodles probably occurs at high levels of HMT-HAS; thus, further studies for improving that aspect are needed.

Table 8. Pasting properties of triticale noodles with different levels of HMT-HAS.

HMT-HAS (%)	Peak Viscosity (cp)	Trough Viscosity (cp)	Breakdown (cp)	Final Viscosity (cp)	Setback (cp)	Peak Time (min)
0	1245.33 ± 45.79 ^{a}	953.33 ± 59.88 ^{a}	292.00 ± 14.18 ^{a}	1870.67 ± 17.79 ^{a}	917.33 ± 75.34 ^{a}	6.18 ± 0.20 ^{a}
10	909.33 ± 39.32 ^{b}	670.00 ± 31.22 ^{b}	220.33 ± 15.01 ^{b}	1523.33 ± 30.62 ^{b}	853.33 ± 13.50 ^{a}	5.84 ± 0.04 ^{b}
20	612.00 ± 23.00 ^{c}	477.00 ± 5.29 ^{c}	144.00 ± 7.55 ^{c}	1112.67 ± 33.84 ^{c}	635.67 ± 29.28 ^{b}	5.55 ± 0.12 ^{c}
30	419.33 ± 3.51 ^{d}	333.67 ± 1.53 ^{d}	83.67 ± 2.31 ^{d}	781.33 ± 4.62 ^{d}	447.67 ± 3.21 ^{c}	5.35 ± 0.03 ^{cd}
40	236.67 ± 2.08 ^{e}	203.00 ± 1.73 ^{e}	35.67 ± 1.53 ^{e}	449.00 ± 6.24 ^{e}	246.00 ± 5.20 ^{d}	5.25 ± 0.07 ^{de}
50	132.00 ± 7.00 ^{f}	129.00 ± 7.94 ^{f}	4.33 ± 1.53 ^{f}	268.67 ± 21.73 ^{f}	139.67 ± 17.62 ^{e}	5.10 ± 0.09 ^{e}

$^{a–f}$: values with different lowercases in the same column are considered significantly different ($p < 0.05$).

3.7. In Vitro Starch Digestibility

As shown in Table 9, the addition of HMT-HAS caused a significant reduction in the contents of RDS and SDS in the noodles, while the RS content increased. Punia [43] believed that the content of amylose may be the main factor influencing almost all the physicochemical properties of barley starch. Similarly, starch digestibility was also influenced by amylose content, and the RS content of starch usually increased with an increase in amylose content [44]. In addition, through optimal HMT conditions, the RS content of HAS increased from 30.98% to 60.23%, so the higher the amount added to the noodles, the higher the RS content. When the addition level reached 50%, the RS content of the noodles was 34.41%, nearly three times higher than that of pure triticale noodles. Similarly, Li et al. [45] reported that the RS content significantly rose by about 6 times when 50% high-amylose wheat flour was added to wheat bread. The results proved that the addition of HMT-HAS could significantly decrease the starch digestibility of triticale noodles, which helps in the development of functional triticale noodles.

GI is a commonly used metric for estimating the postprandial blood glucose response to carbohydrate-containing foods. It is well established that diets low in GI are associated with the prevention and management of diabetes and coronary heart disease [46]. The FAO recommended increasing the intake of low-GI foods, particularly for individuals with diabetes and impaired glucose tolerance [47].

While it is recognized that some inconsistencies exist between GI values predicted by in vitro digestion models and those obtained from human studies, in vitro models are generally considered reliable estimates of GI for foods [48,49]. RS is a type of starch that resists digestion in the gastrointestinal tract and is instead fermented by microorganisms in the colon. There is a strong negative correlation between RS content and GI value, making RS a useful proxy for estimating the potential GI of food products [50].

In our study, we utilized the RS content to infer the starch digestibility of triticale noodles. However, we acknowledge that this is an indirect measure, and the actual GI value of the final triticale noodle product should be validated through human trials to confirm our findings.

Table 9. Starch digestibility of triticale noodles with different levels of HMT-HAS.

HMT-HAS (%)	RDS (%)	SDS (%)	RS (%)
0	64.88 ± 0.10 [a]	23.03 ± 0.08 [a]	12.08 ± 0.08 [f]
10	60.35 ± 0.42 [b]	21.33 ± 0.35 [b]	18.32 ± 0.34 [e]
20	57.12 ± 0.35 [c]	20.29 ± 0.32 [c]	22.59 ± 0.26 [d]
30	53.19 ± 0.07 [d]	19.76 ± 0.26 [cd]	27.05 ± 0.22 [c]
40	49.82 ± 0.35 [e]	19.34 ± 0.45 [d]	30.84 ± 0.33 [b]
50	46.96 ± 0.49 [f]	18.63 ± 0.38 [e]	34.41 ± 0.17 [a]

[a–f]: values with different lowercases in the same column are considered significantly different ($p < 0.05$).

4. Conclusions

To promote the application of HAS in low-GI triticale noodles, we investigated the effects of HMT parameters on the starch digestibility of HAS by response surface methodology. As expected, the RS content was significantly enhanced, but the RDS content was reduced after HMT. The optimal HMT conditions for improving the RS content of HAS were a temperature of 108 °C, a time of 5.8 h, and a moisture content of 25.50%. Under these conditions, the RS content reached up to 60.23%, approximately 1.94 times higher than the untreated counterpart. The results indicated that HMT may be an effective means to attenuate starch digestibility for the development of low-GI foods. The addition of HMT-HAS to the triticale flour significantly affected the physicochemical properties of triticale noodles. With the level of HMT-HAS increasing, the triticale noodles showed a lighter color and a smoother surface, but the gluten network may be weakened. Additionally, the short-range order, relative crystallinity, pasting properties, and starch digestibility of the noodles all decreased. Particularly when the level of HMT-HAS reached 50%, the RS content of triticale noodles greatly increased to 34.41%, approximately three times higher than that of the pure triticale noodles, significantly enhancing the resistance to digestion of triticale noodles. According to the literature and current information, there are no official recommendations for RS intake. Numerous studies suggest that different amounts of RS can positively impact health [51]. Some researchers recommend a beneficial intake of RS to be at least 14% of total starch, or around 20 g per day, or 6 g per meal. It has been shown that consuming 6–12 g of RS per meal can enhance postprandial glucose and insulin responses. However, for additional health benefits such as increased stool volume, a daily intake of approximately 15–20 g is necessary. Therefore, while we cannot provide a precise value or range for the quantity of RS and the level of starch digestibility, we recommend an approximate range (40–80 g/meal) for the consumption of the triticale noodles.

After incorporating 50% HMT-HAS into triticale noodles, the key characteristics of traditional wheat noodles, such as texture, appearance, and processability, were still retained. However, a slight decrease in sensory quality was observed. Further studies for improving that aspect are needed. Based on these findings, we conclude that a 50% HMT-HAS triticale noodle maintains the fundamental qualities of a noodle and can be considered a new variety of noodle. Finally, the principles and methodologies used in this study could be applied to other cereal-based products beyond noodles, such as bread, pasta, and breakfast cereals, to enhance their nutritional profiles and lower their glycemic indices.

Author Contributions: Conceptualization, H.L. (Hua Li 1) and S.S.; methodology, H.L. (Hua Li 1) and S.S.; validation, H.L. (Hua Li 1) and S.S.; formal analysis, H.L. (Hua Li 2), Y.L. and R.L.; investigation, H.L. (Hua Li 2) and Y.L.; resources, H.L. (Hua Li 2) and Y.L.; data curation, H.L. (Hua Li 2), Y.L. and R.L.; writing—original draft, H.L. (Hua Li 1) and H.L. (Hua Li 2); writing—review & editing, H.L. (Hua Li 1) and S.S.; visualization, H.L. (Hua Li 2) and R.L.; supervision, S.S.; project administration, H.L. (Hua Li 1); funding acquisition, S.S. All authors have read and agreed to the published version of the manuscript.

Funding: This research was financially supported by Thailand Science Research and Innovation (TSRI) and Mahasarakham University.

Institutional Review Board Statement: Not applicable.

Informed Consent Statement: Not applicable.

Data Availability Statement: The data presented in this study are available on request from the corresponding authors. The data are not publicly available due to privacy restrictions.

Acknowledgments: This research was financially supported by Thailand Science Research and Innovation (TSRI) and Mahasarakham University. The authors also thank the Testing Center of Yangzhou University for the scientific aids provided.

Conflicts of Interest: The authors declare no conflict of interest.

References

1. Zhong, Y.; Tai, L.; Blennow, A.; Ding, L.; Herburger, K.; Qu, J.; Xin, A.; Guo, D.; Hebelstrup, K.H.; Liu, X. High-amylose starch: Structure, functionality and applications. *Crit. Rev. Food Sci. Nutr.* **2023**, *63*, 8568–8590. [CrossRef] [PubMed]
2. Ang, K.; Bourgy, C.; Fenton, H.; Regina, A.; Newberry, M.; Diepeveen, D.; Lafiandra, D.; Grafenauer, S.; Hunt, W.; Solah, V. Noodles made from high amylose wheat flour attenuate postprandial glycaemia in healthy adults. *Nutrients* **2020**, *12*, 2171. [CrossRef] [PubMed]
3. Singh, R.; Kumar, Y.; Bist, Y.; Pamecha, P.; Shukla, M.; Singh, A. Heat-moisture treatment of starch. In *Starch: Advances in Modifications, Technologies and Applications*; Sharanagat, V.S., Saxena, D.C., Kumar, K., Kumar, Y., Eds.; Springer: Cham, Switzerland, 2023; pp. 173–186, ISBN 978-3-031-35843-2.
4. Chung, H.J.; Liu, Q.; Hoover, R. Impact of annealing and heat-moisture treatment on rapidly digestible, slowly digestible and resistant starch levels in native and gelatinized corn, pea and lentil starches. *Carbohydr. Polym.* **2009**, *75*, 436–447. [CrossRef]
5. Zavareze, E.D.R.; Halal, S.L.M.E.; Santos, D.G.D.L.; Helbig, E.; Pereira, J.M.; Dias, A.R.G. Resistant starch and thermal, morphological and textural properties of heat-moisture treated rice starches with high-, medium- and low-amylose content. *Starch Starke* **2012**, *64*, 45–54. [CrossRef]
6. Ito, T.; Saito, K.; Sugawara, M.; Mochida, K.; Nakakuki, T. Effect of raw and heat-moisture-treated high-amylose corn starches on the process of digestion in the rat digestive tract. *J. Sci. Food Agric.* **1999**, *79*, 1203–1207. [CrossRef]
7. Liu, X.; Ogawa, H.; Ando, R.; Nakakuki, T.; Kishida, T.; Ebihara, K. Heat-moisture treatment of high-amylose corn starch increases dietary fiber content and lowers plasma cholesterol in ovariectomized rats. *J. Food Sci.* **2007**, *72*, S652–S658. [CrossRef]
8. Cervini, M.; Lobuono, C.; Volpe, F.; Curatolo, F.M.; Scazzina, F.; Dall'Asta, M.; Giuberti, G. Replacement of native with malted triticale (x *Triticosecale* Wittmack) flour in dry pasta: Technological and nutritional implications. *Foods* **2024**, *13*, 2315. [CrossRef]
9. Lorenz, K.; Dilsaver, W.; Lough, J. Evaluation of triticale for the manufacture of noodles. *J. Food Sci.* **1972**, *37*, 764–767. [CrossRef]
10. Niu, X. Study on the Quality and Antioxidation of *Triticum aestivum* Flours and Its Noodle Products. Master's Thesis, Henan University of Technology, Zhengzhou, China, 2022.
11. Yan, W.; Zhang, Z.; Ye, A. Nutritional characteristics and processing quality assessment of Heidasui black wheat. *Food Sci.* **2012**, *33*, 146–150.
12. Liu, M.; Fan, M.; Qian, H.; Li, Y.; Wang, L. Effect of different enzymes on thermal and structural properties of gluten, gliadin, and glutenin in triticale whole-wheat dough. *Int. J. Biol. Macromol.* **2023**, *253*, 127384. [CrossRef]
13. Zhu, F. Triticale: Nutritional composition and food uses. *Food Chem.* **2018**, *241*, 468–479. [CrossRef] [PubMed]
14. Lv, Y. Study on the Effects of Heat-Moisture Treatment on the Characteristics of Barley-Wheat Composite Dough and the Quality of Noodles. Master's Thesis, Henan University of Technology, Zhengzhou, China, 2023.
15. Zhang, X. Study on the Wet Heat Treatment of Sweet Potato Flour and Its Application in Noodles. Master's Thesis, Henan Institute of Science and Technology, Xinxiang, China, 2022.
16. Li, H.; Mao, Y.; Ma, D.; Li, H.; Liu, R.; Siriamornpun, S. Impact of cooking methods on phenolic acid composition, antioxidant activity, and starch digestibility of chinese triticale porridges: A comparative study between atmospheric pressure and high pressure boiling. *Foods* **2024**, *13*, 230. [CrossRef]
17. *GB 5009.9-2016*; National Standard for Food Safety—Determination of Starch in Food. Standard Press of China: Beijing, China, 2016.
18. Englyst, H.N.; Kingman, S.M.; Cummings, J.H. Classification and measurement of nutritionally important starch fractions. *Eur. J. Clin. Nutr.* **1992**, *46*, S33–S50. [PubMed]

19. Hoover, R. The impact of heat-moisture treatment on molecular structures and properties of starches isolated from different botanical sources. *Crit. Rev. Food Sci. Nutr.* **2010**, *50*, 835–847. [CrossRef] [PubMed]
20. Chi, C.; Li, X.; Huang, S.; Chen, L.; Zhang, Y.; Li, L.; Miao, S. Basic principles in starch multi-scale structuration to mitigate digestibility: A review. *Trends Food Sci. Technol.* **2021**, *109*, 154–168. [CrossRef]
21. Wang, Y.; Ral, J.-P.; Saulnier, L.; Kansou, K. How does starch structure impact amylolysis? Review of current strategies for starch digestibility study. *Foods* **2022**, *11*, 1223. [CrossRef] [PubMed]
22. Liu, H.; Guo, X.; Li, W.; Wang, X.; Lv, M.; Peng, Q.; Wang, M. Changes in physicochemical properties and in vitro digestibility of common buckwheat starch by heat-moisture treatment and annealing. *Carbohydr. Polym.* **2015**, *132*, 237–244. [CrossRef]
23. Singh, J.; Dartois, A.; Kaur, L. Starch digestibility in food matrix: A review. *Trends Food Sci. Technol.* **2010**, *21*, 168–180. [CrossRef]
24. Wu, C.; Ji, G.; Gao, F.; Qian, J.; Zhang, L.; Li, Q.; Zhang, C. Effect of heat-moisture treatment on the structural and physicochemical characteristics of sand rice (*Agriophyllum squarrosum*) starch. *Food Sci. Nutr.* **2021**, *9*, 6720–6727. [CrossRef]
25. Wang, D.; Guo, M.; Liu, Y.; Guo, Y.; Ye, H.; Guo, Y. The effect of moisture-heat treatment on the glycemic index value and related indicators of glutinous rice flour. *Sci. Technol. Food Ind.* **2023**, *44*, 253–259. Available online: https://link.cnki.net/doi/10.13386/j.issn1002-0306.2022040041 (accessed on 12 June 2024).
26. Mohamed, I.O. Effects of processing and additives on starch physicochemical and digestibility properties. *Carbohydr. Polym. Technol. Appl.* **2021**, *2*, 100039. [CrossRef]
27. Sun, Y.; Qin, R.; Zeng, J.; Li, G. Effect of heat-moisture treatment on the structure and digestibility of sweet potato starch. *Foods* **2023**, *12*, 3076. [CrossRef] [PubMed]
28. Zhang, S.; Xu, X.; Zhang, L.; Meng, L.; Tang, X. The impact of high amylose corn starch on the structure, cooking quality and starch digestibility of extruded buckwheat noodles. *Food Sci.* **2023**, *44*, 116–124. Available online: https://www.spkx.net.cn/CN/10.7506/spkx1002-6630-20220505-044 (accessed on 10 June 2024).
29. Yan, Y.; Feng, L.; Shi, M.; Cui, C.; Liu, Y. Effect of plasma-activated water on the structure and in vitro digestibility of waxy and normal maize starches during heat-moisture treatment. *Food Chem.* **2020**, *306*, 125589. [CrossRef] [PubMed]
30. Pozo, C.; Rodríguez-Llamazares, S.; Bouza, R.; Barral, L.; Castaño, J.; Müller, N.; Restrepo, I. Study of the structural order of native starch granules using combined FTIR and XRD analysis. *J. Polym. Res.* **2018**, *25*, 266. [CrossRef]
31. Lin, L.; Yang, H.; Chi, C.; Ma, X. Effect of protein types on structure and digestibility of starch-protein-lipids complexes. *LWT* **2020**, *134*, 110175. [CrossRef]
32. Shi, P.; Zhao, Y.; Qin, F.; Liu, K.; Wang, H. Understanding the multi-scale structure and physicochemical properties of millet starch with varied amylose content. *Food Chem.* **2023**, *410*, 135422. [CrossRef]
33. Karwasra, B.L.; Gill, B.S.; Kaur, M. Rheological and structural properties of starches from different Indian wheat cultivars and their relationships. *Int. J. Food Prop.* **2017**, *20*, S1093–S1106. [CrossRef]
34. Zobel, H.F. Starch crystal transformations and their industrial importance. *Starch Stärke* **1988**, *40*, 1–7. [CrossRef]
35. Schafranski, K.; Ito, V.C.; Lacerda, L.G. Impacts and potential applications: A review of the modification of starches by heat-moisture treatment (HMT). *Food Hydrocoll.* **2021**, *117*, 106690. [CrossRef]
36. Hormdok, R.; Noomhorm, A. Hydrothermal treatments of rice starch for improvement of rice noodle quality. *LWT* **2007**, *40*, 1723–1731. [CrossRef]
37. Li, C.; Shao, S.; Yang, C.; Li, E. Starch fine molecular structure is the driving factor for starch gelatinization property in both wheat flour and wheat dough. *Starch Stärke* **2023**, *75*, 2200290. [CrossRef]
38. Luo, X.; Cheng, B.; Zhang, W.; Shu, Z.; Wang, P.; Zeng, X. Structural and functional characteristics of *Japonica* rice starches with different amylose contents. *Cyta J. Food* **2021**, *19*, 532–540. [CrossRef]
39. Blazek, J.; Copeland, L. Pasting and swelling properties of wheat flour and starch in relation to amylose content. *Carbohydr. Polym.* **2008**, *71*, 380–387. [CrossRef]
40. Liu, C.; Hong, J.; Zheng, X. Effect of heat-moisture treatment on morphological, structural and functional characteristics of ball-milled wheat starches. *Starch Stärke* **2017**, *69*, 1500141. [CrossRef]
41. He, Z.H.; Yang, J.; Zhang, Y.; Quail, K.J.; Pena, R.J. Pan bread and dry white Chinese noodle quality in Chinese winter wheats. *Euphytica* **2004**, *139*, 257–267. [CrossRef]
42. Zhang, Y.; Nagamine, T.; He, Z.H.; Ge, X.X.; Yoshida, H.; Peña, R.J. Variation in quality traits in common wheat as related to Chinese fresh white noodle quality. *Euphytica* **2005**, *141*, 113–120. [CrossRef]
43. Punia, S. Barley starch: Structure, properties and in vitro digestibility-A review. *Int. J. Biol. Macromol.* **2019**, *155*, 868–875. [CrossRef]
44. Lv, X.; Hong, Y.; Zhou, Q.; Jiang, C. Structural features and digestibility of corn starch with different amylose content. *Front. Nutr.* **2021**, *8*, 692673. [CrossRef]
45. Li, C.; Dhital, S.; Gidley, M.J. High-amylose wheat bread with reduced in vitro digestion rate and enhanced resistant starch content. *Food Hydrocoll.* **2022**, *123*, 107181. [CrossRef]
46. Augustin, L.S.A.; Kendall, C.W.C.; Jenkins, D.J.A.; Willett, W.C.; Astrup, A.; Barclay, A.W.; Björck, I.; Brand-Miller, J.C.; Brighenti, F.; Buyken, A.E.; et al. Glycemic index, glycemic load and glycemic response: An International Scientific Consensus Summit from the International Carbohydrate Quality Consortium (ICQC). *Nutr. Metab. Cardiovasc. Dis.* **2015**, *25*, 795–815. [CrossRef]
47. Raigond, P.; Ezekiel, R.; Raigond, B. Resistant starch in food: A review. *J. Sci. Food Agric.* **2015**, *95*, 1968–1978. [CrossRef] [PubMed]

48. Li, C.; Hu, Y. In vitro and animal models to predict the glycemic index value of carbohydrate-containing foods. *Trends Food Sci. Technol.* **2022**, *120*, 16–24. [CrossRef]
49. Ferrer-Mairal, A.; Peñalva-Lapuente, C.; Iglesia, I.; Urtasun, L.; De Miguel-Etayo, P.; Remón, S.; Cortés, E.; Moreno, L.A. In vitro and in vivo assessment of the glycemic index of bakery products: Influence of the reformulation of ingredients. *Eur. J. Nutr.* **2012**, *51*, 947–954. [CrossRef]
50. Kumar, A.; Sahoo, U.; Baisakha, B.; Okpani, O.A.; Ngangkham, U.; Parameswaran, C.; Basak, N.; Kumar, G.; Sharma, S.G. Resistant starch could be decisive in determining the glycemic index of rice cultivars. *J. Cereal Sci.* **2018**, *79*, 348–353. [CrossRef]
51. Bojarczuk, A.; Skąpska, S.; Khaneghah, A.M.; Marszałek, K. Health benefits of resistant starch: A review of the literature. *J. Funct. Foods* **2022**, *93*, 105094. [CrossRef]

MDPI AG
Grosspeteranlage 5
4052 Basel
Switzerland
Tel.: +41 61 683 77 34

Foods Editorial Office
E-mail: foods@mdpi.com
www.mdpi.com/journal/foods

www.ingramcontent.com/pod-product-compliance
Lightning Source LLC
Chambersburg PA
CBHW061440210926
51850CB00008B/1565
* 9 7 8 3 7 2 5 8 7 0 1 4 1 *